UNDERSTANDING BUILDINGS

UNDERSTANDING BUILDINGS

A MULTIDISCIPLINARY APPROACH

ESMOND REID

Illustrations by the author

Longman
Scientific &
Technical

Longman Scientific & Technical
Longman Group UK Limited
Longman House, Burnt Mill, Harlow,
Essex CM20 2JE, England
and Associated Companies throughout the world

First published 1984 by Construction Press
Reprinted by Longman Scientific & Technical 1988, 1989, 1990, 1991,
1992, 1993, 1994

British Library Cataloguing in Publication Data
Reid, Esmond
    Understanding buildings,
    1. Building   2. Architecture
    I Title
    690   TH145

ISBN 0-582-00971-5

Set in 9/10½pt Linotron 202 Times Roman
Produced by Longman Singapore Publishers Pte Ltd.
Printed in Singapore

to Elma and Trisha

CONTENTS

Contents

PREFACE

A few years ago, when working as a technical editor at the Architectural Press in London, it occurred to me, and to those who first put me in mind to try and do this book, that there appeared to be no general, introductory text on the technical subjects in building design. To be sure, there was the ever-increasing wealth of excellent, specialised texts on structure, services, lighting and so on, but nothing on the whole technical context at a simpler level. It seemed so curious. Surely, such a book would be useful?

So this is intended as an introductory text on how buildings work, constructionally and technically. Building planning is covered only in so far as it is involved with the technical function. The chapter headings on the cover show the scope. The book is primarily for the student, though, perhaps, without defining 'student' too closely – goodness knows, I discovered my areas of ignorance in writing it! But, for the architectural student, it tries to explain the practical and technical constraints in building design, and to give an early insight into the problems confronting the other specialisations in what has come to be called the 'design team'. For students in the other allied courses that have to do with buildings, it may help as a primer in their particular course where appropriate but, more than that, hopes to offer reciprocal insight between the specialisations and, not least, into the problems confronting the architect. In short, the book is multidisciplinary.

The approach is to explain building technology in the light of the basic functions and practical physics from which it springs, on the basis that the underlying principles are the key to understanding what is done in practice – practice being only the symptom of the principles. And, generally, the text rests at giving one good, typical example of what is done in practice: for one example, in Chapter 2 *Enclosure*, the wall constructions shown are to illustrate how the needs for stability, water exclusion, insulation and so on, are met – there is no intention to offer an anthology of constructions. Also, the text usually stops short of doing the sums, of quantifying what is done in terms of calculations and sizes, except where this seems to be the clearest way to express the principles themselves. In Chapter 1 *Structure*, it is one thing to describe how the various internal stresses influence the shape of a beam but it would have been quite another to go on and show how beam sizes are calculated. Usually, the text leaves off where there are already more specialised texts in the particular subject better fitted to go further and, wherever possible, the bibliography refers the reader to them.

The book was mostly written and researched in the UK, but not exclusively, and it is not exclusively written as a UK book. Again there must be some bias, but the very fact of its being introductory and on basic principles should help avoid tying it down nationally, in as much as the principles do not change markedly from one country to the next. Gravity is gravity and rain is rain. Of course, some principles do change – Chapter 2 *Enclosure* and Chapter 3 *Climate services* assume the reasonably temperate climate found in much of Europe and the USA, but make sallies to the colder and hotter extremes. Where other practice does happen to differ, the text, at points, compares what is done in different countries, notably the European mainland, the UK and the USA.

But apart from the book itself, or in further explanation of its intention, a few words may be allowed here on the place of technology in architectural teaching – albeit my own angle will already be plain enough. Few would deny that architectural design has found it hard to keep pace with the fast-changing technology over the past few decades nor that, among the members of the design team, the architect has been the most criticised on the technical performance of modern buildings. This is all well known and needs no elaboration. But, accepting there is a problem in keeping pace with innovation in materials and methods, and for that matter with the increased sophistication we have come to expect from the performance of buildings, then one would have expected this to reflect in the way, and extent to which, architectural technology is taught.

Some while ago, I was asked to give a course of lectures – on fire safety, as it happened – at a leading school of architecture in London. Discussing the time-table, the tutor in charge of technical teaching warned me that technical lectures were poorly attended and not compulsory. But it did not matter I was told. Building technology was only common sense and did not lend itself to formal teaching anyway – far better the student years be left unfettered by technology, allowing a freer design expression to develop at the drawing board. Anything lacking in technical knowledge would soon be made up, once in practice. Well, I never did give those lectures and I tell the story only because it so aptly expresses the opposite of what I believe – when told to practising architects, engineers and builders, it has been received with almost universal horror. True, the view it expresses is too extreme to be typical but one hears more modest versions of it and I submit there may be some confused thinking on the matter?

We had better be quite clear about this. There is no argument here that technology and function should control building form, that would be doctrinal and im-poverishing; of course, architecture has wider social, cultural and aesthetic contexts than that. Nor is there any argument that free design expression is an un-worthy aim, only a question as to how that freedom be achieved. The essential thing, surely, is that useful free-dom does not come, and never has come, through ab-sence of discipline. Good design solutions, even in terms of aesthetics alone, were it possible to speak of them in isolation, do not spring fully armed from the mind as if by magic. To think they can is mistaken, irresponsible,

even arrogant. There is simply no historical precedent for it, not anywhere, not in engineering, not in architecture.

As the absolute bottom line one can say that the practical and technical constraints are there, like it or not, and that it is therefore prudent to appreciate their nature if only as a matter of competence. To ignore them even in the introductory stages of 'learning' de-sign is self-deceiving, rendering solutions less meaning-ful, and missing out on lessons to be learnt. It's cheat-ing! There is a fear sometimes expressed that technolo-gy might come to control and inhibit design, but surely not understanding it would make this unhappy outcome more likely rather than less, with designers surrender-ing technical responsibilities to third parties and sub-ordinating themselves in consequence? In any case, a weak technical grasp must in itself be inhibiting, tend-ing to restrict and polarise design around existing solu-tions or, worse, tending to polarise it around misunder-stood and often ill-grounded technical 'fashions'. And this is saying nothing of the risks of straight, technical failure.

However, the fair and balanced view can afford to be less defensive in its tone. Yes, over-emphasising tech-nology in design is impoverishing, but denying it its sensible place is equally so. Properly understood the physical 'constraints' are of course not only constraints but also *inputs*. Consciously or subconciously in the mind of the designer this cannot fail to help generate good design – whether one is talking about the absolute responsibility of making buildings work properly or ab-out their whole spirit and form.

ACKNOWLEDGEMENTS

To say I did not know it all at the start would be a gross understatement. I am indebted to those who helped, who, despite their own obvious cares in specialist practice and authorship, yet found the time to give their advice so freely and thoughtfully – their contact and encouragement was one of the most rewarding parts of preparing this book.

Allan Hodgkinson rigorously checked *Structure* and the constructional parts of *Enclosure*; David Adler advised on *Structure* and was continually enthusiastic in wider technical discussion; Eddie Buckle (Captress Ltd) stubbornly maintained the builder's point of view; Cecil Handisyde, to use his own words, was a 'fierce critic' of *Enclosure*; Peter Kennedy gave early advice on services; David Clark (Building Services Copartnership) undertook the lengthy task of checking the chapters on services and, at same firm, Ian White helped on electrics and Barry Jones on water services; Fred Hall, also, gave valuable advice on the chapters on services; David Loe gave early advice on *Lighting* and, much later, Joe Lynes (Thorn Lighting) offered further invaluable advice and was scrupulous in checking the chapter; John Miller (Bickerdike Allen Partners) was equally scrupulous in checking *Acoustics*; at the Architectural Press, London, Anthony Ferguson was kind enough to check *Fire safety* and, also at the AP – without which the book might never have got started – my thanks to The Architects' Journal editor, Leslie Fairweather, for putting me in touch with many sources and allowing me the use of the library, Dorothy Pontin for putting up with my use of the library, Barrie Evans for advice on the final chapter and Louis Dezart for patiently reminding me how to draw.

Many organisations and firms were helpful, including in the UK: the libraries at The Architectural Association, at The Royal Institute of British Architects and at Edinburgh University; The British Steel Corporation; The Building Research Station and The Fire Research Station of The Building Research Establishment; The Cement and Concrete Association; The Mastic Asphalt Council; Arup Associates; Bison Concrete; Cape Universal Claddings; Chubb Fire Security; Colt International; Conder Services; Crittall Construction; Dowling Design and Development; Evode; Greenwood Airvac Ventilation; Honeywell Control Systems; Imperial Chemical Industries; Otis Elevator Co; Philips Electronics; Pilkington Glass; Potterton International; Rainham Timber Engineering; and Ruberoid Building Products.

And in the USA: The National Bureau of Standards, Washington DC; New York City Fire Department; Skidmore, Owings and Merrill, San Francisco; Libby-Owens-Ford, Ohio; The College of Environmental Design, University of California.

My thanks, also, to my publishers, Construction Press of Longman Group Ltd., and to Andrew Best of Curtis Brown Academic Ltd.

1

STRUCTURE

INTRODUCTION

The first thing that any building has to do is stand up. The building is hardly there to satisfy the needs of structure but, whatever its purpose and plan, structural needs will have had a vital hand in shaping its form. And yet, we are so used to buildings not falling down that we hardly stop to wonder why they do not. In fact, 'structure' is often thought of as a rather rarefied subject obscured by tedious sums, but it need not be so, *for sums do not in themselves shape structural design and nor are they essential to a basic understanding of the principles from which structural solutions spring*.

Why does, say, a simple stone tower stand? With no special knowledge, we can say 'because the walls are vertical and thick and because the floors and roof have strong enough beams to span'. True, this may not take us far in comprehending the structure of cathedrals and high-rise buildings, but it is a start. Our everyday experience brings us an intuitive grasp of structure, a feeling for what is structurally right. We have all experienced how materials behave. Observation has told us that stone and wood are relatively strong and bananas and blotting paper weak, and we expect tree branches to thicken towards the trunk and need no sums to tell if a fence is strong enough to sit on.

Certainly, our intuition is fallible. At the risk of irritating readers more familiar with structures, which of the spans in 1.1 is the stronger? Be assured, you are in numerous company if you pick the left one at first glance, where the struts and wires are doing nothing useful at all. But then, again, it only takes a moment's extra thought to understand the arrangement and set the matter right – thought, that is, coupled with quite ordinary experience?

This kind of intuition, and experience, often from hard trial and error, must have schooled the early builders. They did no sums in the modern way but their skilled traditions grew. The mediaeval master masons were delicately balancing the outward thrust of cathedral arches with the counter-thrust of external buttresses, long before the beginnings of modern theory when men like Galileo in the seventeenth century came to wonder about the mathematics of bending, and Newton about the action and reaction of forces, and Hooke about how materials deform under load.

Where, then, does calculation come in? Well, for one thing stress calculations enable engineers to check whether their assumptions about the likely behaviour of their structural proposal are correct. The intuitive, or empirical, can be mathematically verified. In this respect, the early builders were blindfold and, despite their achievements, they sometimes did get it wrong – in quantity, if not in concept. On Beauvais cathedral the roof fell twice and the tower once. Winchester collapsed. Wells, propped too much by the stone buttresses outside, had to be counter-propped by buttresses inside. Many mishaps were due to men who were uncertain of the stresses their structure had to take and dared too much. They could only follow what modern jargon has called the PPI rule – the 'put plenty in' rule – making elements more massive and hoping to err on the safe side: at least they had ample labour force and ample stone. In the advancing iron technology late last century, it took a whole series of collapses – about 25 bridges a year in America – before it was finally accepted that, as structures got larger, commonsense rules-of-thumb simply had to be backed up by calculations.

More specifically, there is the need to minimise self-weight. Just 'putting plenty in' is hopeless in terms of

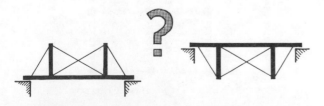

1.1 Which of the two spans is the stronger?

1

structural economy. Having the structural parts bigger than they need to be wastes material and demands that the structure be even bigger, to carry its own superfluous weight – a vicious circle that is ever more critical as scale increases. For instance, if you spanned a wooden metre rule across a gap it would be stiff and could carry a load much greater than its own weight. But if you increased all the dimensions a hundredfold, creating a plank spanning 100 m, the plank would sag and break under its weight alone. This is called *scale effect*. Suppose a 1 cm cube is increased in size to a 2 cm cube, and then to a 3 cm cube. The area of a cross-sectional slice goes up in the ratio 1 : 4 : 9, but the volume goes up in the ratio 1 : 8 : 27. Volume increases faster than cross-section, so to speak. As we increase the plank size, the volume and, hence, weight, soon outstrip the ability of the cross-sectional area, i.e. the plank's 'thickness', to cope. Weight overtakes strength. In nature, it is the reason why spiders have thin legs and elephants have fat ones. In building structure, it is the reason why a room can be economically spanned by a wooden joist, while the factory roof-span needs, say, an open-web truss, and a long river span, the ultimate efficiency of a lightweight suspension bridge.

But stress calculations do not define the shape of structure, they simply allow an exactness of sizing so that available strength is tailored to the loads with maximum economy, always allowing a known margin of safety. Exactness checks and trims the idea, it gets it right the first time, but we do not need the same exactness to understand the idea. The vaulted roofs of Sydney Opera House needed computers to solve the structural problems they posed and yet their final shape reflects the first idea of an architect watching the sails of yachts in a bay.

THE ARCHITECTURE OF COMPRESSION

As soon as man learnt to pile up stones for shelter, he started to play the compression game. Stone is strong under compression but pulls apart fairly easily under tension and, moreover, stones in a wall lock together under compression but under tension can pull apart at the joints. The need to keep stones or bricks, or other blocks, under compression has influenced the shape of every masonry building ever built.

HOW MATERIALS RESIST LOAD – THE WALL

Think of just one stone at the base of a wall. It must carry its share of the weight of the stones above – the wall's *deadload* – and so must 'press up' as much as that weight is pressing down. If it pressed too little, the wall would squash, if it pressed too much, the stones above would presumably fly away. But it presses just the right amount and the system is stable. And, if someone sits on the wall, a minute part of their added *liveload* must reach the stone and the stone obliges by pushing up that fraction more. The system is still stable. But how does the stone and, for that matter, the earth below it, always know how much to push up? The answer to this apparently silly question has only been known for the past few decades, in fact: it lies at the microscopic scale of the atom and sheds light on the whole nature of structural strength.

The molecular 'grid'

The component molecules of every substance are connected by forces which, for our purposes, can be regarded as magnetic springs. The rather notional three-dimensional grid (1.2) gives the general idea. It is these springs which compress or extend when a material is squeezed or stretched – a tiny deformation often, but it is there, for nothing is ever truly rigid. Even a marble slab is like a very stiff internally-sprung mattress and will compress locally as an ant walks over it, as will the ant's feet. The world is a springy place. So, in fact, structure – materials – do not actively 'push back', but they deflect and, in doing so, find their passive reaction to loads placed upon them. You can see the deflection of a tree branch when you swing on it, and the flapping of an aircraft's wings when it hits air turbulence, and you can just about feel the sway in some tall steel-frame buildings when the wind blows strongly. Wood and steel are relatively elastic. You will not notice the sag in a stone-arch bridge when a car crosses it, nor the compression in the wall stone, but they are there.

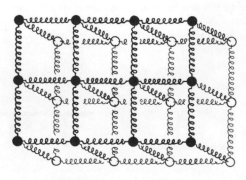

1.2 All matter is 'springy'

Stiffness and strength

However, just because one material gives more, i.e. is more elastic than another, it is not necessarily weaker. A brittle dry twig will snap under load long before a green springy one, although the green one may, at first, have seemed the weaker. So *stiffness* is not the same as *strength*, although both qualities are needed in the materials we use for building. We will come back to this distinction later, particularly when talking of beams and frames.

Stress and strain

If the load on a stone is two tonnes, the stone must react with two tonnes' force, but without knowing the bearing area of the stone, we cannot tell how hard the material is having to work (1.3). Instead, we must think in terms of *stress*. If the stone's bearing area is 1000 square centimetres (cm^2), then the compressive stress in the material is 2 tonnes (or 2000 kilograms) force per 1000 cm^2 – i.e. 2 kgf/cm^2. The units do not matter. The same information could be given in pounds force per square inch or dynes per square centimetre, for stress is simply a ratio of load to bearing area – much more useful. Strain is the linear distortion in response to

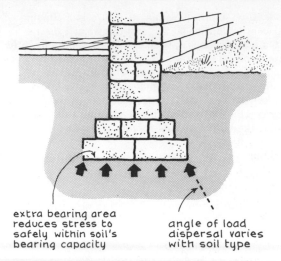

extra bearing area reduces stress to safely within soil's bearing capacity

angle of load dispersal varies with soil type

1.4 Foundation, i.e. increased bearing, reduces soil stress

Shear

It may, at first, be hard to see how compressive stress could ever crush stone, instead of just compressing the atoms more tightly together. In fact, what happens at the point of failure is that the stone squashes out sideways, rather as clay will under the weight of a modeller's hand, except that stone, being more brittle than clay, fails by cracking or even a virtually explosive release (1.5). The compressed material is actually failing

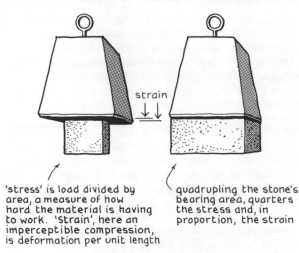

'stress' is load divided by area, a measure of how hard the material is having to work. 'Strain', here an imperceptible compression, is deformation per unit length

quadrupling the stone's bearing area, quarters the stress and, in proportion, the strain

1.3 Load and bearing area, stress and strain

1.5 Compression induces inner tensions and eventual failure by shear

stress. Nails are pointed so that the impact load of the hammer is concentrated on a tiny area of the material pierced – maximum stress and strain to yield point, for minimum load. Obviously, the same load that produces a high stress in a thin wall will produce a safer, lower stress in a thick one. Foundations are made wider than walls to reduce compressive stresses to a level that weaker subsoils can bear. In contrast to the nail, the foundation (1.4) spreads the load, preventing the wall from punching through the ground, which would cause settlement and eventual building collapse.

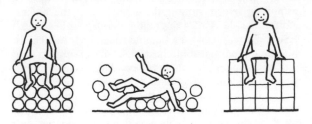

1.6 Masonry is more stable and strong if well fitted . . .

by *tensions* induced within it. There is a slipping of adjacent molecular layers, and this we call *shear*.

On a larger scale, a wall can fail at the joints if the stones roll off one another (1.6), sliding into a heap, as a pile of cannon balls would. This is an extreme way of illustrating that a wall made of irregular boulders is weaker than one built with accurately cut rectangular blocks – the strength of early cathedrals owes much to the workmanship of long-forgotten masons.

Bonding

A masonry wall is also stronger if the horizontal courses of stones or bricks are staggered, so that the vertical joints are not continuous (1.7). If they were continuous,

1.7 . . . and more stable and strong still with an interlocking bond

the wall would act like a series of adjacent, vertical piers, rather than as a cohesive unit. The interlocking bond has the further merit of dispersing point loads evenly.

THE NEED FOR STABILITY

But readers more familiar with structures may be getting worried by this focus on direct, compressive stress, knowing that masonry walls are usually too strong to fail by compression alone. In truth, the real danger is instability, overturning and buckling, and the higher and thinner the wall, the greater the danger becomes.

Suppose you lean on a flexible cane (1.8). As long as the cane remains absolutely straight, it can support you but, as soon as the slightest bending occurs, things become suddenly worse and over you go. If only you could be sure of keeping the direction line of your weight acting exactly through the centre of the cane down to the ground, you would be all right but, invariably, some slight twist of your hand on the handle,

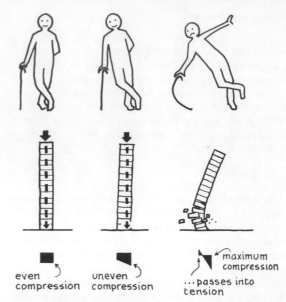

1.8 Instability owing to buckling

or some irregularity in the cane, will get the buckle started. We call this situation *dynamically unstable*.

So, either the wall can get out of line with its load or the load can get out of line with the wall. Take the first case. A vertically loaded wall is all nicely in compression – the stress distribution is shown diagrammatically by the shaded even band across its base. But, suppose the foundation is on poor subsoil and the wall starts to tilt. The load line acting vertically down from the centre of gravity – the centre of mass – of the wall moves towards the outward tilting face. The stress distribution alters until either the overstressed edge crumbles, making a bad situation worse, or the load-line passes outside the wall causing, horror, the face on the other side to go into tension, so that the joints open and the wall buckles and fails.

The second case, where the loads get out of the vertical, happens all the time. Horizontal wind forces or secondary, sideways, thrusts from roofs and floors are threats, again tending to tilt the main load-line. In other words, as far as the wall is concerned, the *component* forces on it, its own vertical loading and any secondary, sideways, loading, have the combined effect of a single force, the *resultant*. For example, imagine a 3 kg brick stood on end and subjected to a sideways thrust of half a kilogram force from wind (1.9). The brick weight and wind force can be represented in magnitude and direction by appropriately scaled arrows – it does not matter what units are chosen as long as the proportions are right. The magnitude and direction of the overall resultant can then be found by simply connecting up the remaining side of the force triangle. Combining component forces into a resultant is a structural shorthand.

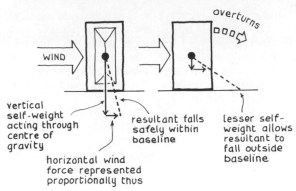

vertical self-weight acting through centre of gravity

horizontal wind force represented proportionally thus

resultant falls safely within baseline

lesser self-weight allows resultant to fall outside baseline

1.9 Triangle of forces. Self-load and wind-load components and their resultant, for 'heavy' brick and 'light' box

The stronger the wind, the more the resultant tilts away from the vertical: the heavier the brick, the less it tilts. As long as the resultant falls within the bearing area, i.e. the base, the brick will stand. Well, it would take a gale to blow over a heavy brick, but a cardboard box the same size with its low weight component would blow over in a breeze. The double benefit of having a thick, low wall is, therefore, that the thickness adds weight, keeping the resultant near the vertical, and that the geometry is such that the resultant has further to tilt, in any case, before it falls outside the wall width.

Thus, the tower wall (1.10) is stepped, not only to increase the bearing area to match the downward increasing loading but, more important, to tailor the wall face to contain any load-line tilting. The original builders may not have known all this, but their craft and native wit must have told them that, if they built this way, their walls would stand.

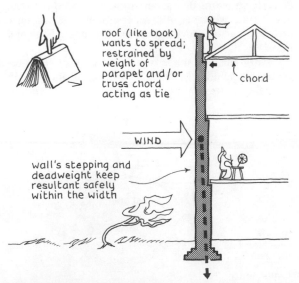

roof (like book) wants to spread; restrained by weight of parapet and/or truss chord acting as tie

chord

WIND

wall's stepping and deadweight keep resultant safely within the width

1.10 Wall is stepped for stability as well as to increase bearing area . . .

Effective thickness

Further, the fact that the walls tend to act in concert and not in isolation, adds to the overall stability. Just as the book stood on end is more stable open than closed, so a building is stabilised by the right-angled 'cellular' interconnection of its walls on plan (1.11). Any

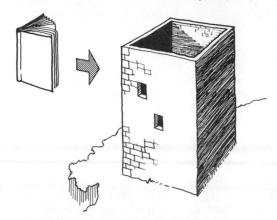

1.11 . . . wall plan arrangement adds stability – book is stable, so is tower . . .

wall of the tower will be helped in resisting wind load or other forces by the side walls flanking it. The flanking walls restrain the first wall, giving it a greater *effective thickness*. The garden wall (1.12) has greater effective thickness if recurring piers are built at intervals along its length and the retaining wall is stronger if it zig-zags on plan.

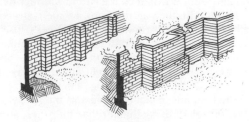

1.12 . . . and so is wall with piers or stepped on plan

THE COLUMN

Given that any self-respecting wall has strength and to spare to resist downward loading, it can easily cope with reductions in its bearing area caused by door and window openings. The openings are simply topped with lintels, short beams, and the loads shed round them and down to the ground. Logically, we can continue this opening-up process until the loads are concentrated into a row of separate supports, so that solid construc-

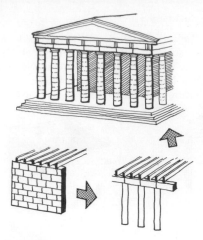

1.13 The Parthenon

tion gives way to the skeletal or trabeated construction of column and beam (1.13). Columns cannot be built as high as walls of the same thickness owing to the greater internal stresses and because buckling coming from any direction becomes that much more of a problem. But they bring interesting new possibilities to the compression game, allowing larger open spaces inside a building, i.e. uninterrupted by intermediate support walls and, on the façade, allowing a lighter façade of recurring columns and load-free, open bays.

Greek architecture, in all its surpassing elegance, exalted the column. There is no real occasion for discussing aesthetics here – and nor, some may add, the scholarship – but it would be meagre to consider buildings like the Parthenon in terms of their structure alone. One can but wonder at the Greeks' use of entasis, an almost imperceptible swelling of columns, to improve apparent proportion. A straight beam would appear to sag, they thought and therefore, the Parthenon's architraves, running along the façades above the columns, silhouetted against the sky, would give the illusion of being thinner than the others and so were made slightly thicker. And all the columns, to correct the illusion of their falling outwards, had their axes inclined fractionally inwards – the axes projected would meet at a single hypothetical point some three and a half kilometres above the temple, a rare refinement in a building started some 450 years before Christ. It is also interesting how a tradition that was to influence those that followed it, itself visibly reflects an earlier tradition of building in timber. There is no functional reason for the stone capitals above the column heads, they simply echo a time when a hardwood insert prevented a timber post from crushing into the side grain of a softwood beam above. The stone column bases found in some temples derive from hardwood inserts, preventing ground moisture and insects from attacking the vulnerable timber above.

But a cardinal point is how closely the columns were spaced – at barely more than 4 m intervals. We have said that stone is weak in tension, and beams mean tension: exactly why, we shall see later. If a stone lintel were too long or too heavily loaded, it would crack in its underside and collapse. The Greeks could span only the shortest distances in stone, and this placed a fundamental restriction on the planning of their buildings. So how can one span sideways in masonry, but all in compression? The next development in the compression game is not so much a step as a gigantic leap.

THE ARCH

1.14 Hardly a difficult problem

The solution to the support problem in the diagram (1.14) seems so obvious that it is curious no one thought of it earlier. Of course, the same might be said of something like the wheel and, certainly, when the Etruscans in the mid-west part of Italy started to use the arch – around the time the Greeks across the Ionian sea were building the Parthenon – it was thought such an unlikely structure that superstition attributed it to the devil! But, whatever the superstitions, the masonry arch is a practical device capable of spanning 50 m and more.

Clearly, the arch shape is such that each stone is prevented from falling by the stones on either side of it. In 1.15, the centre keystone, in trying to slip down into

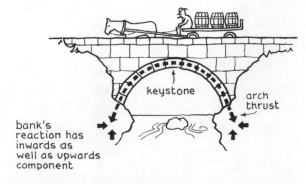

keystone

arch thrust

bank's reaction has inwards as well as upwards component

1.15 The arch's basic mechanics

the river, comes up solid against its immediate neighbours and, therefore, resolves its load sideways onto them. This load resolution continues to the banks, effectively forming a compressive line that locks the system together.

But the arch has an Achilles' heel. Like the propped book, it tends to develop an outwards thrust at its base, which must be contained. If the river bank did not push inwards, the arch would spread outwards, unlock and fall. The flatter the arch, the greater the outward thrust becomes. With rock river banks, a low-rise arch is all right but, where the support is softer, or where an arch drops onto the top of a wall, it needs a higher rise to help bring the thrust closer to the vertical.

The arch can also fail by buckling. A heavy load on one side of the bridge could, conceivably, cause the other side to lift. In practice, unless the arch is too flat, its geometry and deadload should keep everything safely in compression. The horizontal parapet helps by stiffening the arch with an action similar to that of the tower's vertical flanking walls. Most compression arches are *built-up* arches, that is to say, are stiffened by the walling around them.

Not surprisingly, the Roman engineers made the arch their own. It was a practical device for practical minds and recurred across Europe in their enduring structures, in buildings, bridges and aqueducts. The Pont du Gard aqueduct in southern France (1.16) was completed in AD 14 as a link in a 40 km water-supply channel to Nimes. Notice how the loads progressively shed down through the tiers of arches to the river level some 48 m below. Lateral stability comes mainly from the increased width of each tier towards ground level and, in the centre span, from the heavy side buttresses. Occasional projecting stones are a remaining clue as to how the aqueduct was built. An arch must be complete if it is to work at all and, therefore, cannot be built up from the ground like a wall but needs some temporary support until the keystone is in place. The stone nibs are the points where the original timber support framework, the centring, came to bear.

laterally stable since wider and buttressed towards base

1.16 Pont du Gard aqueduct, southern France

But, so far, we are still limited to two dimensions, building vertically with the wall and column and, horizontally, with the short beam or arch and this, inevitably, limits the size of the space enclosed. In other words, the distance between the opposite walls of a building can still only be as great as the maximum span of the timber roofing or flooring between. How can a three-dimensional enclosure be achieved all in compression?

THE DOME AND THE VAULT

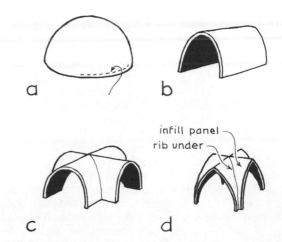

1.17 The arch as enclosure, (a) Dome (b) Wagon vault (c) Groin vault (d) Gothic ribbed vault

The dome is really only an arch rotated on plan (1.17). The Romans used domes but, moving on to the sixth century, the great Santa Sophia in Constantinople – Istanbul – took the compression dome near its practical limits. Domes, like arches, tend to spread at their bases. Much later, architects like Wren at St Paul's, London, used restraining chains around dome bases to avoid the need for side buttressing. At Santa Sophia, the Byzantines used auxiliary half domes and massive side arches to restrain the main dome over the nave. The result was a column-free enclosure 69 m × 33 m on plan and 55 m high, as large as anything until the steel-framed industrial buildings and railway stations of the late eighteenth century.

And then, there is the vault, as shown. The simplest type is the *wagon vault* but it has the disadvantage of needing continuous support walls, curtailing the internal planning freedom. The *groin vault* is an intermediate solution, but the *rib vault* is the cleverest – masonry infill panels arching onto pointed arch ribs which, in turn, carry the load to the single corner supports or columns. The rib vault was the more significant in that it freed the planning from the restrictions imposed by the circular arch and dome, allowing extended rectangular plans to be freely roofed.

THE GOTHIC CATHEDRAL

The stone vault completed the Gothic master masons' kit of parts. Now, they could not only build high buttressed towers, and walls pierced with pointed arched windows but, also, could extend the roofs of the long cathedral nave and intersecting transepts as a series of vaulted stone bays, repeating over the rows of columns inside. The cathedral could reach upwards and across, a spectacular three-dimensional stone skeleton held together by gravity alone (1.18).

The structural pattern is, again, clear. The loads shed downwards from vault panel to rib, to column, to foundation. The outward thrusts from the high nave roof are contained by flying buttresses, which thrust inwards like massive arms from the main buttresses straddling the building. The main buttresses have pinnacles on their shoulders, not just as ornaments but for useful deadweight and, again, have stepped increases in thickness towards their base, all serving to keep the resultant forces near vertical and safely within the structure. The skeleton, therefore, concentrates the main load-carrying elements to regular intervals along the plan, allowing the bays between to be opened up. The nave space can flow transversely through the arched colonnade to the side aisles, and the external wall bays can be relatively light and pierced with large, traceried windows.

These arrangements characterise virtually all the great European cathedrals, products of a highly developed building tradition practised by master masons who travelled widely, offering their skills. It was through this kind of daring and applied tradition that the Gothic magic was achieved: soaring structures that remain the highwater mark of building in stone – of the compression game.

For it is a point of history that the architecture of the Renaissance which followed acknowledged that, structurally, there was little more that could be achieved in stone. This was one reason for the return to the classical Greek and Roman styles. There were profound intellectual and social changes, as castle and cathedral gave way to country house, but structural advance awaited materials strong in tension as well as compression. Timber was only moderately strong and, being difficult to joint well, could not be built up into large structural members. Modern glues have changed all that but, historically, the story was one of short spans until the arrival of steel, which was strong and could be bolted and welded into large trusses and frames.

Summary

- Masonry structures are primarily compressive.
- All structure is elastic: stiffness is not the same as strength.
- The load on e.g. a stone induces compressive stress the magnitude of which is inversely proportional to the bearing area. Stress, internal force per unit area, is a measure of how hard a structural material is having to work. Strain, deformation, is the result of stress.

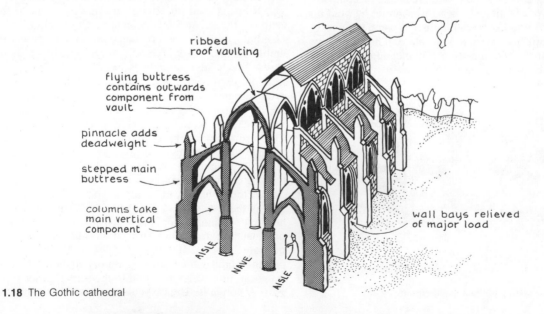

ribbed
roof vaulting

flying buttress
contains outwards
component from
vault

pinnacle adds
deadweight

stepped main
buttress

columns take
main vertical
component

wall bays relieved
of major load

AISLE
NAVE
AISLE

1.18 The Gothic cathedral

- Masonry in compression eventually fails, owing to shear, a slipping between the material layers.
- Bonding integrates masonry and spreads loads.
- Buckling is caused by resultant load-line passing outside effective wall width: prevention is by adequate masonry mass to swamp horizontal thrusts (the force triangle) and by adequate wall width.
- Piers, buttresses and plan articulation increase effective wall width.
- Columns free the masonry building's floor plan.
- The compression arch increases the lateral span – and leads to the enclosing dome and vault.

BEAMS AND FRAMES

BEAM THEORY

If we stand on a beam (1.19a), we cause it to bend and, remembering that the beam, like everything else, is just one vast molecular grid, it follows that the grid is being distorted against its will. In fact, it distorts until the beam's internal protesting forces balance our external

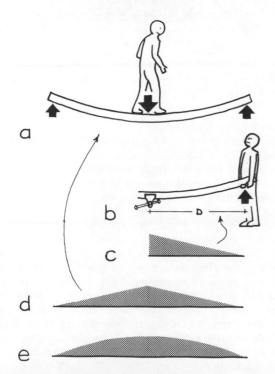

a

b

c

d

e

1.19 Beam deflection and bending moment

load. The beam deflects in order to support and the way it diverts the load sideways – in this case to the supports – is what beam theory is all about.

The externally applied bending moment

We can, similarly, bend the beam by clamping the middle and pulling up on the end (1.19b). If we could pull hard enough, it would break, the break occurring at the vice and not near our hands. It is a matter of leverage, the principle of the crow-bar, really. The bending severity at any point in a beam is called the *bending moment* and is proportional to the externally applied loads and their leverage. For example, in the clamped beam, the maximum bending moment occurs at the vice, being equivalent to our upward pull multiplied by the whole lever arm length of the beam. Moving away from the vice, the bending moment reduces as the lever arm reduces, a state of affairs represented graphically in the shaded bending moment diagram (1.19c). The beam works hardest at the vice and least hard at the end. It follows that the bending moment pattern for the original centre-loaded beam (1.19a) is minimal at the supports and at its maximum in the centre – in fact, as shown (1.19d). If the load were too heavy or the span too long, the beam would eventually break, probably at the centre. Increasing load, or span, increases the maximum centre bending moment that the beam has to resist. Incidentally, if the beam is uniformly loaded along its length instead of point-loaded at its centre, the bending moment pattern becomes the gradual curve (1.19e).

The internal moment of resistance

The beam, for its part, supplies an *internal moment of resistance* which is the product of the internal stresses and *their* lever arms. Take the stresses first.

Bending stresses

The most critical stresses are longitudinal, running along the span. Here, it helps to think of the deflected shape. In bending, the top of the beam, i.e. the inside of the curve, must have got shorter, and the bottom, the outside, must have got longer (1.20), so the molecular grid must have been compacted at the top and

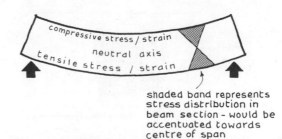

1.20 Longitudinal stresses in deflected beam

9

stretched at the bottom. This creates a maximum horizontal compression in the upper fibres, which reduces through a zero value in the centre or *neutral axis*, where the fibres are effectively unaltered in length, and increases to a maximum tension in the bottom. It is this bottom tension that bothered the early stone builders.

Shear forces

There are also secondary stresses which protest against shear which, as we have said, is the tendency of the adjacent molecular layers to slide over one another like a pack of cards. The effect is complex. The engineer tends to see shear forces through calculus. The architect is normally content with merely disapproving of them.

Basically, shear has two complementary components, *horizontal shear* and *vertical shear*. First, take horizontal shear. If we sliced the beam as shown (1.21a), its separate layers would have less strength than the original whole, because they could slide relative to each other like a rather unstructural ham sandwich. In the same way, a dozen or so pages of this book are less stiff than a single sheet of cardboard of the same thickness. The horizontal shear stress that builds up in the beam is the result of the wood's protest against this tendency

to slide. Then, there is vertical shear stress, which is the wood's protest against the beam's natural tendency to escape its loads by slicing down through its section (1.21b).

Vertical and horizontal shear have a combined effect best explained by looking at how a random square chunk in one side of the beam is being distorted (1.21c). The horizontal and vertical sliding tendencies, represented by the slip arrows, try to distort the square into a parallelogram, which means that the diagonals of the original square are respectively stretched and compressed. This is the key. The combined effects of shear are setting up secondary compressive and tensile stresses *diagonal* to the beam's main axis. Unlike bending moments, which are maximum in the centre, the shear stresses are maximum towards the ends of the beam, which is logical if you visualise the way the slippage and distortion occur.

So this gives us a beam picture (1.22) – rather simplified – of longitudinal stresses doing most of the work and secondary shear stresses criss-crossing between. But we are not quite there yet. Hold this picture in mind.

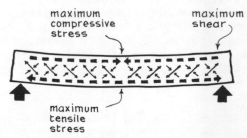

1.22 Simplified pattern of resultant stresses from longitudinal and shear stresses combined

Internal leverage in beams

The beam's internal moment of resistance, and that is what we are ultimately interested in, depends not only on these stresses but also on *their* leverage. Think how much easier a wooden ruler is to bend with its section lying flat than with the section on edge, although the cross-sectional area of wood resisting the bending is the same in each case. The diagram (1.23) shows why. Think of the ruler as being in two halves. The applied load and its leverage would tend to rotate the halves as shown, but they do not, so we can take it that the internal stresses (shown simplified), and their leverage, are producing a balancing counter-rotation. A deeper beam means greater internal leverage and, hence, greater strength. However, there has to be enough width to provide torsional stability, i.e. keeping the sectional depth upright and, therefore, structurally in play. Sectional twisting towards the span centre is a form of buckling which can cause a beam to fail.

1.21 (a) Horizontal shear (b) Vertical shear and (c) Their combined effect

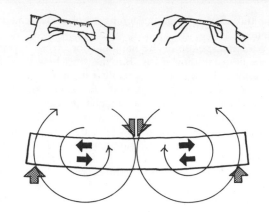

1.23 External bending moment is balanced by internal moment of resistance

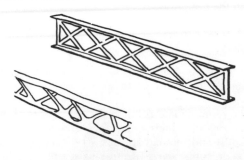

1.24 Anticipated stress pattern, and the need for sectional depth to lever, reflect in shape of truss and wing bone

The reasons for the shape of the lightweight metal truss (1.24) or, to pluck an example from nature, the metacarpal bone of the vulture's wing, now become clear. Top and bottom chords take the main longitudinal compression and tension, and the lighter diagonal members, the secondary shear. And the diagonals ensure that the main chords act together while yet allowing them to be as far apart as possible so that they can lever.

Triangulation
Now, before we look at other beam types, two points in passing. First, the truss diagonals raise the important concept of *triangulation*. A triangle is the only multisided geometric figure that is automatically rigid, and it crops up everywhere in structure. You can distort a square by pushing on one of its sides, but not a triangle and, indeed, the easiest way to stiffen a square is to place a diagonal brace between opposite corners, effectively turning it into a pair of triangles. In the truss and wing bone, triangulation and shear resistance are related – it is no accident that the necessary shear resistance between the chords can be most efficiently developed if the secondary members are triangulated.

The distinction between stiffness and strength again
Second, we come back to the distinction between stiffness and strength. Understandably, people tend to think of things as being made strong enough not to break. But, in fact, when architects or engineers are sizing things in relatively elastic materials like timber, steel or even reinforced concrete, they are usually far more concerned with limiting the deflections. For example, a timber-joist floor strong enough to support its loads perfectly safely would still be insufficiently stiff if it sagged so much that the plaster ceiling underneath it cracked and fell. Steel-frame buildings must only deflect a limited amount if they are to be regarded as stable. A steel bridge must not sway unduly in wind. It is perfectly all right to talk about strength as long as we remember that, in design, *stiffness* and not ultimate breaking strength is the main concern.

Beam types

Moving from left to right in the diagram (1.25), we start with the simple timber joist. It has a sensible, maximum span of 5 m or so and does not bother with the efficient weight-saving refinements of the truss, although the stress patterns it has to cope with are similar. The next step is to shift material away from the neutral axis to where it can better lever. The timber box beam does this by having main top and bottom chords connected by a plywood side webbing, making it much lighter than the solid joist of equivalent strength. In the I-section universal beam, the molten steel ingot is passed through rollers in the steel mill to give it heavy top and bottom flanges connected by a lighter web. Hollow steel tubing resists bending in all directions. Really, it is like an I-section rotated, and we see similar tube shapes in nature, in grass or bamboo shoots, or in our bones, which have the structural part around the outside and the soft marrow inside. The castellated beam is developed from the I-section – the web is sliced longitudinally in the factory with the zig-zag cut, and the resulting peaks on the cut edge of the lower half are welded to the peaks on the top half, i.e. giving a deeper

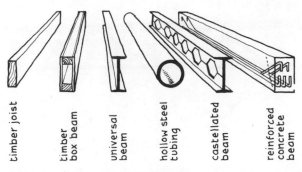

timber joist

timber box beam

universal beam

hollow steel tubing

castellated beam

reinforced concrete beam

1.25 Some beam types in building

11

beam for the same weight of material. The very shape of the resulting holes usefully distributes steel away from the neutral axis. The reinforced concrete beam is a composite. Concrete, like stone, is strong in compression but weak in tension and so the beam has steel rods embedded in its lower section to supplement its tension-resisting deficiencies. In a typical casting process, reinforcing steel rods are laid along the bottom of a temporary timber or steel 'shuttering', i.e. mould, and the wet concrete poured in. The mixture usually comprises a graded stone 'aggregate', sand, cement and water, which hardens in anything upwards of a few hours to grip the rods and form the strong composite-action beam. The shuttering is later removed. At the end of the beam, where the bending moment and, hence, bottom tension is less, some of the rods can be diverted upwards to shoulder the new role of resisting the increased diagonal shear tension there – otherwise, the concrete might crack along the dotted line as shown.

And to digress for a moment, the reinforcement in concrete can also be *pre-stressed*. Clearly, ordinarily reinforced concrete requires some deflection (often with slight cracking) before the tensile contribution of the steel comes effectively into play. But limiting the deflection is generally the primary task of useful strength and, if the reinforcement can be already under tension with the element in the unloaded condition, then the 'strength' against deflection will be ahead of the game and enhanced. Common ways of pre-stressing are to heat and, hence, expand the steel and clamp it either end before it contracts on cooling, or to tension the steel by jacking it between end anchorages – all before pouring the concrete into the surrounding mould. *Post-tensioning* is where the reinforcement is tensioned after the concrete has hardened, for example, by having the end projecting from the concrete,

and threaded so that tensioning turnbuckles can be applied.

We have spoken of the parallel-chord truss, typical in medium building spans. But the steel bow truss of the railway bridge (1.26) is really a beam theory summary. As before, top and bottom chords take the main longitudinal stresses and a diagonal lattice connects them, triangulating the whole thing and resisting shear. But, as an additional rationalisation, the profile is made to follow the anticipated bending moment distribution for uniformly loaded beams. The chords are furthest apart, i.e. have maximum leverage, where the bending is most severe. And notice how the lattice gives way to a solid web at the ends, where the shear forces are most severe. Obviously, the bridge is relatively complicated to make but, in a long span, the structural efficency and saving in self-weight which results makes this worthwhile – it would plainly be ludicrous to make a floor joist in the same way.

Summary

- A beam is subjected to an external bending moment proportional to span (i.e. leverage) and load.
- The resulting internal moment of resistance is proportional to internal leverage (related to beam depth) and internal stresses – primary stresses of compression above neutral axis and of tension below, and secondary stresses resisting horizontal and vertical shear.
- Beam design, ideally, directs most material towards top and bottom chords, with lighter web between: web has secondary related role of resisting shear – truss diagonals reflect this and, also, reflect concept of triangulation.
- In practice, beams have to be stiff enough to avoid excessive deflection rather than strong enough not to break.

MODERN STRUCTURE AT DOMESTIC SCALE

We can now turn for a while from theory to practice and see how the foregoing compression game and bending principles apply to the commoner building types. It will also give us a chance to touch on some building construction.

The brick–joisted, 'traditional' house

The two-storey house (1.27) is sometimes described as 'traditionally' constructed in the UK, though the US term 'brick–joisted construction' is rather better, today's construction being increasingly rationalised,

triangulated lattice unites top and bottom chords, resisting shear and, hence, allowing their respective compression and tension to lever

overall profile reflects anticipated externally-applied bending moment. Maximum leverage and, hence, internal moment of resistance at centre

webbing reinforces against maximum shear at truss supports

1.26 Bow truss – virtually a beam theory summary

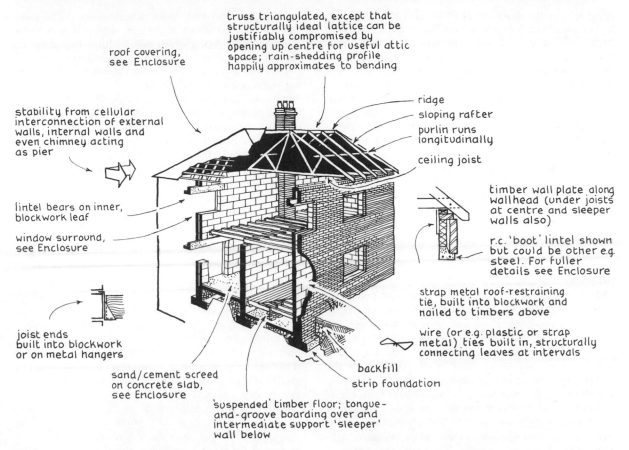

truss triangulated, except that
structurally ideal lattice can be
justifiably compromised by
opening up centre for useful attic
space; rain-shedding profile
happily approximates to bending

roof covering,
see Enclosure

stability from cellular
interconnection of external
walls, internal walls and
even chimney acting
as pier

lintel bears on inner,
blockwork leaf

window surround,
see Enclosure

ridge
sloping rafter
purlin runs
longitudinally

ceiling joist

timber wall plate along
wallhead (under joists
at centre and sleeper
walls also)

r.c. 'boot' lintel shown
but could be other e.g.
steel. For fuller
details see Enclosure

strap metal roof-restraining
tie, built into blockwork and
nailed to timbers above

wire (or e.g. plastic or strap
metal) ties built in, structurally
connecting leaves at intervals

joist ends
built into blockwork
or on metal hangers

sand/cement screed
on concrete slab,
see Enclosure

'suspended' timber floor; tongue-
and-groove boarding over and
intermediate support 'sleeper'
wall below

backfill

strip foundation

1.27 'Traditional' brick-joisted house – some typical construction features

using standard, factory-made, lintels, roof trusses and other components.

The foundation, formerly a widened masonry footing, is now a *strip* foundation of poured concrete. The excavation trench acts as its permanent shuttering. Strip foundations at domestic scale are not normally reinforced, since the stresses are mainly compressive – and steel is expensive. The only case for reinforcement is where the soil has uneven bearing strength and where the reinforced strip's ability to act as a beam adds valuable stability.

Then, there is the concrete oversite slab, a 100–150 mm layer of concrete sitting directly on the ground or, rather, on a made-up base of *hardcore*, i.e. stone rubble, possibly topped by a more finely graded *blinding*. Sometimes, the slab acts as the ground flooring itself and, sometimes, there is the traditional timber suspended floor above it. If the slab area is small, say, where it is confined to the kitchen and wet utility areas as shown here, it is not normally reinforced but, where it comprises the whole ground floor, as is increasingly

the case nowadays, a steel mesh reinforcement adds stability and has the advantage that internal block walls can be built directly off it rather than having to descend through it to their own strip foundations. There is a topping layer of *screed* – sand and cement but no aggregate – poured as a fine level floor finish.

The suspended ground floor joists can reduce in size if they are relieved at mid-span, say, by masonry piers or a honeycomb *sleeper* wall – 'honeycomb' meaning that gaps are left between adjacent bricks or blocks along each course, primarily to save material, though the resulting voids must also allow freer ventilation through the underfloor space. (See Chapter 2 *Enclosure*). The sleeper wall will probably have a timber *wall plate* (plank) running along its head, offering an even bearing and possible nail fixing for the joists.

The external walls are *cavity construction*, two leaves and an air gap between. The main purpose of this is to keep water out and heat in but, again, more on this in Chapter 2 *Enclosure*. The outer leaf is shown as brick – reasonably dense, water-resistant and durable. The inner leaf is blockwork. The typical block is of mineral composition, burnt clay or foamed concrete, and may be hollow. It is larger and lighter than brick, and cheaper – particularly, since it is faster to lay. Being less dense, it is a better insulator, but is still adequately strong for domestic loads (in fact, there are high-strength types for other applications). The leaves are braced together by metal ties built in at intervals of a metre or so. This gives the wall greater effective thickness and, hence, stability than it would have if the leaves acted independently. The window and door openings are topped by lintels, small beams, in effect. Common types are galvanised steel and reinforced, factory-precast, concrete. (Incidentally, concrete components can be cast in place, often known as *in situ*, or they may be factory-*precast*).

The lateral stability of the whole house comes from each wall's weight and width and, again, by all the walls' cellular interconnection on plan. For example, no matter which face the wind hits, there will be walls at right angles, i.e. parallel to the wind direction, to stabilise. The internal walls will also help, and even the chimney, acting as a heavy pier.

The external and internal walls which support the horizontal floor and roof spans are said to be 'structural' or 'supporting'. You can normally tell which are the supporting walls in a traditional building by lifting the carpets and seeing which way the floor boards run – the joists must be at right angles underneath the boards, spanning to the supporting walls at either side. The house illustrated has a simple, central spine wall.

The floor joists are, of course, shaped deep in section to maximise their strength. As already explained, it is not the intention to include detailed calculation methods in this book but, in any case, structural sizing at domestic scale is as much by accepted good practice and rule-of-thumb. The typical timber floor has joists sized around 225×50 mm and at 500 mm intervals, *centres*, on plan. The maximum centring is not only governed by the need to get adequate structural timber into the floor but, also, by the obvious need to limit the span intervals of the floor boards going the other way.

The floor boards are tongue-and-groove, interconnecting along their edges so that, if one of them deflects, its neighbours deflect also. This helps spread impact loads, like people jumping, and, of course, the heavy point loads of the grand piano legs or similar articles. That said, various types of reconstituted timber sheet floorings are increasingly used.

The pitched roof is framed in timber, hipped to give the familiar sloping triangular sections at each end. A row of triangular trusses shapes the pitched section between, with longitudinal purlins connecting them. The pitched shape may be mainly for drainage but is useful in creating space for water tanks and, provided the trusses' web layout permits it, an open attic storage space as well. The maximum truss depth in the centre helpfully corresponds to the maximum bending moment there. The bottom chord is both ceiling joist and tie – a tie because, otherwise, the roof construction

could fall by spreading outwards at its base. The truss ends are nailed down to timber wall plates running along the wall heads which are, in turn, tied down to the wall by metal straps. For the roof to lift would require the top few brick courses to lift as well which, barring hurricanes, is unlikely.

People, not least finance companies, like 'brick-joisted' construction and, perhaps rather unfairly to other construction types, it has the reputation of being particularly durable – you know the kind of thing, 'a house built to last'. But there are practical advantages. The brick and timber components being small in scale are easily erected. In fact, a brick has always been sized so that a man can hold it in one hand while trowelling mortar with the other – and, as such, is indisputably a 'building system', albeit the modern 'building system' is taken to include very much larger components. There is greater design flexibility with small components – virtually any plan shape is possible in brick. Brick and timber can be easily cut on the job so that design tolerances do not have to be absolutely exact for things on the drawing board to fit together on site. Also, there is the ease of making later alterations: we need different things of our houses from time to time and a brick partition or timber floor framing is more easily removed, or pierced to accommodate new plumbing or electrical runs, than is the equivalent construction in, say, precast concrete. Certainly, there are heavy-component house building systems, for example with whole floor and wall panels either factory-precast, or cast in situ, using large steel shutterings which can be moved on from house to house. But they lack flexibility and need long runs to make the setting up of the process worthwhile. At the very least, they seem to have fallen short of their post-war promise.

The timber-frame house

Naturally enough, timber construction (1.28) is most common where the timber supply is plentiful, for example, North America and Scandinavia. Generally, there are two main types of timber frame, the *platform frame* and the *balloon frame*. The platform frame shown treats each storey as a separate construction, whereas the balloon frame runs the vertical wall studs the whole two-storey height from sole plate to roof plate, and nails the first-floor joists directly to the studs. In contrast to masonry walls, timber construction has little deadweight to stabilise the construction and, anyway, a whippy frame in a wind would soon lean far enough for its downward deadweight component to pass outside the base. Instead, stability comes from the frame's action as a stiff box. The sole plate is bolted down to the foundation, the walls act as stiff panels owing to the bending resistance of the members, and the whole is then stable from all the walls' interaction on plan. If, as is now common, the walls are sheathed

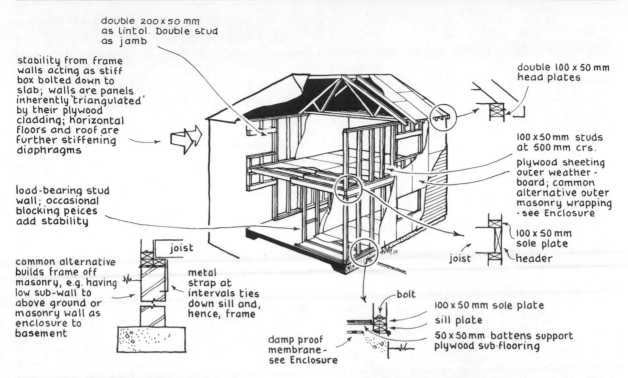

double 200 x 50 mm
as lintol. Double stud
as jamb

stability from frame
walls acting as stiff
box bolted down to
slab; walls are panels
inherently 'triangulated'
by their plywood
cladding; horizontal
floors and roof are
further stiffening
diaphragms

double 100 x 50 mm
head plates

100 x 50 mm studs
at 500 mm crs.

plywood sheeting
outer weather-
board; common
alternative outer
masonry wrapping
- see Enclosure

load-bearing stud
wall; occasional
blocking peices
add stability

100 x 50 mm
sole plate

joist   header

joist

common alternative
builds frame off
masonry, e.g. having
low sub-wall to
above ground or
masonry wall as
enclosure to
basement

metal
strap at
intervals ties
down sill and,
hence, frame

bolt

100 x 50 mm sole plate

sill plate

50 x 50 mm battens support
plywood sub-flooring

damp proof
membrane-
see Enclosure

1.28 Timber 'platform' frame

in plywood or other sheeting, this can act as a stiffening skin, inherently 'triangulating' the frame. Temporary diagonals would then be needed during construction, though, before the skin was added.

The frame is shown springing off an overall raft foundation. This is a simple construction method, although it requires the raft to be reinforced. The more traditional alternative is to have an ordinary strip foundation with masonry walling to just above ground level and to build the timber frame off that. The oversite slab inside is then structurally separate from the main frame loads and is unreinforced – and, in the persisting view of some, there is the added advantage that the foundation concrete is then safely below the ground frost level. In some countries, for example, the USA and Scandinavia, basements are common, i.e. with masonry or concrete walling below ground and the timber frame springing off above. Another construction gaining popularity is a timber frame as the inner stable structure with a single-leaf masonry external wrapping.

The terraced house

The modern terraced house (1.29), also known as the 'double-aspect house' or 'row house' in the USA, is a familiar form, with a long and respected pedigree. It works well from so many points of view. It is an increase on the small-house scale, but not on the scale

of the components used – large terraces are, generally, still brick-joisted. Running houses together with common party walls is economic in terms of structure, land use and – as we shall see in Chapter 2 *Enclosure* – thermal performance. Yet the form retains considerable design flexibility. In the terrace illustrated, each bay could comprise four flats, or two two-storey maisonettes, or a four-storey house. Sloping or irregularly shaped sites can be coped with by stepping the terrace on elevation or on plan, slipping each bay vertically or horizontally, relative to its neighbour on the other side of the party wall.

In modern terraces, the party walls are usually the supporting ones. This is called *cross-wall construction*, recurring parallel supports along the terrace length. They have to be thick enough, anyway, for stability, and for noise suppression and compartmentation against fire spread between neighbouring occupancies, so it makes sense to utilise their ample potential for support. Cross-walls keep spans short and regular, and fit in with the idea of the deep plan and correspondingly narrower frontage inevitably attractive to the developer anxious to achieve an economic number of units along a street. Often, the windowed front and back elevations, free of significant load, will be of lightweight construction. The diagram shows some typical constructional details.

Where there is more than one occupancy in a vertical bay – say, two two-storey maisonettes, an adequate noise and fire barrier is, again, needed between them.

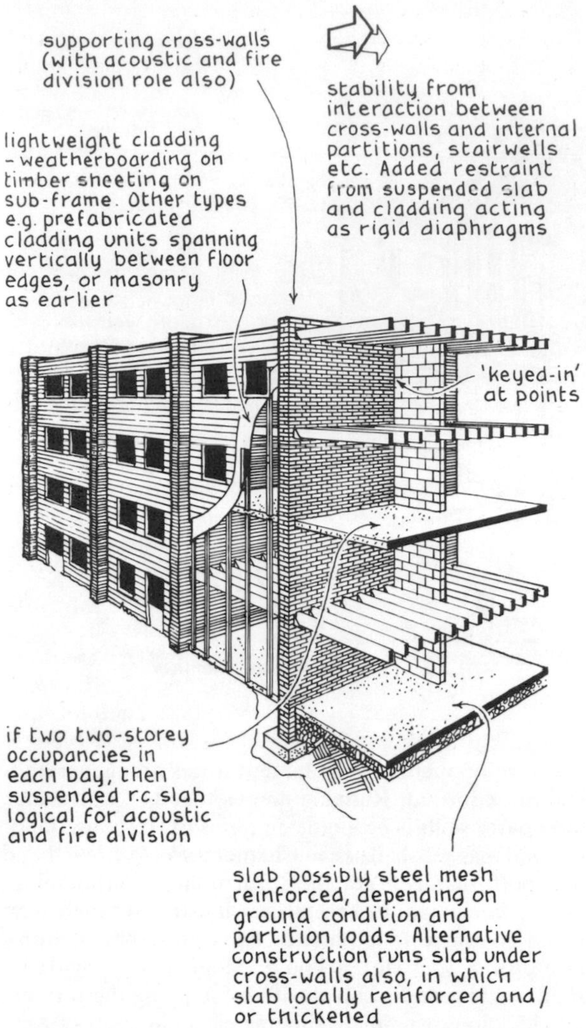

supporting cross-walls (with acoustic and fire division role also)

stability from interaction between cross-walls and internal partitions, stairwells etc. Added restraint from suspended slab and cladding acting as rigid diaphragms

lightweight cladding – weatherboarding on timber sheeting on sub-frame. Other types e.g. prefabricated cladding units spanning vertically between floor edges, or masonry as earlier

'keyed-in' at points

if two two-storey occupancies in each bay, then suspended r.c. slab logical for acoustic and fire division

slab possibly steel mesh reinforced, depending on ground condition and partition loads. Alternative construction runs slab under cross-walls also, in which slab locally reinforced and / or thickened

1.29 Terraced or 'row house' construction

Timber construction with sound deadening between the joists may satisfy both counts, but the concrete floor shown is the surer separation. The shuttering onto which the slab is poured would probably be made up of plywood sheet or other sections small enough to be manhandled into position over scaffolding propped from the ground floor. For variety, the illustration shows timber floors within the occupancies themselves though, in practice, there is the argument that with concrete at one level, there may as well be concrete at all levels and, if anything, this is now the trend. In large projects, it might pay to streamline the operation further by having a floor-sized 'table-top' shutter craned into position for each pouring and then moved onto the next.

The terrace is inherently stabilised by the right-angled interaction of the party walls with the elevational walls and/or the internal block partitions. Mind

you, it is often said that the joy of terraced houses is that they hold each other up, and there is truth in it!

INCREASING THE BUILDING SCALE LATERALLY

Longer simple spans

Larger single-enclosure buildings, factories, sports halls, call for longer spans. The simplest spanning system is, in fact, called the *simple span* (1.30), beams freely resting on supports under each end. The garage roof is a simple span. Ordinary steel I-section steel joists, may be used up to 15 m or so, but the greater efficiency of castellated and truss forms makes these a more likely choice as spans increase.

Granted one might expect the profile of the truss illustrated to echo the bending moment distribution for, surely, the parallel chords imply a superfluous depth, i.e. strength, at the ends? But, in fact, the roof is a compromise. A truss with parallel chords and regularly sized lattice members is far easier to make. It allows a convenient flat roof as opposed to a curved one – and, in one application, facilitates the familiar *northlight* roof shown. The resulting slight loss of efficiency, strength to weight, is not critical. Also, of course, the loading might not always be uniform: the trusses might easily have electric hoists or block and tackle attached at any point along their length, to lift car engines and the like, and this would move the maximum bending moment off centre. The trusses are shown stiffened against torsion by lateral strutting triangulating between them, by their end fixation at the walls, and by the secondary system of lateral T-section purlins: alternatively, there are proprietary triangular-section trusses inherently stable against torsion. The concentrated load at the truss ends is borne by concrete bearing pads set into the brick. Another construction is to bear the trusses on steel columns, stanchions, which carry the load independently of the enclosing wall and down to their own foundations.

Achieving more even bending-moment distribution

Beams can be shaped to suit the bending moment, but might it not be possible to shape the bending moment to suit the beam? Actually, this is not as structurally arrogant as it sounds. There are ways of arranging things so that the bending moments are more evenly distributed, ironing out that troublesome peak value in the centre.

The encastré beam

When a simple-span beam deflects, the ends must rotate ever so slightly at the supports. This is the clue

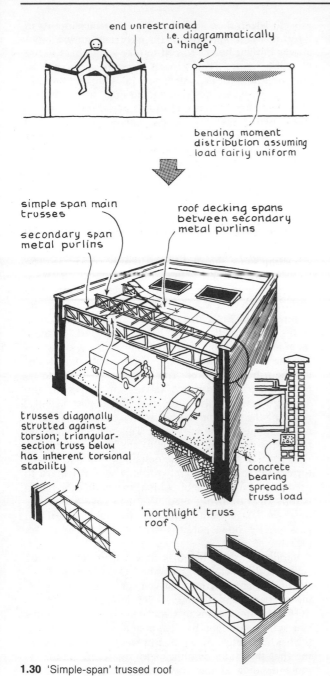

end unrestrained
i.e. diagrammatically
a 'hinge'

bending moment
distribution assuming
load fairly uniform

simple span main
trusses

secondary span
metal purlins

roof decking spans
between secondary
metal purlins

trusses diagonally
strutted against
torsion; triangular-
section truss below
has inherent torsional
stability

concrete
bearing
spreads
truss load

'northlight' truss
roof

1.30 'Simple-span' trussed roof

able weight of masonry to clamp a beam – for example, roof beams could never be clamped by the weight of a parapet alone, they would just revert to being simple spans and, possibly, crack the wall – and the ends of an open-web truss would, like as not, collapse under the high local shear and bending the clamping induces. So the method is really confined to solid beams or concrete floor slabs and, as such, to shorter spans.

The cantilever frame

There are more subtle solutions. The garage canopy (1.31) is a column and beam construction, a frame, the cantilevered parts of which are counteracting the centre portion's end rotation tendency with a counter-rotation of their own, a balancing act which effectively freezes the beam at the supports. As far as the centre portion is concerned, its ends are restrained as if they were encastré. There are now three bending-moment peaks but none is as high as would be the single peak in the equivalent simple span supported at the canopy edges. The bending has been redistributed and the peaks ironed out, allowing a lighter frame in consequence. The optimum position for the supports under a uniformly loaded beam is 1/6 of the beam length in from each end, at which point the three peaks are equal. Incidentally, notice how the points of *contraflexure*, where the beam is momentarily straight as it changes its bending from one direction to the other, are logically matched by the zero values on the bending moment diagram.

So, the overhanging first-floor gable of the mediaeval timber house is often helping to relieve the bedroom floor beams although, again, the builders would hardly have known it – or, if they did know it, they surely

for, if some way can be found to hinder the rotation, then the beam will deflect less. *It will have been effectively strengthened.* Looking at it another way, if the ends are fixed, they will automatically attract some of the bending to them.

The most obvious method of restraint is to build the ends of the beam into the support walls. They are then said to be *encastré* (fitted-in). But the applications of encastré construction are limited. It takes a consider-

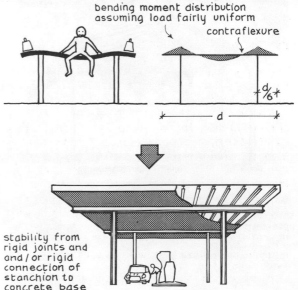

bending moment distribution
assuming load fairly uniform

contraflexure

$\frac{d}{6}$

d

stability from
rigid joints and
and/or rigid
connection of
stanchion to
concrete base

1.31 Cantilever frame relieves centre bending

would not have known why. Today, cantilever frames are often used in those buildings where the resulting parallel rows of columns set inside the side walls will not interfere with the inside activities – in a library, perhaps, or in a large dining hall. And the external walls, freed of primary load, then take only horizontal wind forces.

The portal frame

Rigid portals. The portal frame (1.32) is really rather clever. It frees the centre space of columns, and yet, still manages to redistribute some of the bending away from that top span. The essential thing about a portal is that the corners, or knees, are rigid. This means that any deflection of the top member must distort the legs

rigid 'knees', i.e. no hinge, so legs obliged to share some of the bending. If feet can be restrained against rotation, legs take further bending – if not then portal 'two hinged'

contraflexure

hinge

knees and centre ridge attract highest bending so locally strengthened

portals inherently stable in span direction; triangulating ties in wall for longitudinal stability

e.g. sheet metal cladding

screed

slab reinforced and possibly strengthened for heavy floor loads by rib-grid cast integrally under – see foundations later

stanchion welded to plate bolted to slab

cement 'grouting'

1.32 Two-hinged portal frame passes some of the bending to the legs

as well, which then act not only as vertical supports in direct compression – that is the easy part anyway – but also as upright beams forced to take some share in the overall bending. Their protest helps the top's protest, and the whole frame is stronger. Looking at it another way, the legs are restraining the ends of the centre span, and vice versa, more or less as did the cantilever frame overhangs. The portals bending moments are similar, almost as if you had simply taken the cantilever overhangs and turned them down to the ground. As you can see, the peak values are at the centre and at the knees, which explains why portal sections are thickened at these points for extra strength.

Hinged portals. The bending moment diagram for the first portal assumes that the feet are restrained against rotation – encastré as it were – and, therefore, attracting some share of the overall bending too. The factory portals could be restrained, for example, were they embedded in the concrete foundations supporting them, though that would call for pads heavier than the ordinary slab. More usually, they will be bolted down to the slab directly, or to a steel sole plate, locating them and preventing the legs from springing outwards like the base of an unrestrained arch – but not restraining them from rotation. The portal is then said to be *two-hinged*. We are not talking of hinges in the accepted sense, only joints that can deflect slightly, releasing the bending stresses so that the internal bending moments there are zero. The bending moment peaks then inevitably move to the more obstinately rigid parts of the frame, the parts that refuse to be distorted. Often, the amount of foot restraint is hard to predict, making it a moot point as to whether the portal is fully rigid or two-hinged. We then have a *statically indeterminate* structure (just to mention the term in passing), i.e. one where the stresses are hard to predict exactly.

Suppose a third hinge is now introduced into the centre of the portal's top member. One can be forgiven the immediate reaction 'but you can't, there's a bending-moment peak there', and yet, looking at the drawing (1.33), one sees instinctively that a three-hinge frame can stand. All that has happened is that the bending has been released in the centre and at the feet, meaning that the frame is obliged to take correspondingly more bending at the knees – the bending has to go somewhere, and this is why the knees of a three-hinged portal are so massive. The system has its advantages. It is *statically determinate*, meaning that the stress patterns can be foretold in design; it readily adjusts to any differential settlement of the feet; and, consisting of two halves, it is less cumbersome to manufacture, handle and erect, especially where the frame is factory-prefabricated or precast and then transported by lorry to the building site. Such portals would typically be in reinforced concrete or laminated timber – timber built

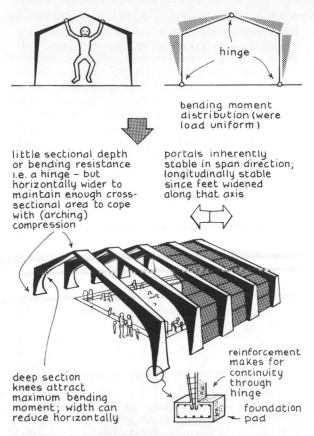

bending moment distribution (were load uniform)

little sectional depth or bending resistance i.e. a hinge – but horizontally wider to maintain enough cross-sectional area to cope with (arching) compression

portals inherently stable in span direction; longitudinally stable since feet widened along that axis

deep section knees attract maximum bending moment; width can reduce horizontally

reinforcement makes for continuity through hinge

foundation pad

1.33 Three-hinged portal frame

up in layers glued together – because these materials offer a good surface finish and, more important, can be easily tailored to the frame's exacting shape requirements.

The portal's centre and feet are hinges, in as much as their reduced profile depth can offer no significant bending resistance. But notice how the section widens horizontally. Why is this? One immediate benefit is that feet widened in this way stabilise the portals sideways. But there is more. The widening is in compensation for the reduced sectional depth, maintaining a sufficient cross-sectional area to resist the compressive thrusts that continue, regardless of what the bending may be doing – 'thrusts', because a portal is really a kind of arch, albeit one in which bending rather than compression prevails. The two halves would fall inwards if you chopped a bit out of the centre. In the portal shape, it is as if you had *frozen an arch in the act of buckling, by suddenly making it able to resist bending stresses.*

So we see that any argument as to whether a portal is an arch or some arches are portals is partly provoked by the labels we are obliged to hang on to the different structural types. In truth, types often overlap and it is best, sometimes, just to look at structures and see how their shape relates to the job they are trying to do.

The arch in bending

The masonry arch bridge resisted its loads almost entirely by compression, and we saw that the only way it could avoid the potential tensions and buckling from moving loads was by possessing enough self-weight and built-up spandrel restraint to contain them. The concrete or steel arch is in compression, also. The difference is that it can take bending from wind or other non-uniform loads as a bonus and, hence, can be made much lighter over the same span. But, in contrast to the portal, it still invokes compression rather than bending to resist its loads – and compression is structurally easier to resist.

In an arch uniformly loaded around its length, i.e. loaded only by its own self-weight, compression prevails if the shape is a *catenary*, that is, an upright version of the shape naturally adopted by a chain hung between two horizontal points (1.34). This is clear when you think about it. A chain *cannot* take bending and so adopts that shape which allows it to resist its uniform self-weight entirely by tension. Logically, if the arch is this shape inverted, its uniform load resolves entirely into compression. A catenary describes the arch's ideal

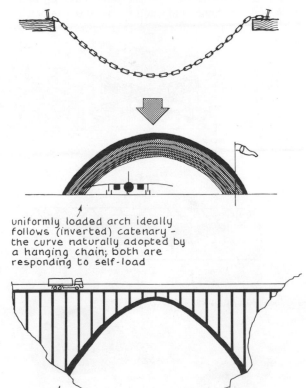

uniformly loaded arch ideally follows (inverted) catenary – the curve naturally adopted by a hanging chain; both are responding to self-load

non-uniform loading since frequency of road supports per unit length of arch intensifies towards crown – so curvature intensifies also

1.34 Arch in bending – catenary and parabolic

curvature and the rise can vary as a chain's sag can vary.

However, the bridge arch is not uniformly loaded. You can see from the frequency of the roadway supports that the loading/unit length around the arch is more concentrated towards the crown. The crown is, therefore, required to curve more sharply there, giving a curve nearer to a *parabola* than a simple catenary – just a point in passing.

In the light of all this, the portal now appears as a kind of compromise. It distorts the ideal arch shape into something squarer, giving a more convenient space inside and offering flat wall and roof surfaces. After all, if the structural purity were the only consideration, we would all live in domes.

Two-way spanning – the space frame

So far, we have confined ourselves to main spans going one way across a building with a secondary roof system between. The space frame (1.35) follows the logic that, where the area to be roofed over is square or near square on plan, in other words, where it makes as much sense to span the main frames one way as the other, it makes more sense to span both ways at once. Compression domes and vaults do it and so can beams.

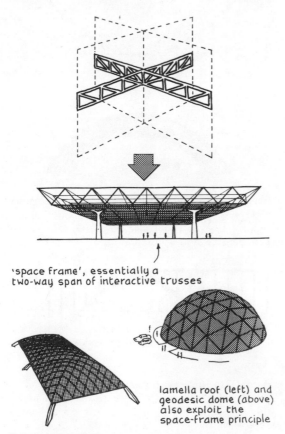

'space frame', essentially a two-way span of interactive trusses

lamella roof (left) and geodesic dome (above) also exploit the space-frame principle

1.35 Two-way spanning – the space frame

In effect, one is increasing the depth and stiffness of the 'secondary' system to that of the 'main' system, and making the systems interactive. Thus, the simple truss, which was rigid in one vertical plane only, is converted into the space frame which is rigid in two mutually perpendicular planes, utilising all the members to shed the loads radially towards the perimeter. The space frame can be lighter and more efficient than the equivalent one-way system and is better at distributing point loads placed at random. Its long potential span allows large column-free areas, often spectacularly so. But it is a departure from the ordinary, is complicated to fabricate and is seldom economic on small buildings, especially where the grid is not square. For example, where the span one way is twice that of the other, the shorter span takes 90 per cent of the loading in any case – having a space frame would then be a waste of time.

Just in passing, the lamella roof illustrated is a kind of space structure where arched spans interact at right angles, and so is Buckminster Fuller's celebrated geodesic dome. But they are also verging on surface, form-resistant structures, about which more in the last section.

Summary

- The simple span has beams resting on supports either end.
- The encastré beam has ends clamped at the supports (by building-in) to prevent rotation. This reduces deflection, better distributes the bending moment and effectively adds strength.
- Beam ends cantilevered beyond the supports act as counter balances, similarly reducing rotation at the supports and attracting some of the bending moment away from the centre span.
- Portal frames can be fully rigid or two- or three-hinged; design intention is again better to distribute the bending moment throughout the material available; and typical feature is strengthened 'knee' where maximum bending moment occurs.
- The steel or reinforced concrete arch, like the masonry arch, invokes mainly compression to resist its loads, but with a tensile capability to resist buckling.
- The space-frame roof spans both ways on plan: span grid is interactive and sheds loads radially towards the perimeter.

INCREASING THE BUILDING SCALE VERTICALLY

Now, we must look at the structural implications of *vertical* scale increase, at the form of the modern multi-storey building. Essentially, there are two structural choices in going upwards, cellular construction using supporting walls and skeleton-frame construction using columns and beams.

Cellular construction – higher supporting walls

In high-rise cellular construction, floor loads are directly passed to the supporting walls and the lateral stability derives from the walls' cellular interaction on plan. Think of the terraced-house type extended vertically. But there are two governing factors.

The first, obviously, is that this solid-wall solution for multi-storey buildings is only suited to types whose plan organisation is vertically repetitive and requiring small rooms rather than large uninterrupted areas. Hotel bedroom blocks and blocks of flats are examples. Walls there are needed, in any case, for separation between occupancies – and solid-block walls rather than lightweight partitions, the better to provide acoustic and fire compartmentation – so, again, it makes sense to use them for support.

The second factor is height. Masonry, with its high self-weight and minimal tensile resistance, has height limits in terms of direct compressive stress and the related risk of buckling. So the practicable limit is around 15 storeys or so, albeit this is influenced, in turn, by how far cellular interaction is achieved. Walls can be stepped, widening them towards their bases to get further height safely, but this approach soon becomes uneconomic with height. Modern cavity brick or hollow-concrete-block walls are sometimes stabilised by vertical steel rods inside, anchored in the foundation and passing up within the cavity and bolted to a metal wall plate on top of the wall head. The bolts can then be tightened so that the rods go into tension and exert a stabilising, vertical compression within the wall line – it is rather like the way a string of beads becomes rigid when the beads are pushed hard against the string's knotted end. The masonry would have to stretch the rods to topple. But, in fact, this is verging on the reinforced concrete wall, the integral reinforcement of which imparts buckling resistance – and, more immediate to our discussion, it brings us to the skeleton frame.

The multi-storey skeleton frame

The term 'skyscraper' was coined in the 1880s in the USA, when buildings of 20 storeys and more started to appear on the Manhatten and Chicago skylines. These were masonry structures with an internal stiffening, stabilising, frame of steel – 'skeleton cage' construction, it was called. The growing pressure for city space pushed heights upwards and the compressive and tensile strength of steel provided the way. Today, cage and masonry have long since given way to the steel frame, or the steel-reinforced concrete frame, enclosed in a skin of lightweight cladding panels (1.36).

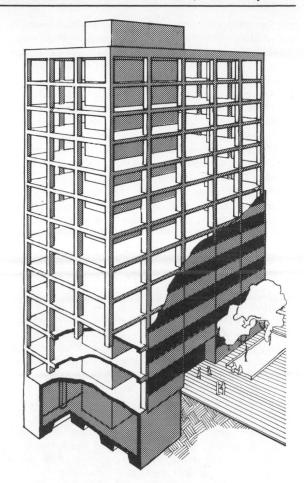

1.36 Multi-storey skeleton frame

The skeleton frame is appropriate to multi-storey buildings of any height – low-rise included – where there is a need for floor areas uninterrupted by walls. It is also appropriate, owing to its lower self-weight and vastly superior buckling resistance, to all buildings rising above the practicable masonry limits. In fact, this means most multi-storey buildings, but the familiar office block will serve as our central example.

We will look at the frame in two ways: first, at its design as a *total* structure, seeing how its planning organisation is made to fit in with the building planning as a whole, and how structurally it achieves its strength and stability; and second, at its individual components – foundations, columns, beams and floors.

Outline shape

Naturally, the office's outline shape and size will be influenced by the initial brief, i.e. by the client's accommodation requirements and by the finance available, and by the innumerable, wider influences applicable to any building, including site configuration, vehicle and pedestrian access points, outlook, surrounding buildings and so on. But our immediate interest is the structural bones.

Reinforced concrete or steel?

Whether to use reinforced concrete or steel is an early question, of course. There is more than a hint of the doctrinaire when outlining the pros and cons in design decisions, but here are some pointers. Reinforced concrete has lower strength to weight. To date, the highest concrete-frame building is the 218 m One-Shell Plaza building, Texas, by Skidmore, Owings & Merrill. Certainly, that is high, but only about half as high as the present ultra-high possibilities in steel. Concrete construction is slower: erecting the shuttering, laying the reinforcing steel, pouring the concrete and allowing it to set and strengthen, removing the shuttering, all this consumes time. The nature of shuttering construction makes it hard to build a concrete skeleton to fine tolerances. On the other hand, concrete can be cast to virtually any shape.

In contrast, steel erection is simple, fast, immediately strong and accurate. It rusts, but traditional maintenance problems reduce with modern weather coatings, and are eliminated if weathering steel is used, an alloy of steel and copper. If that were the end of the story, steel would win hands down, but there is a big snag. As temperature rises through 550° C or so, and that is quite modest where building fires are concerned, the strength of ordinary steel starts to reduce sharply and soon falls away completely – witness the grotesquely twisted shapes of steelwork after a fire. So all the frame members have to be protectively encased, i.e. with insulation. Concrete is a common encasement and there are other materials and methods described in Chapter 7 *Fire safety*. But, whatever the method, it must add to the cost and bring concrete right back into the running. That apart, the choice may as much be influenced by local costs and practices, or other valid design intentions.

Whole frame action

Still looking at the frame as a whole, it is immediately apparent that a grid of reasonably stiff columns and beams can stand – always provided that joint stiffness or other restraint prevents its flopping over. We will come back to stability in a moment. What is, perhaps, less immediately apparent is that the connection stiffness of the joints and their consequent ability to transmit stresses between members inherently adds strength. It is as if the frame were acting as a series of interconnecting portals, with consequent advantages in stress sharing and stiffness overall. A beam in a frame must be strengthened if it is restrained at its ends by its interaction both with the columns flanking it and with the beams beyond the columns. Equally, a column is stiffened against buckling by vertical continuity

through the floors and by the beams flanking it. In practice, stiff joints are rather easier to achieve with reinforced concrete than with steel. Concrete joints can be made *monolithic*, that is to say, cast with the reinforcing bars running continuously through. Steel members offer smaller cross-sectional area to the joint, reducing its stiffness in consequence. Continuity there can be improved by welding and/or additionally plating the joint, as opposed to simply bolting it – but only up to a point.

Frame stability

The main ways a frame can be stabilised against wind forces are shown in 1.37. *Joint stiffness* (a) is one of

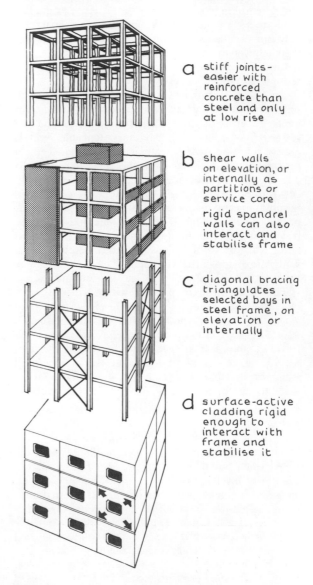

a   stiff joints-
easier with
reinforced
concrete than
steel and only
at low rise

b   shear walls
on elevation, or
internally as
partitions or
service core

rigid spandrel
walls can also
interact and
stabilise frame

c   diagonal bracing
triangulates
selected bays in
steel frame, on
elevation or
internally

d   surface-active
cladding rigid
enough to
interact with
frame and
stabilise it

1.37 Common ways of stabilising the frame

course but, except perhaps in a low-rise concrete building, other measures are needed. By geometry, height increase must increase both the total wind thrust and the thrust leverage on the windward façade, producing strong overturning moments in relation to the frame depth available behind to resist them. And steelwork design is often obliged to regard joints as being pinned rather than rigid, assuming them to have no stiffness at all. *Shear* walls can be added (1.37b). Solid partitions can act as rigid diaphragms within the column to beam angles, as shown. There is also, usually, the central service core running up the building, housing lifts and other utilities on each floor, and its capacity for shear wall action can be exploited – there are shades of cellular action here, though if the walls are of reinforced concrete, they will be acting as much like a stiff cantilever springing from the foundations. The concrete floor edge turned up to form a rigid 'spandrel' wall – or 'apron' wall to use the US term – is also a kind of shear wall. *Diagonal bracing* (1.37c) can triangulate selected bays in the steel frame, forming trusses keeping the whole frame from serious deflection. *Surface-active cladding* (1.37d) can stiffen the frame more or less as a floppy cupboard frame is stiffened by its plywood skin.

Returning to an earlier point, buildings have to be stable during construction – obviously. There was an illustrative case in Scotland some years ago, in Aberdeen, when a half completed steel-frame university building collapsed. The cladding was intended to stiffen the frame but it happened that, for the wind direction on the day in question, more cladding had been erected on the windward elevation than on the sides, making for much more sail area than stiffening. The frame deflected considerably. It also happened that no proper allowance had been made for the added overturning moment that any deflected structure suffers from the consequent eccentricity of its own off-centre weight. The situation was dynamically unstable. The harder the wind gusted, the greater the deflection and the greater the deflection, the greater the eccentric moment – it was too much and the frame fell. Closer attention to the erection sequence and the use of temporary bracing would have saved the building and, more important, the number of the work force who were killed. On all projects, the service cores, shear walls and other bracing have to proceed apace with the rest of the structure.

The structural grid

So a regular grid of supporting columns rises through the stack of floors. We call this the structural grid, the (it is to be hoped) orderly arrangement of the columns on plan. From the structural point of view, it is undesirable to break grid, i.e. to shift columns off-line on successive floors (1.38). It means that the collected vertical loads channelling down through the column on the floor above have to be diverted sideways, breaking the

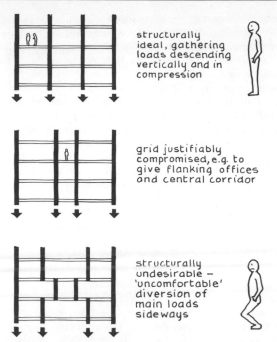

structurally ideal, gathering loads descending vertically and in compression

grid justifiably compromised, e.g. to give flanking offices and central corridor

structurally undesirable – 'uncomfortable' diversion of main loads sideways

1.38 Ideally, the column grid should be regular on plan and uninterrupted through the floors

structural rhythm and *involving heavy costly floor construction to resist the extra bending*. Bending, using internal leverage to resist far greater external leverage, is far harder to cope with than the mainly compressive stresses in columns. Each floor bay should carry only its own load horizontally, vertical members should take the vertical collective carrying – which, when you think about it, is why it is more comfortable to stand straight than with your knees bent.

But you cannot drop columns indiscriminately. Many an obstructive column later excused as 'a feature' was doubtless an unfortunate oversight in early design! In a well-run project, the clients, architects, structural and services engineers and, not least, the contractors who have to build the building, in fact, the whole design team, get together from the start to ensure that internal planning, the structural grid and all the other parts mesh together. Picture the designers working with a series of tracing-paper overlays, one showing the structural grid, above it the proposed partition layout and, perhaps, above that, another with the air-conditioning runs, adjusting and compromising towards the preferred solution.

The grid dimensions must make structural sense – too wide, and the floor spans become heavy and costly, too narrow, and there is a clutter of columns, tedious to build and not allowing a floor with only the minimum thickness needed for constructional convenience to realise its spanning potential. But the suitable dimensions will ultimately depend on the building use, of course. Generally, a narrow grid of 5 m or so in each direction is acceptable where there are to be many small spaces internally. Partition and column lines can be sensibly aligned across the floor plan: also, the closer column lines occurring vertically at the building perimeter, possibly with only one window unit between, offer more possibilities for locating partitions which can really only hit the outside wall where a column (or cladding mullion) occurs, not at a window. Open-plan office layouts or buildings requiring larger unobstructed areas favour a wider grid.

Often, the preferred grid dimensions are explicitly dictated by the planning. In a multi-storey car park, the grid spacing would have to be a multiple of a car-bay width one way and car-bay lengths and circulation alleys the other. In a library (very high floor loads), records office or supermarket, the spacing would have to be a multiple of the shelving banks and access aisles between. As you will realise, these despotic planning requirements can cause conflict where the multi-storey building has more than one occupancy or function in its height, each occupancy wanting its preferred grid for floor-planning economy but the structure wanting a single grid through the floors for building economy. Changing grid costs money but then, so does wasted floor space. So, although a grid change is always structurally undersirable, it is not always wrong.

The module
A related point, and it is one which applies to virtually all building design, and would certainly apply to the design of the frame building from its earliest stages, is the practice of *modular co-ordination*. The grid orders the structure but of course the rest of the fabric must be ordered as well – internal partitions, suspended ceiling tile grid, service ducts and related components, doors, external cladding – a considerable jig-saw. A module is simply a basic control dimension in relation to which the component parts of the building are sized – a rationalisation ensuring the jig-saw fits. It is a moot point though whether modular dimensions have most commonly sprung from the larger scale of structure or the smaller scale of building components. Rayner Banham has it that the dimunitive, acoustic tile long exerted a tyrannical influence on suspended ceiling grids and by result on overall structural grids, without anyone really noticing!

Summary
- Cellular construction masonry is limited in height and to close-walled, vertically repetitive, floor plans.
- Steel or reinforced-concrete skeleton frame is most common at larger scale.
- Various factors, e.g. fire-protection, affect steel/concrete choice: very high-rise demands steel.

- Structural continuity, especially through monolithic joints, strengthens members and helps stabilise the whole frame.
- Other stability measures include shear walls, diagonal bracing and surface-active cladding.
- Structural and planning needs intermesh – column grid and overall module order the design from the start.

Frame foundations
Turning to the individual components, we can start from the bottom and work up. Admittedly, the foundations are not a frame part as such but, with work below ground level sometimes amounting to as much as a third of the structure cost, they merit some coverage!

Principal foundation types are shown in 1.39. Reinforced concrete *pads* (a) are a logical base to the frame columns. However, high loads or weak subsoil may call for an overall *raft* (b) to spread further the loads and keep the stresses within the soil's bearing capacity. A close grid also argues a raft rather than pads, possibly using more concrete but allowing the contractor to sweep the whole excavation through with mechanical diggers rather than messing around with closely spaced holes. Rafts can be thickened and additionally reinforced at points where the column loads arrive. They may be stiffened by integral beams underneath, particularly to resist any tendency for them to dish downwards at their centres, where column loads are heavier.

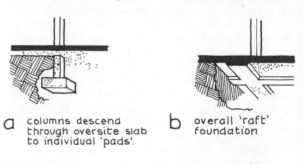

a columns descend through oversite slab to individual 'pads'.

b overall 'raft' foundation

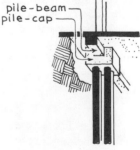

pile-beam
pile-cap

c basement construction can act as 'buoyancy raft'

d piled foundation

1.39 Typical frame foundations

We will explain the heavier centre columns in a moment. A basement below ground level may be built with the floor as the load-spreading, supporting raft, cast monolithically, with the retaining side walls cantilevering upwards. But, with weak unstable subsoils especially, the construction may act more as a *buoyancy raft* (1.39c). Here, the volume of basement underground is built as a reinforced chamber and, broadly speaking, acts like a ship's hull, resisting the inward pressures of soil and static water, and effectively floating the building by displacement. Finally, there are piled foundations (1.39d), where concrete or steel piles reach far down to solid strata or are held up by their friction with the soil packed around them. As shown, there may be multiple rather than single piles at a point, 'thrusting up' under the pile-cap, and there may be a pile-beam above that. Piles may be 'replacement', where concrete is cast into a pre-drilled hole, or 'displacement', where they are forced down into the ground by jacking, screwing or hammer-driving.

Differential settlement. Incidentally, where a building has high- and low-rise parts adjoining, some differential settlement is virtually inevitable. The solution is then 'if you can't beat it, don't join it'. The respective foundations and superstructures are made independent, so that what looks like one building is really two and small relative movements over the years will do no harm. Such breaks, *movement joints*, are similar to, and often synonymous with, the thermal expansion joints to be described in Chapter 2, *Enclosure*.

Translating from foundation to frame
Turning to the frame, as said, the ideal thing structurally is to have the column grid springing from the foundations and rising unchanged to the top of the building. However, the grid on the lower levels, typically the entrance level and, in consequence, the basement

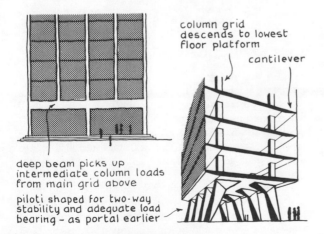

deep beam picks up intermediate column loads from main grid above

piloti shaped for two-way stability and adequate load bearing - as portal earlier

column grid descends to lowest floor platform

cantilever

1.40 Altering the grid at entrance levels

below if there is one, is sometimes widened to suit the particular planning requirements there – for example, to provide a more open podium entrance area or shops under an office development, or a more open foyer and public areas under, say, a hotel bedroom block. 1.40 shows two ways in which this justifiable grid translation can be achieved. The deep first-floor beam is common. Also, the concrete first-floor platform on piloti is interesting in the way it tailors its shape to the loads it receives, but it is arguably making a bit of a meal of things structurally and is, perhaps, more redolent of the Corbusian 'brutalist' forms of the 1950s than of today.

The columns
The column stresses vary enormously. The vertical loadings increase down the building, that is obvious. So will the lesser but, depending on the construction, quite significant, stresses from wind – the wind is a uniformly distributed load on the upright frame and the resulting overturning moments must increase towards the base.

Further, the vertical loadings increase in the plan centre. By symmetry (1.41), it is easy to see that a corner

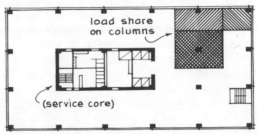

load share on columns

(service core)

1.41 Column loadings vary across the plan as well as with height

column takes only a quarter share of the floor load in a bay, a perimeter column twice as much and a central column four times as much. So, one way and another, there is a very great difference in the responsibilities of a corner column under the roof and a central column in the basement. But, in practice, the advantage of paring down structural self-weight and saving material is balanced with the constructional advantages of sticking to standard sizes. Uniform concrete column sizes allow the same shutters to be used on each floor level and, in both concrete and steel, uniform sizes ease the fitting of the structural jig-saw. So the typical frame compromises. The perimeter columns are smaller than those in the plan centre but reductions up the building are at limited intervals only, not at each floor. And, incidentally, where perimeter columns are exposed on the elevation, sectional reductions will normally be confined to the dimension running back into the building. A change on the elevation side would complicate the cladding sizes and appear visually awkward, however structurally justified. The corner columns are usually constant. They are lightly loaded anyway and having a face exposed on two elevations makes a sectional change hard to hide.

On the other hand, the amount of reinforcing steel in concrete columns can be varied, progressively reducing higher in the building. The steel distribution within each column reflects the stresses it is expected to take. It has to resist buckling as well as direct compression – if anything, buckling is how it would fail – and therefore the vertical bars are located just inside each face so that, no matter how the buckling or bending may come, there will be steel available to take the tension. Thickening a column therefore strengthens it in two ways, by increasing its cross-sectional bearing area and also increasing its 'depth' as a vertical beam.

And just a point in passing, the perimeter columns, albeit more lightly loaded, may have to resist some bending owing to the eccentricity of the load placed upon them by the dishing tendency of the more heavily loaded frame centre. One can visualise this by thinking of the frame action as a whole.

Again, structural continuity is desirable. In construction, the columns on each floor are cast with the vertical reinforcing bars left sticking out at the top, perhaps a metre or so, some sideways, some vertically. These protrusions are then *starter bars* for the next floor and columns above, the reinforcement of which can be made to overlap, ensuring solid joints and reinforcement continuity.

The floors

A variety of floor constructions is shown in 1.42. The first four concrete types are all cast in situ, with the consequent structural continuity advantages this brings. 1.42a is the 'simple slab'. It is shown here one-way spanning between the main beams, but can equally be a two-way span, i.e. where the main beams occur across both axes of the floor. It needs only flat table-top shuttering underneath, with troughs to cast the beams and, as such, is relatively easy and cheap to construct. It is common over the more modest grid spans. The reinforcement takes the form of a heavy steel mesh. However, there will inevitably be an increase of stress around the column head: 'punching shear' this is called, the tendency of the column to punch up through the floor. Sometimes, it is enough to beef up the slab reinforcement there but, alternatively, column capitals can be used, acting like inverted foundations to spread the load (1.42b). 1.42c is the one-way 'ribbed' or 'troughed' slab. The underside shuttering is more complicated but, in terms of strength to weight, it scores in having greater effective depth as a beam by distributing more material away from the neutral axis. Broadly speaking, it starts to come into its own over longer grid spans and/or with heavy floor loads. The two-way ribbed slab (1.42d), severally known as the 'coffered', 'honey-pot'

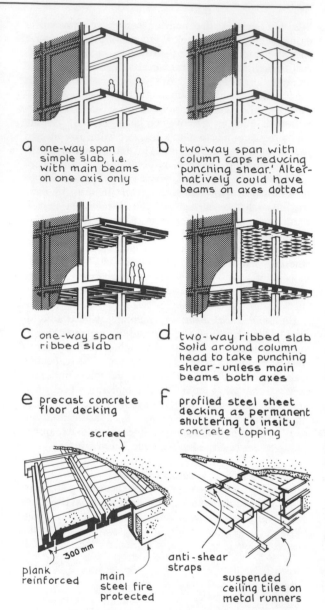

a one-way span simple slab, i.e. with main beams on one axis only

b two-way span with column caps reducing 'punching shear.' Alternatively could have beams on axes dotted

c one-way span ribbed slab

d two-way ribbed slab Solid around column head to take punching shear - unless main beams both axes

e precast concrete floor decking

f profiled steel sheet decking as permanent shuttering to insitu concrete topping

screed

300 mm

plank reinforced

main steel fire protected

anti-shear straps

suspended ceiling tiles on metal runners

1.42 Typical frame floor constructions

and, rather quixotically, 'waffle' slab, takes the argument further again – maximum shuttering complication but maximum efficiency. You get what you pay for. It is suited to the square grid, as was the space frame (in fact, it is something of a space frame) and also where longer spans and high loadings make the effort worthwhile.

Having said all that, where spans and loads are modest, the advantages of structural continuity, as far as the floors themselves are concerned, can be completely traded for simple and quick erection. 1.42e is the typical precast floor, precast concrete units topped by a light topping screed. Steel frames can use similar pre-cast units or steel decking (1.42f). Note the anti-shear straps at intervals on the steel decking – they improve the steel-to-concrete composite bond and the cohesive beam action in consequence.

The cantilever slab and curtain wall

The floor construction can be projected beyond the perimeter columns (1.43), with the structural logic that the cantilevering slab then helps relieve the flooring in the internal bays behind. Moreover, there is a fairer division of load-carrying responsibility in that all the columns across the plan are now carrying the same load and the load eccentricity and, hence, bending on the perimeter columns disappear. The cladding is supported at the floor edges, which is constructionally rather simple – this is the familiar curtain wall.

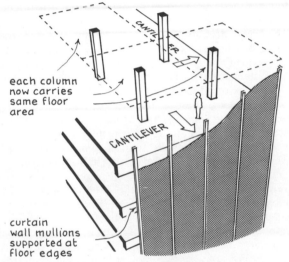

each column now carries same floor area

curtain wall mullions supported at floor edges

1.43 Cantilever slab and curtain wall

Towards the limits

The ultra-high-rise building (1.44) is hardly recent. The American skeleton cage construction had produced the 235 m high Woolworth building in New York by the First World War and, by 1932, the 380 m Empire State, the height of which has only recently been surpassed. What continues to be innovatory is the economy and sophistication with which ultra-high frames (indeed, frames generally) are achieved. As explained, ultra-high-rise demands steel. Lateral stability can be provided by braced bays, either selectively located or vertically repeated on each floor to form a building-height, interactive truss (left). But stability needs can breed other interesting solutions. For example, the 60-storey US Steel Building, Pittsburg (centre), by Harrison Abramovitz & Abbe, has a triangular plan with braced steel core, together with intermediate and 'top-hat' structural levels, whole floors given over to structural

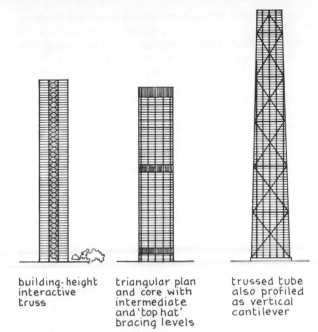

building-height interactive truss

triangular plan and core with intermediate and 'top hat' bracing levels

trussed tube also profiled as vertical cantilever

1.44 Stability at ultra-high-rise

bracing, interacting with the vertical frame passing through them and so reducing its lateral deflections. In the 100-storey John Hancock Center, Chicago (right) by Skidmore Owings & Merrill, stability needs have been allowed to dictate the whole form. It is a complete, trussed tube, an Eiffel-Tower-like vertical cantilever, 'logically' widening towards its base and with cross-bracing on the elevations. The twin 412 m towers of the World Trade Center, New York (Minoro Yamasaki with engineers Worthington, Skilling, Helle & Jackson), have rigid, close-grid framing linked to their central cores by prefabricated, lattice steel floor units. The 483 m Sears Roebuck Tower, Chicago – to date, the world's tallest – uses the innovatory 'bundle tube' concept, having nine 25 m square units rising to different heights. There is no stabilising core, but the façades are stiffly connected to the concrete/steel composite floor decking and, like the US Steel Building, there is added bracing from intermediate structural levels.

However, such buildings stretch the limits of this introductory text. It is one thing to outline the main rules that shape a concrete or steel skeleton, as we have done, and quite another to analyse exactly the three-dimensional interaction of floors, beams and columns, and the complex stress patterns within them. For example, how far is the office skeleton acting like a huge column? Or is it a cantilever? Or is it like a space frame? Even engineers must analyse structures in a simplified way and then add safety factors to allow for the area of uncertainty. In a simple structure, it is often easier to over-design for strength (the PPI rule again) than to spend an enormous amount of time on calculation. On the other hand, we have the design team of the high-rise building using *global* analysis, running the grid arrangement, the loadings and other variables through a computer to find the optimum layout and structural sizes. And, as a closing point in passing, there are other limits to scale and height than those imposed by structure. The services chapters will show how, technically, ultra-high-rise brings increasing problems in achieving water supply, air-conditioning and mechanical transport. (Happily, the social desirability of increased scale is not part of our present discussion!)

Summary

- Foundations are pad, raft, buoyancy raft, pile.
- Lower floors sometimes have wider grids, requiring deep beams to pick up loads from the narrower grid above.
- Column loads vary across the plan and with height, but sectional changes are limited by constructional practicability.
- In floor construction, the advantages of continuity and underside ribbing are similarly balanced against constructional practicability.
- Cantilever slab (with curtain wall) evens up the column loads across the plan.
- Ultra-high steel buildings are stabilised by interactive vertical trusses, and/or braced cores, 'structural' levels or interactive facades.

CABLE AND SURFACE

The architecture of compression followed its logical path to the forms of the Gothic cathedral, and the added possibility of resisting tension gave us beams and bending, and all the applications of building frames from house to modern high-rise skeleton – the scope, by this time, having touched most of the buildings around us. Now we come closer than ever to purely structural form, the form of suspension systems, inflated and tension membranes, and shells inherently strong owing to their curved or folded shape.

SUSPENSION STRUCTURES

Bridges

The suspended chain discussed earlier was a purely structural form, efficient in automatically adopting the

1.45 Suspension bridge

catenary curve, with the tension everywhere equal and, therefore, minimal. The modern suspension bridge (1.45) is a similarly efficient form, accounting for all the world's longest spans. It exploits the enormously favourable strength-to-weight ratio of high tensile steel cable. The structural principle is clear. The support towers are primarily in compression, while the cables, all in tension, rise over them and drop to massive reinforced concrete ground anchorages behind. The bridge load is balanced by the resisting pull of the anchors. The bridge deck itself is supported at intervals by hangers from the main cables and, therefore, develops little bending and can be extremely light. The beauty of the system is that each element – cable, tower and deck – is ideally fitted to its task.

We can look at the significance of cable sag another way. Take the hanging chain again. The further up the curve at either side, the greater is the portion of chain being supported. Yet the tension remains the same? This is possible because the curvature brings the direction of the tensile force, the pull, ever closer to the vertical and, therefore, better able to oppose the increasing downward self-weight. The tension is constant but the upward component of the tension improves. This is why the catenary is the shape it is. No matter how hard the chain were stretched, it would always have some sag, i.e. upwards inclination towards the ends, to give the upward tension component to balance the downward self-weight – only if the chain were weightless, could it be stretched straight. So, in a suspension bridge, the sag enables the cables to provide the support, and the greater the sag, the less the tension in the cables need be. In practice, the sag-to-span ratio is limited by the height to which the towers can be practicably built. They have their own problems in sway and buckling.

The technique has seen significant improvements in recent years. Sophisticated model testing and computer analysis enlighten the designer on the complex stress patterns. Higher-strength steel construction stiffens the towers. The steel box-deck principle, whereby the road deck is constructed as a hollow beam, stiff against winds and the moving imposed loads from traffic, leads to spans over ten times lighter than, say, the 1937 Golden Gate. The 1981 UK Humber Bridge spans 1396 m, and the theoretical limits are nowhere near reached.

Suspended buildings

The architectural applications of suspension systems vary. In the suspended office building (1.46) – the illustration is based on a Dutch building at Eindhoven – the

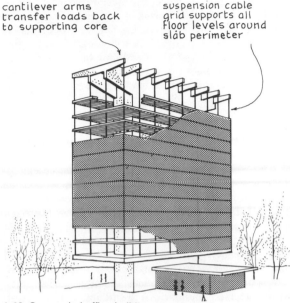

cantilever arms transfer loads back to supporting core

suspension cable grid supports all floor levels around slab perimeter

1.46 Suspended office building

central core is the compression element, and the vertical cable grid hanging from it, the tension element supporting the floors around their perimeters. Of course, the roof works are much more expensive than in a conventional structure but, on the advantage side, the cables are much thinner than would be the equivalent columns and this saves structural self-weight and floor space. Regarding stability, the extra compression on the core ensures that it never gets into tension on the windward side. Additionally, the cables in such buildings can be carried down to ground anchorages, stabilising as guy ropes stabilise a tent.

Suspended roofs

Then there is the possibility of entirely suspending a roof, for example, by having an intersecting grid of horizontal cables running between the outside walls. The roof is then thin and can only be supporting itself by tension. There is no bending, not a truss or beam in sight. Now, there are two considerations here. First, the span has to be sufficiently great to exploit the potential of the suspension principle and make its use worthwhile, so we are really talking of buildings like large exhibition halls and indoor stadia. Second, there is, again, the important question of stiffness. A sagging, simply-hung roof would be liable to what is called *flutter*, flapping about like a stalled yacht sail in even moderate winds. Unlike the suspension bridge, there is no

stiff deck to hold the cables down. Systems have been designed where suspended roofs are stiffened by trusses or are tied down, but this kind of fussiness tends to miss the potential simplicity of the suspended span.

A building that pointed the way was the livestock arena in Raleigh, North Carolina (1.47a), by the engineers Severud-Elstad-Krueger. The structure comprises two inclined, parabolic, compression arches in reinforced concrete, intersecting one another and with a grid of suspended cables between. The cables support a light roof decking. The weight of the arches stresses the cable roof and the cable roof restrains the arches, so that compression and tension elements balance. The walls are load-free. But what is important here is that the arched shape of the boundary gives the cable roof an opposed *double curvature*, like a horse's saddle. Viewed from above, the cables in the concave (sagging) direction, i.e. between the opposite arches, are stressed by the presence of the cables in the convex (hogging) direction – and vice versa. On any point of this three-dimensional surface, the curvature one way is opposite to the curvature the other, a mutually opposed bracing, a continuous tug-of-war, that stiffens the roof against flutter.

By a quirk, the suspension roof of Eero Saarinen's 1959 Yale University hockey stadium (1.47b) usefully demonstrates the point about double curvature. The compression element there is a 100 m span reinforced concrete arch, a kind of dinosaur spine. The roof consists of a first set of cables dropping to the perimeter

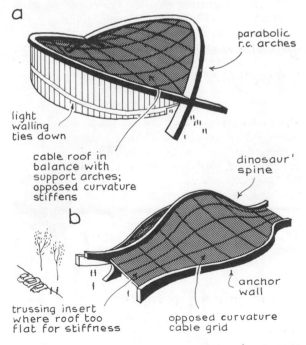

a

parabolic r.c. arches

light walling ties down

cable roof in balance with support arches; opposed curvature stiffens

dinosaur spine

b

anchor wall

trussing insert where roof too flat for stiffness

opposed curvature cable grid

1.47 Opposed-curvature cable roofs in USA, (a) Raleigh Arena and (b) Yale University Hockey Stadium

walls and a second set superimposed longitudinally, again giving a taut double curvature. But, at the very ends of the building, the opposed curvature eases. Viewed from above, the roof there appears totally concave, sagging in both directions as if it were loosely hung. So we can deduce that either it must be free to lift and flutter or it must have something else stiffening it – and we are not deceived. Steel trussing lying flat within the plane of the roof ends, primarily intended to anchor the longitudinal cables, is also supplying the stiffness.

Although Raleigh and Yale are cable roofs, they act as if they were continuous tension membranes. In architecture, a *membrane* is a thin surface the internal stresses of which act tangentially to, i.e. along, the surface plane. There can be no bending resistance. If a man stands on a cable roof, it is just distorted imperceptibly until the upward component of its marginally increased tensile stresses balances his added downward load. Membranes do not necessarily have to be tensile. For example, a masonry wall is a kind of compression membrane, no bending and all the forces lying within the surface plane.

Lightweight tension structures – LTS

An ordinary tent is a tension membrane, fabric in tension, poles in compression and the whole system in equilibrium. In recent years, the tent principle has been successfully applied to the larger scale of structural engineering, notably in Frei Otto's cable-net structures erected for the 1974 Munich Olympics 1.48.

Cable-net membranes are tensioned from their boundaries to initiate the stiffening. That is rather obvious, perhaps. After all, it is only when you tension a drum skin or umbrella fabric that it develops strength in the useful way you want. In the cable-net, the structural value of this tensioning is maximised by the use

cable-net membrane

boundary cable

steel support masts

adjustable 'guys' impart tension

ground anchor

1.48 Minimal surface, cable-net membranes. Munich Olympics 1974

of double curvatures and the more pronounced the curvatures, the greater the stiffness becomes.

Tension, double curvature . . . the next criterion is that the membrane tension be everywhere the same, since it would be a poor use of material were the membrane slack in one place and near breaking point in another. Now, equal, wrinkle-free tensioning is only possible if the membrane is pre-shaped from smaller, specially tailored, pieces. You will not get it with a single, large piece. And the membrane has to be a minimal surface – has to have the minimum surface area possible given the shape of the boundary and supporting mast system.

The nature of the minimal surface becomes much clearer when one looks at the ingenious scale-model techniques used to find the correct topography for these membranes. Briefly, the layout of support masts and ground boundaries needed to give the required enclosure and sufficient membrane curvatures is decided (with experience and a few sums), and scaled down into a model. In one method, the model is passed through a patent soap solution so that a soap film attaches itself between masts and boundaries to represent the membrane. Now, the surface tension present in any liquid makes the film want to contract and, in fact, automatically pulls it into the smallest possible surface area within the fixed boundaries – the minimal surface. And, since the surface tension is a constant value and acts equally in all directions, it follows that this must be the surface shape that will allow even tensioning in the larger cable-net. Moreover, as a bonus, on any part of a minimal surface, the surface curvature one way is automatically equal and opposite to the curvature the other way, optimising the double-curvature advantages described earlier. It only remains to transfer this desirable shape to full scale. Soap solutions have been developed that set after a time, so that the contours can be fairly easily measured, the film acting as a miniature tailor's dummy for the real membrane, allowing a 'dressmaking' pattern for the net pieces to be worked out. There are techniques which use thin rubber films, where elasticity rather than surface tension does the contracting, but the principle is the same, namely, that nature arrives at the optimum membrane shape, model film leading to cable net with only a few modifications in the real thing to cope with local wind uplift, snow-loading and so on.

Pneumatic structures

Another way to tension a membrane is, quite simply, to blow it up like a balloon. The plastic, pneumatic structure is a cheap and light enclosure (1.49). It is still a minimal surface tension membrane, and still a kind of suspended structure, but it is supported by a slight excess of internal air pressure rather than by compression masts. The excess pressure in the 'bubble' is main-

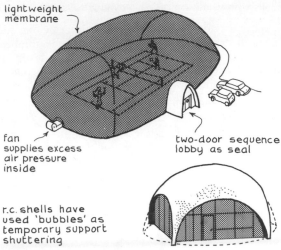

lightweight
membrane

fan
supplies excess
air pressure
inside

two-door sequence
lobby as seal

r.c. shells have
used 'bubbles' as
temporary support
shuttering

1.49 Air-supported, 'pneumatic' structure

- Suspended buildings, similarly, exploit cable strength.
- Suspended cable roofs are lightweight, but have to meet the flutter problem with opposed-curvature or other bracing.
- Lightweight tension structures are opposed-curvature, minimal-surface, equal-stress roof membranes.
- Pneumatic structures are also minimal-surface tension membranes, supported by slight excess air pressure inside.

tained by the type of fan typical of mechanical ventilation systems, blowing air through a hole in the side wall. An entrance lobby with a two-door sequence helps maintain the air seal so that large numbers of people can come and go without the structure deflating – one is conscious both of the push needed to open the doors and of the air escaping as one passes.

The actual tension in the membrane depends on the excess air pressure and on the bubble's radius of curvature. For a given pressure, a small radius (i.e. a more pronounced curvature) will reduce the tension and, in some larger spans, the membrane tension has been usefully reduced by composing it of many bubbles of smaller radii rather than a single hemisphere.

The air-bubble idea was proposed for field hospital tents towards the end of the First World War but was not so used until Vietnam. Happier and more familiar applications include exhibition pavilions, indoor market gardens (where the pumped air can be additionally used for climate control) and medium-sized sports facilities, like tennis courts and swimming pools.

Bubbles have also been used as temporary shuttering for reinforced concrete shells. This is significant. It shows, as we might have come to expect by now, that the minimal surface tension shape can yield a membrane mainly in *compression*. The very thin concrete shell roof also shown is supporting itself with virtually no bending, partly because it was formed on a tension bubble – which leads us, remarkably conveniently, to surface structures.

Summary

- The suspension bridge primarily exploits the very high tensile strength of woven steel cable – the system balances cable tension, tower thrust and deck weight.

SURFACE STRUCTURES

Folded plates

The flat piece of paper (1.50) is unable to span the gap because it is too thin to develop significant bending resistance, to give the internal stresses sufficient lever-

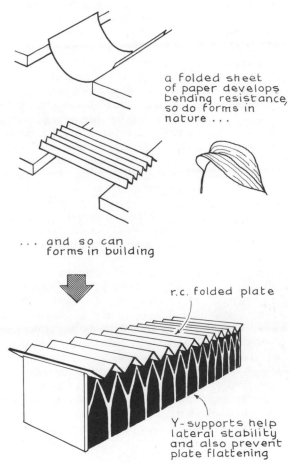

a folded sheet
of paper develops
bending resistance,
so do forms in
nature ...

... and so can
forms in building

r.c. folded plate

Y-supports help
lateral stability
and also prevent
plate flattening

1.50 Folded plates – folds stiffening in paper model, in nature and in building

age against the external bending moment. It is a rather feeble beam with no real thickness. But, if the paper is given a series of folds, it can easily support itself because it has been given a stiffer form. The amount of supporting material remains the same but it has mostly been moved away from the neutral axis.

The folded paper works in two ways. First, it is developing conventional bending resistance because the longitudinal folds tend to act as stiffening ribs. The flat inclined sections of paper merely span the short distance between the folds. It is rather like a system of main beams and thin secondary slabs. Second, the paper is also showing some of the properties of a membrane. Paper is perfectly capable of resisting compression and shear, as well as tension, provided these stresses lie within the plane of its surface. And so it is here. The narrow inclined planes have a longitudinal function as membranes helping to resist the anticipated downward deflection. They have depth but no real thickness so they are more than longitudinal beams. Any tendency to buckle is counteracted by the small width they do have and by the presence of the flanking folds. In short, the folded paper is part beam and part *form-resistant surface*, a rib-stiffened membrane.

In nature, we find leaves and grass blades folded along their centres for stiffness. In man-made things, the examples are countless – corrugated paper, corrugated metal decking and so on. The folds in motorcar body panels are for strength as well as styling.

In building, folded plates can be made up of laminated timber or metal sheeting but, in longer spans, are more usually of concrete. Concrete shuttering is easy to construct from straight timber planks on scaffolding supports. Alternatively, the individual roof planes can be cast flat on the ground, all from the same mould, with the transverse reinforcing bars left projecting. Each plane is then lifted to roof level by crane and aligned with the plane before, so that the transverse bars can be welded together and the small seam filled in with concrete poured in situ – the monolithic joint again.

In the building also illustrated, notice the Y-shaped supports. The triangulation that results obviously helps lateral stability. But there is another function. Think of the folded paper again. If it were progressively loaded, the folds would soon flatten – that is the way it would fail. But paper strips glued across the zig-zag ends would tend to preserve the folded form and so increase the overall strength. On the larger scale, this is what the Y-supports are doing for the concrete roof, providing a row of fixed points that prevent any sideways movement of the folds. Other ways of locking the folded form include casting thin concrete diaphragms across the folds underneath, or thickening the zig-zag edge into a stiffening beam.

Folded-plate construction can be used in walls or, indeed, for the whole building envelope.

Shells with single curvature – cylindrical shells

Suppose we now pick up a piece of paper by one end (1.51). It hangs limply, as before, because its thinness prevents bending resistance. But if we crimp it slightly to give it curvature at right angles to the direction in which it is supposed to be cantilevering, it becomes stiff.

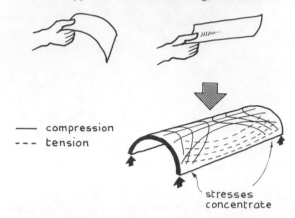

—— compression
--- tension

stresses concentrate

1.51 Cylindrical shell – single curvature imparting stiffness

We have created at model scale what, in architecture, is called a *shell*, a membrane or resistant surface that is curved and takes compression and shear as well as tension. *The stresses are still all in the surface plane*, but the paper has been so arranged that its inherent stiffness in that plane (in other words, along its surface, the only direction in which it can be regarded as 'thick') is available to resist the deflection. Again, material has been moved away from the neutral axis – the paper takes mainly tension along the upper parts of its sides and compression in the valley and, to that extent, follows beam theory. But there is virtually no bending. The paper's surface form does the resisting and membrane action prevails. The surface, as long as it is thin, just distorts a trifle in response to the bending moments, to the point where the in-plane stresses alone balance the load.

As we shall see, shells in building can have many shapes and curvatures. In all cases, the main things are their thinness, lightness and strength. Shells' limitations are that they are weakened by openings in their surface, and tend to dislike point loads which inevitably introduce the possibility of local buckling.

Construction is nearly always concrete, exploiting the material's tremendous mouldability. Interestingly enough, when reinforced concrete first became common in building, it naturally took the forms of materials then in use. A concrete wall looked much like a masonry wall and a concrete frame followed the same lines as a wooden or steel frame. This in no way diminishes concrete's importance in these roles but it is intriguing how the first use of a new material or technique so often follows existing traditions. Greek masonry temples echoed the earlier timber tradition – awaiting Gothic to exploit stone. In curved shells and other flowing forms, reinforced concrete finds full expression, doing the job that, for the moment, only it can do.

The cylindrical concrete shell shown is shaped like a slice of a cylinder, curved the short way and straight the long way. Unlike the masonry wagon vault which arched the short way only, the cylindrical shell spans the long way as well, and so only need be supported at its corners and not all the way along its sides, consequently freeing the planning below. It is lighter than its masonry predecessor. The dotted lines show the approximate stress pattern – there would be a general reinforcing mesh inside the surface throughout, supplemented by curved heavier bars, logically following the anticipated tension flow.

The cantilever canopy of the sports stand (1.52) is

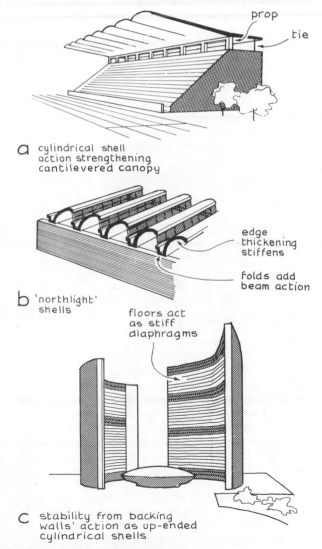

a cylindrical shell action strengthening cantilevered canopy

b 'northlight' shells

edge thickening stiffens

folds add beam action

floors act as stiff diaphragms

c stability from backing walls' action as up-ended cylindrical shells

1.52 Cylindrical shells in building

made up of a series of thin reinforced concrete, cylindrical shells. Typically, the repeating shells would all be cast from a single mould in the factory or on site, and then erected and joined, as described for the folded plate. As a further comparison with the folded plate, the junctions between the shells must inevitably act as stiffening folds, taking some bending in addition to the membrane action of the shells themselves. The same is true of the northlight shells (b), shell versions of the northlight trusses mentioned earlier. The valleys and the thickening along the edges stiffens them, giving a complex stress pattern that is part beam and part membrane. Trussed northlights with a steel decking might be cheaper but, of course, strict utility is not always the only determinant in design. Shell action on a very large scale is seen in the novel structure of the Toronto City Hall complex (c) (architect, Viljo Revell). The backing walls of the two towers are cylindrical shells up-ended, vertical cantilevers that are inherently stiff against wind forces. The 36 floors cantilever out from the shells and, important, stiffen them. Therein lies a tale.

Shells with only single curvature have the advantage of being easy to construct, the shuttering being composed of straight planks set on curved cross supports, or of curved timber or steel sheeting. But, like folded plates, they have the disadvantage of being *developable*. That is to say, they can fail by flattening or unrolling into the flat sheets from which they theoretically came, without even having to buckle or tear. This is clear from the original paper analogy. The possibility of developable, single-curvature shells losing the form that gives them their stiffness is their potential weakness. In some cases, the shell's slight thickness does, in fact, offer enough bending resistance to hold the shape but, usually, the curved edges have to be thickened so that they can resist the unrolling tendency, or there have to be stiffening diaphragms cast across the shell laterally. The Toronto floors are diaphragms.

Shells with opposed double curvature

Double-curved shells have the enormous structural advantage of being *non-developable*. They are inherently rigid. There are two distinct categories of double curvature. Shells like the hyperbolic paraboloid have *opposed* double curvature, convex one way along the surface and concave the other – a saddleback shape like that of the Raleigh tension membrane. Shells like the simple hemisphere have allied double curvature. It is impossible to model a saddleback or hemisphere from an uncut single sheet of paper and, by the same token, if you did cut and tailor the paper so that it could be glued into these shapes, it could not then be flattened out without buckling or tearing – *membrane action locks the shape.*

The hyperbolic paraboloid
The hyperbolic paraboloid, or HP, is a simpler surface

than its geometric title suggests. We have talked of the parabolic curve. Geometrically, the HP surface (1.53) is derived by suspending a series of identical 'sagging' parabolas between two parallel 'arching' parabolas. Although there is double curvature, the HP saddle can be defined by two sets of intersecting straight lines, as shown – in technical parlance, the surface has straight-line generators – in other words, at any point on the saddle there are two directions in which the surface is straight. Any sectional shape extracted from the saddle will, like the saddle itself, be non-developable and, moreover, can be made up of a series of straight sections – this is a great constructional bonus. For example, the most familiar HP extraction, used for the roof illustrated, could be simply constructed from a series of straight timber planks, spanning one way between the opposite edges, with planks or curved sheet material overlaid the other way. Alternatively, an easily assembled straight-plank timber shuttering could be used to form the reinforced concrete version of the shell.

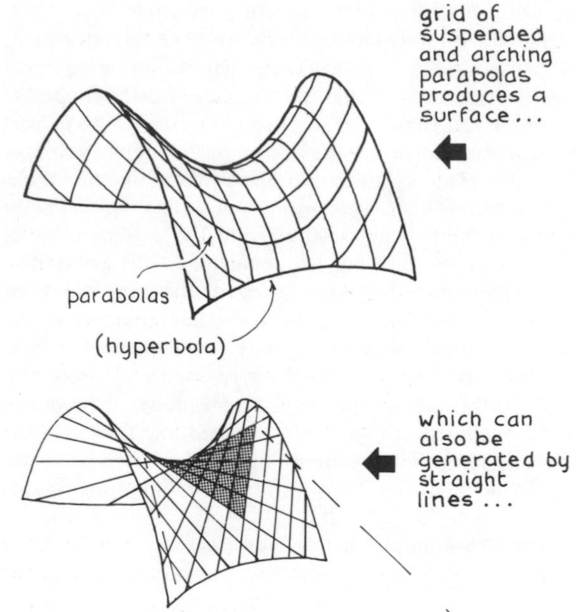

Incidentally, the roof is shaped as if you had taken a four-sided frame and twisted it and, for this reason, HP surfaces are sometimes known as *warped* surfaces.

Another advantage is that the HP shape is readily analysable. The stresses can be easily calculated and this goes a long way towards explaining the shells' popularity with engineers since the war. This is in no way intended facetiously because, as we shall see, there are any number of shapes that would be structurally sound if only there were simple mathematics or model techniques to reveal the likely stress patterns within them.

The stress flow in the HP roof, as one might intuitively feel, is predominantly compressive along the convex parabolas down either side towards the two supports, and tensile at right angles along the concave parabolas, i.e. tending to hold back the high points. Also, you can view the roof in profile as two cantilevering shells tied back-to-back, with a resulting stress tendency for tension along the top and compression towards the bottom. The convex parabola 'push' and concave parabola 'pull' collects at the roof edges, causing a resultant compressive flow that grows towards the supports. This is why the edges have a correspondingly thickening section as they descend.

The nightclub at Xochimilco, Mexico (1.54) – by the noted HP pioneer, Felix Candela – combines eight HP

grid of
suspended
and arching
parabolas
produces a
surface...

parabolas

(hyperbola)

which can
also be
generated by
straight
lines...

.. and
from this
the HP
roof shape
derives

thickening edge
passes gathering
stresses to
support

1.53 Hyperbolic paraboloid – opposed double curvature

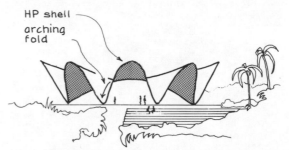

HP shell
arching
fold

1.54 HP shells in combination

elements around a common centre. You will understand this if you imagine each radiating element as one half of the simple HP roof we saw earlier, only with the pointed high parts cut back to form a curved edge. The main stress flow from the non-developable elements channels to the valleys which, acting rather like the folds in the folded plate, span as stiffening arches across the plan. The folds follow the ideal arching pressure line.

The hyperboloid of revolution

The hyperboloid is another opposed curvature. At model scale, it could be obtained by connecting two circular hoops with straight threads and then turning one of the hoops relative to the other (1.55). In the

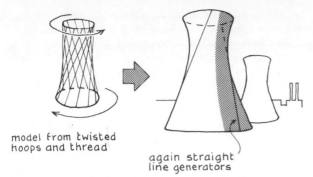

model from twisted hoops and thread

again straight line generators

1.55 Hyperboloid of revolution – opposed double curvature in power-station cooling towers

large scale, it is a non-developable surface which can, again, be cast in concrete with a straight-plank shuttering, sandwiching the concrete wall inside – witness the familiar shape of the power-station cooling tower.

Shells with allied double curvature

We have seen the form-resistant membrane as folded plate, as single-curvature shell and as opposed double-curvature shell. Obviously, a last important category remains. The sea shell, the egg shell, the skull, or the light bulb or the monocoque car body, are all shells with *allied double curvature*, spherical or near-spherical surfaces that are incredibly strong. In building, we see the surface at its simplest in the thin concrete hemisphere (1.56), a first use being for the domes of planetaria in the 1920s and 1930s, where the spherically concave surface inside was ideal for the projection of the artifical sky. As you can imagine, concrete spherical shells are difficult to construct because they need a complicated shuttering that curves both ways, and placing curved reinforcing steel is harder. But they have vital merits. For one thing, the alternative warped shell, like the hyperbolic paraboloid with its powerfully

compression ———
tension - - - -

compression cap, i.e. both around horizontal parallels and down arching meridians

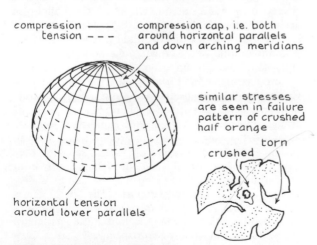

similar stresses are seen in failure pattern of crushed half orange

torn

crushed

horizontal tension around lower parallels

1.56 Dome acting as a membrane – allied double curvature

undulating, structurally dominated surface, only loosely relates to the useful space it encloses. Often, this will not matter and the strong structural expression may fit the building's purpose there, be it nightclub or church. But the spherical shell, itself a pure structural form, is also the form that trims the internal volume and minimises the area of the enclosing surface. For example, in practical terms, the minimised enclosure means lower heating costs (see Chapter 3 *Climate services*). Structurally, it saves material and cuts self-weight and, as spans increase, that is more and more important. An engineer would hardly use an opposed-curvature shell to roof a large sports stadium, he would more likely use a spherical shell because, strength-for-weight, it is the most efficient shell there is.

Stresses distribute around a shell, recruiting material into resisting them by membrane action. You will not break an egg by squeezing it in the palm of your hand – if you do not believe this, try it. You can break it if you cheat and press it, but that is because you are then putting pressure on one point and introducing bending stresses against which the shell is weak. Continuing this kind of analogy, suppose you took half an orange, laid it igloo-like flat on a table, and squashed it. Granted this would be an unlikely thing to do and one that would be at least as messy as any misfire on the egg experiment, but the point is that the skin would fail by crushing in at its crown and splitting around its lower edge, as shown. This test to destruction gives an indication of the normal stress patterns in the concrete shell. In the upper part, there is compression both horizontally around the parallels and in the arching direction of the meridians, i.e. there is a compression cap. In the lower part, the arching meridians are still in compression, as we might expect, but the parallels go into tension which corresponds to the splits in the orange skin and, indeed, to the spreading of the masonry dome, which needed chains around its circular base to restrain it.

The essential difference between the concrete hemisphere and the masonry dome is rather similar to the difference described between the concrete and masonry arch. The masonry dome can be regarded as a series of independent arched strips, all with a common crown. To avoid buckling, each strip has to be thick enough to contain any deviations in the compression line that results from wind, snow or settlement loads, or from inaccuracies in erection. But the shell can take tension and, because it is thin, it can adjust slightly until membrane action prevails. The arch strips are acting cohesively to form a three-dimensionally stiff surface. Shells have a structural continuity if you like, and can be built with a thinness impossible in masonry. The span-to-thickness ratio of the massive masonry dome over St Peter's, Rome, is only 1:14. For an egg, it is in the order of 1:50. In a concrete shell, it can be 1:1000 and more, slashing the self-weight. Spherical shells only

70 mm thick can span 100 m and, often, the thinness achievable is limited by constructional rather than structural considerations. For example, the thinner the shell, the more difficult it becomes to locate the internal reinforcement accurately and it is often worth accepting a slight increase in thickness in order to avoid the higher labour costs in building to very fine tolerances.

But we are not quite done with stresses yet. As in the HP roof, but for slightly different reasons, there are some problems at the edge of a hemispherical shell. There is, again, some unavoidable bending. Because the edge is fixed, there is bound to be some local distortion at the foot of the meridians when the rest of the shell is distorted, as it must be, to cope not only with its own load but, also, in response to thermal expansion and contraction of its surface, and to wind, snow and settlement loads. Also, the upward reaction from the support may not exactly align with the descending surface – well it should in a hemispherical shell, where the surface immediately adjacent to the ground is vertical but, in a flatter shell, the surface might be slightly inclined to the direction of the support reaction and cause out-of-plane forces. This unavoidable bending around the base of any spherical shell is called *boundary disturbance* and some edge thickening is needed to cope with it. Fortunately the disturbance cannot reach far into the shell proper because the troublesome meridian distortion is soon damped down by the stresses it induces in the direction of the parallels, in other words, by the added protest and, hence, resistance of the horizontal hoop of material around the shell base.

The Palazzetto dello Sport (1.57), designed by the engineer P. L. Nervi, was built for the 1960 Rome Olympics. Its 68 m span is quite modest amongst modern shell roofs, not least, compared with its big sister, the Palazzo dello Sport, also by Nervi. But the Palazzetto has an easily-read simplicity which has tended to make it something of a text-book classic. Like most modern shells, it is flatter than the full hemisphere, again, to reduce volume, weight and height. This results in a considerable outward thrust at the shell edge which, as you can see, is counteracted by a sur-

rounding ring of Y-shaped concrete buttresses – flying buttresses, actually – which, in turn, thrust back on a 2.5 m wide foundation ring running right round the building. The buttresses and, hence, their thrust, align with the angle of the shell perimeter and thereby keep out-of-plane forces and edge disturbance to a minimum. In any case, the edge is stiffened against bending by its slight corrugation – nature scallops sea shells in this way. Again, the Y-heads are triangulated for overall stability. The shell proper has a lattice of intersecting ribs in its underside, strengthening it against buckling and channelling the loads to the buttress prongs. Structurally, the effect of these ribs is the most complicated part of the design, because it is impossible to say how far the dome is acting as a membrane and how far as a rib latticework like a space frame – to that extent, Nervi's design must have been intuitive rather than precisely calculated. Details neatly follow the building function. For example, the Y-shape of the buttresses allows freer crowd access to the stadium entrances than would V supports springing directly from the ground, and the roof corrugation helps admit more light to the interior.

In passing, there are many ways of constructing spherical shells. The Palazzetto shuttering was an overall honeycomb, each cell being an inverted, precast concrete pot. The pots were supported on scaffolding like the pots under the two-way ribbed floor slab. When the membrane was poured on top, shuttering and skin were united into a permanent ribbed shell. If there are ribs, these could be cast first, to act like arches, and the membrane then made up of a reinforcing mesh fixed between them and 'gunned', i.e. pressure-sprayed, with wet concrete. Or the membrane could be of precast sections, the projecting reinforcement of which is welded to that of the ribs, and the joints filled. For the small, simple shell, there is the pneumatic structure method we mentioned: an inflated balloon shutter is covered with reinforcing mesh and the concrete skin is poured or gunned on top. Once set, the balloon is deflated and withdrawn. Simple shells can also be cast in stages as a series of horizontal hoops working up from the ground to the crown, a method that is inherently stable and self-supporting at all stages in erection.

Eero Saarinen's auditorium shell at the MIT campus, Cambridge, Massachusetts (1.58), is a spherical extraction, an exact eighth part of a sphere, like a slice of our orange skin. It springs from three equidistant supports 49 m apart, giving a triangular ground plan that well fits the auditorium function, allowing a speaking position at one apex, with the audience fanning away from it. The shell's tendency to spread is checked by underground ties between the supports. The shell edge is only lightly thickened, that is, except towards the supports, where the somewhat uncompromising geometry of the surface inevitably leads to massive stress concentrations.

shell membrane – scalloping adds strenthening folds at boundary

flying buttresses carry thrusts down to foundation. Y-heads add lateral stability

restraint from foundation ring surrounding building

1.57 Palazzetto dello Sport. Rome Olympics 1960

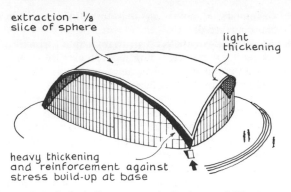

1.58 Spherical shell segment. Auditorium at MIT campus, USA, 1955

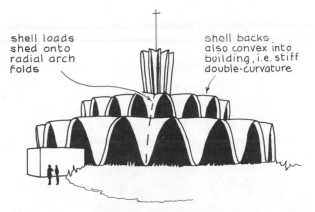

1.59 Convex shells in combination. Church at St Louis, USA

The church (1.59) outside St Louis, Missouri, by Hellmuth, Obata & Kassabaum, is an elegant example of convex shells in combination, two tiers of segments and lantern all radiating from a common centre. This form reflects the way the church works, centrally planned with the congregation surrounding the altar (the Reformation plan really) and side chapels on the outer wall. Structurally, the first thing to notice is the stiffening folds which occur naturally between the segments. The folds of the lower tier are continuous with those of the upper tier so that the whole building is acting like a radially ribbed dome. The shells themselves are parabolic at their open edges and, obviously, shed their loads to these ribs. The ribs, by flanking the shells, help them hold their stiff form. But there is something else. Although the shells appear to have single curvature only, there is, in fact, a slight arching along their backs, i.e. into the building, giving them allied double curvature, so that they are non-developable and much stronger. The church is an interplay of surface light and form, a design that could only have been achieved in concrete shell construction.

In Saarinen's TWA terminal building at Kennedy Airport (1.60), the cantilevering roof shells are like out-

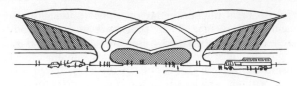

1.60 TWA terminal, Kennedy Airport, New York – part cantilever, part shell. A free form, defying exact stress analysis

spread wings. This is really an example of free form, form neither derived from geometry nor lending itself to exact analysis. Intuition and experience must have led the thinking. Here, again, was a case for modelling techniques, exhaustive scaled-down tests to supplement the blank spots in the mathematical picture.

Correct structural form, or workable structural form, need not necessarily spring from the shapes of geometry or mathematics – who would try and stress-analyse a thigh bone? Often, structural things are regular only because they are easier to make and assemble that way and, to be fair, because we prefer square interiors to irregular ones. Improving methods of stress analysis may see a growing use of free form in coming years.

In closing, we come back to the spectacular, massive 'shells' of Sydney Opera House (1.61), completed in 1974

1.61 Sydney Opera House, 1974. In truth, these are not really shells at all – rather tied-back, arching vaults

– and, in fact, not really composed of shells at all. The architect B. Utzon's early resignation and other politics are now history. Suffice it to say, it was left to the engineers, Ove Arup & Partners, to translate the competition-winning concept into a reality that worked. It was confirmed that the roofs could never be built as the simple shell membranes intended because the shape would inevitably attract large bending moments. But, briefed to preserve the envisaged form, the engineers investigated a series of structural solutions, including parabolic shells with stiffening ribs, and steel space-frame shells sandwiched inside concrete skins. The computers whirred away and nothing worked. Curiously, the final solution (and it took 20 000 man-hours, we are told, to find it) is one that happens to bring us full circle, for the roofs ended up as the largest compression masonry vaults in the world. The 'stones' are of precast concrete, arching across each vault and tied back at right angles by post-tensioned steel wires running down inside the masonry thickness to the supports. It has been said that Sydney was too much a preconceived form and too little a form derived from structural or internal planning needs. But, at $A100 000 000, was it not an inspired folly in the grandest tradition, surprising, and in many ways gratifying, to see built at all?

Summary

- Form-resistant surfaces – plates and shells – are membranes. They are stiff, owing to their shape, developing compression, tension and shear within their surface planes.
- Folded plates develop both beam action (primarily owing to the folds acting as stiffening ribs) and membrane action: they are developable and so need restraint against flattening out.
- Single-curvature shells have similar characteristics.
- Double-curvature shells are non-developable – membrane action locks their shape: they can have opposed double curvature, e.g. the hyperbolic paraboloid, or allied double curvature, e.g. the hemisphere.
- Opposed-curvature shells are produceable from straight 'plank' shuttering: allied-curvature shells are harder to construct but achieve minimal enclosure and are very strong.

2

ENCLOSURE

INTRODUCTION

Having a roof over our heads means more than to be dry. It includes our most basic ideas of comfort, well-being and security. Man operates comfortably within quite narrow temperature limits only, about 18 – 30° C and, historically, his growing inventiveness in shelter must have gone hand in hand with extending his territory and the things he could do. Obviously, shelter is sharply influenced by climate and the materials to hand. Desert mud dwellings with thick walls and small windows, lightweight jungle leaf huts, insulating log cabins, igloos, all are responding to climate and place.

From early times, we find the enclosure acting as a quite complicated multi-purpose mechanism. In pitched roofs, with tiles or thatch, and chimneys through, and in masonry or timber walls with small openable windows and shutters, the building skin is acting as both barrier and filter. It is barrier against the effects of cold or extreme heat, wind, rain, moderating the conditions outside to those needed within. It is also a barrier against intruders. The windows and doors are filters, maintaining the barriers but allowing controlled ventilation, daylighting, views out and friendly access.

Modern expectations of comfort place ever greater demands on the enclosure, yet scale requires it to be lighter and there are associated needs for standardisation, fast erection and low labour content. *In theory*, at least, technology has matched the need. Materials are stronger and the physical performance of individual components has improved enormously. The mediaeval builder would be amazed at the insulating efficiency of expanded polystyrene and double glazing, and at the weather resistance of synthetic membranes, claddings and joints but, considering the performance of some modern enclosures, he might occasionally be forgiven for wondering that technology had bothered at all.

Complexity has been the main problem. The traditional wall usually supported its loads and acted as shelter all in the same material, but the modern wall is as often a structural frame with cladding skin, comprising separate layers each doing a different job, waterproofing, insulating, vapour checking, interior surfacing, all presenting the architect with a host of decisions and chances for error. Claddings, windows and other elements have to co-ordinate with the supporting structure, offering further pitfalls in dimensioning, erection and jointing. Established building practice has inevitably lagged, sometimes, behind the pace of change.

But it is in thermal performance that the building enclosure has its most urgent need of improvement by far. There is now sharp awareness of what has come to be called the 'energy crisis' and yet, energy-wise, many a modern enclosure still performs worse than its traditional counterpart. Earlier this century, as enclosures lightened and windows became larger, and when central heating and cooling systems improved, energy was still cheap and, hardly surprisingly, there came a tendency to underemphasise the enclosure's thermal role and rely on environmental services to put things right. Now, suddenly, energy is no longer cheap and practice has been badly caught out. Insulation standards are rising in a rush but, as we shall see, there are other things crucial to thermal performance that must strengthen as basic influences on the drawing board.

The physical theory behind the enclosure's successful performance is less amenable to intuitive understanding than was structural theory and, whereas the theory and practice in the previous chapter progressed rather happily together, the design of the enclosure is subject to so many different and often conflicting influences – climate, building-type and materials – that the permutations are countless and it is harder to say 'this is the theory and this is how it's done'. So we must start by concentrating on the *principles* – heat kept in, rain kept out and so on – and we will do so by having a strictly *hypothetical* house, a 'small' house, rather than one resolved into a finished design.

THE 'SMALL' HOUSE

The shape of the small house enclosure is, of course, fundamentally influenced by the internal planning arrangements it contains, but our essential interest here is in how it works, how it responds to the variety of shelter needs and other practical influences (2.1). We can break these down into the following headings.

- Structural performance
- Thermal performance
- Water exclusion
- Daylighting, views and ventilation, i.e. windows
- Access, i.e. doors
- Security
- Noise exclusion
- Cost
- Durability
- Appearance

Of these, structure, lighting and noise exclusion have their own chapters, coverage here being on the ways they specifically affect the enclosure itself.

STRUCTURAL PERFORMANCE

Constructional types

Broadly, the enclosure elements, the floor, walls and roof, fall into three constructional categories –

masonry, monolithic (which really means concrete, since plastics and so on can be discounted as serious options as yet) and supporting frame with cladding.

For the floor, the masonry choice, brick or stone, is well in the past. The monolithic choice is the in-situ concrete slab with the underlying ground as its permanent shuttering. The frame and panel choice is the suspended joist floor. For the walls, monolithic concrete would rarely be economic at small scale where its potential strength would be largely redundant, and the shuttering and other erection costs disproportionately expensive. The common masonry option is the cavity wall, probably with brick outer leaf and insulating-block inner leaf. The frame choice is timber, traditionally with weather boarding as cladding but now, also, sheathed with factory-made panels, plastics or tile-clad sheet material – or, as a hybrid, the masonry wrapping mentioned in the last chapter. The roof choice, pitched or flat, is usually frame and cladding – the pitched-truss roof, tiled or slated, or the flat roof with horizontal joists and top decking. Clearly, we will be looking at these constructions in more detail, but these central examples are a convenient peg on which to hang our discussion.

Loads

Structure is all about resisting loads. In the last chapter, we were concerned with primary loads, here, we are concerned with secondary loads, playing on the enclosure itself, against which it has to retain its stiffness and integrity. Admittedly, the functions in resisting primary and secondary loads are sometimes hard to distinguish

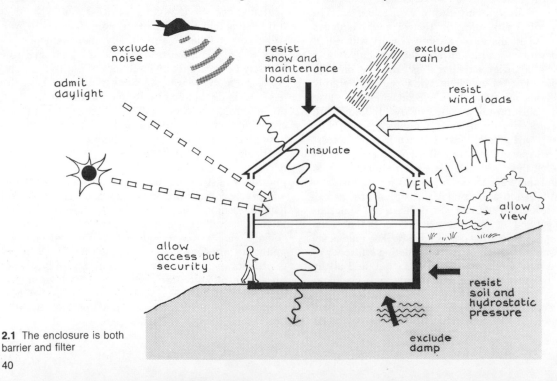

2.1 The enclosure is both barrier and filter

– the masonry wall strong enough for vertical support and lateral stability will usually have little trouble withstanding wind-buffeting or ladders leaning up against it. With a frame and panel wall, the distinction is clearer, though here again there will be times when the panels may be stiffening the frame in the primary sense as well as being secondarily robust in themselves.

Wind load

Here, site location is a first factor, of course. Design for wind will have added significance with the house on an exposed seaboard site, as compared with a sheltered city one. Chapter 1 *Structure* treated wind as a simple sideways force to be resisted by the building's lateral stability but, actually, the forces can be more complex and threatening than that. They can come from all angles. Buildings are hardly streamlined shapes and the simple house will smash the surrounding airflow into buffeting turbulence. And there may be irregular gusting, owing to the related effects of ground contours or adjacent buildings. In a rain-bearing gale, water in the air-stream adds greatly to the mass hitting the building.

There is suction as well as positive pressure. In the first shape (2.2), things are much as might be expected, pressure to windward and suction to leeward. Just as a car at speed develops pressure build-up in front and a vacuum behind, owing to the air's inability to rush in instantly and fill the space just vacated, so the house at rest creates suction and pressure in the moving airflow. But, on the flatter roof of the middle house, one might be surprised, at first, to find suction on the windward pitch as well. This is partly explained by air 'piling up' ahead of the windward wall and itself deflecting the incoming air upwards before it hits the roof. But, also, it is a fundamental law of aerodynamics that a gas in motion parallel to a surface exerts less pressure on the surface than it does at rest – the opposite of what might be expected. Air flowing into a room past a half-open door exerts reduced pressure on the face adjacent to it, so that the excess pressure from the still air inside slams the door shut. Wind blowing over a roof can cause a net uplift because it exerts less pressure than the air underneath. This is particularly apparent in the flat roof shape.

Wind gusting can bring further problems. With a constant wind speed, the pressure inside the enclosure construction is able to equal that outside, as air passes through the tiles or cladding – assuming we are not talking about a continuous 'windproof' membrane. A gusting wind allows no time for this equalisation and can cause sudden, damaging stresses.

The important thing is to appreciate the variety of wind forces, and to bear their nature in mind when we come to look at the enclosure elements – windows, claddings and so on – in more detail. Fairly obviously, if the simple house has a pitched roof, then the trusses will need tying down to the wall head and the tiles will need secure fastening to the battens below them. But it is less obvious, perhaps, that windows or wall claddings are as likely to blow out as in, and that a flat roof membrane needs to be firmly bonded to the underlying screed or decking and, perhaps, weighted on top with tiles or stone chips.

Snow and maintenance loads

The roof also has to contend with snow loads and with the weight of maintenance personnel. The design of the structural members must allow for these additions and, on the smaller scale, flat roof coverings and, ideally, the tiles or the like on pitched roofs, should resist the concentrated load of a footfall. Wall coverings have to be able to support ladders leaning on them and, near ground level, have to resist minor impacts. Masonry copes easily here but a lighter panel wall may have to have impact strength consciously designed-in.

Soil and water-pressure loads

Basements pose their own special problems. The small house has no basement as such, but is shown set into the hill, creating a basement-type wall on the uphill side (2.3). Incidentally, noting the original ground line, it is easy to see how the cut-and-fill technique minimises excavation.

The vertical wall is most tested, having to act as a retaining wall against the horizontal thrusts. Now, these thrusts come not only from the surrounding soil but, also, from the ground water it contains. In fact, it is water pressures that are the most likely cause of trouble. As shown, the static head of pressure at any

2.2 Loads from wind. Apart from buffeting turbulence, the house profile induces pressure and suction in the moving airflow

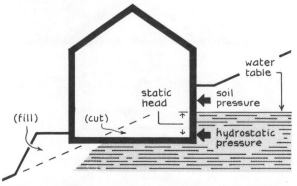

2.3 Loads below ground from soil and hydrostatic pressure

point is dependent on the height to which the *water table*, the level to which the soil is saturated, rises above it, and this pressure can be greater than anything soil itself can produce. Anticipating the likely water table is not always so easy. In porous sand or gravel soils, it is usually fairly constant but, in clay soils, it is notoriously variable and a table level found to be well below the proposed building depth, by a trial bore in the initial site survey, may rise sharply after the building is built. Often, the safe thing is to assume saturation to ground level and ensure that the structure and water impermeability are designed adequate to cope.

New construction can, itself, dam the soil's existing natural drainage, for example, on a sloping site, and cause locally high water tables. Water tables rising during construction have floated concrete basements, as yet unloaded by the intended building above, out of the ground.

But, at the small house scale, the masonry wall can usually resist these pressures by its inherent mass – it can have an added course width at basement level – and by its cellular shape on plan. We shall see later how it is waterproofed. Reinforced concrete is a viable alternative, since the whole basement then acts structurally as a monolithic, rigid box.

Loads from movement

The enclosure fabric is bound to suffer intrinsic movements owing to changes in its moisture content and temperature, and may suffer ground movements as well. The heading 'loads from movement' is literal enough but, although movements can be sometimes restrained, the defence is generally to design so that movements are minimised and/or enabled to occur freely without damaging stresses developing.

Ground movement. Here, again, the inevitable overlap with the last chapter – good siting and adequate

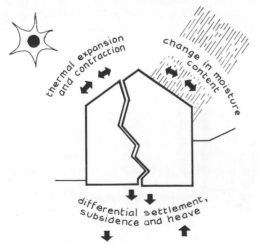

2.4 'Loads' from movement

strip foundation or overall slab foundations are defence against the movement damage to walls and slab itself from uneven settlement, and ground subsidence and heave.

Moisture movement. Moisture movement occurs in porous building materials, such as brick, concrete, some plastics and, of course, timber – expansion on wetting and shrinkage on drying. In masonry and concrete, the movements should be small enough to be accommodated within the construction. Masonry, especially, has the inherent leeway of being able to move imperceptibly within its joints, and concrete derives some restraint from its reinforcement. Concrete, though, and mortar and plasterwork, can suffer damaging shrinkage as they dry out after construction – a rather extreme analogy is a cracked, dried-out, river bed. The main prevention is to have the initial mixes as dry as is possible, often with additives to improve the otherwise reduced workability which would result.

Timber is a notorious moisture-mover. In the living tree, the moisture content is around 50 per cent by weight, depending on type, and this drops to 10 or 15 per cent for timber used indoors and, perhaps, 20 per cent outdoors, i.e. as the moisture content stabilises with the relative humidity of the surrounding air. A lot of water has to be lost between forest and building site and, for the better grades of timber, the drying out, 'seasoning', process has to be slow and even, often over a period of months or more, to avoid unequal contraction and warping. By the time of construction, timber should already have reached a moisture content at least closer to the 10% or lower value to which it may be subject in use, thereby minimising movement later.

Thermal movement. Finally, there is thermal movement. Put simply, the molecules in any material are in a state of continuous, vibratory motion, the inherent energy we call heat – an ice block has heat, there being no such thing as cold, only relative degrees of 'hotness'. When we heat a brick or a steel rod, or the sun shines on the small house wall, the vibrations are excited further. There is an increased 'jostling' and it causes expansion. Cooling causes contraction. Different materials move by different amounts. The usual basis for comparison is the *coefficient of linear expansion,* which is simply the unit increase, per unit length, per unit increase in temperature – for steel and concrete it is around 0.000 012 per $^\circ$ C, a lucky similarity since, otherwise, temperature change would tend to break the steel-to-concrete bond in reinforced concrete.

Actually, in domestic buildings, the relatively small scale of the enclosure elements will help prevent thermal movements from building up to damaging proportions. And, remember, masonry has some inherent slack in its joints, and timber is flexible and, in contrast to its moisture movement characteristics, hardly moves thermally at all. But, as we shall see, the designer ignoring the chance of thermal movements in large buildings is asking for trouble.

Summary

- Structural choices can be related to the three categories, masonry, monolithic and supporting frame with cladding.
- Often, enclosure is also a main structural element, resisting primary loads, e.g. where there is a supporting wall. Generally, there has to be integrity against secondary loads, including wind, snow, maintenance and impact loads above ground, and soil and water-table pressures below ground.
- Movement loads, especially thermal, are more significant in large buildings: design aim is normally to reduce movements and/or allow for their free occurrence.

THERMAL PERFORMANCE

Good thermal protection from the enclosure means greater comfort and, ever more important, less energy consumption in heating and cooling. The extremely loose equations

$$
\begin{aligned}
\text{clothing} + \text{body heat} &= \text{comfort} \\
\text{building enclosure} + \text{services input} &= \text{comfort}
\end{aligned}
$$

are analogous and, just as warm clothing in winter, and light, reflective clothing in summer, mean less for our body mechanisms to do, so does an efficient enclosure mean less for the environmental services to do – and less energy consumed. Really deficient enclosures may pose problems the services can never meet. Insulation is, of course, a main factor in good thermal enclosure, but it is by no means the only one.

Ways in which heat is transmitted

Thermal performance has mainly to do with reducing heat transmission outwards – or inwards – through the enclosure. Where there is a temperature difference between two places, heat tends to flow from the higher temperature to the lower, nature always hating imbalance, and the transmission can occur in three ways, conduction, convection and radiation (2.5).

Conduction is where heat passes through a solid. The process is easy to visualise. If one face of a brick wall is heated, then the vibrations of the atomic particles on the surface will intensify, pass their added excitement to the particles behind them and so on as a jostling chain-reaction through the wall. The energy moves but the matter does not – picture someone pushed hard at the head of a queue.

In *convection*, the matter does move since it is heat

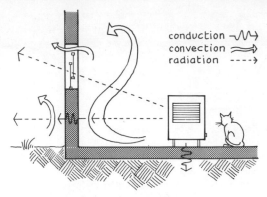

2.5 Heat transmission and, hence, loss, is by conduction, convection and radiation

transmission by the flow of a liquid or gas. Air currents collect heat from warmer surfaces and impart it to cooler ones. In fact, it is the local temperature differences that cause the currents – thus, air warmed by the stove shown expands, becomes less dense and starts to float upwards over cooler, denser air flowing in to replace it. This is natural convection, as opposed to forced convection by mechanical fans. (Wind is natural convection on the grandest scale, powered by the heat of the sun.)

Radiation involves no matter at all in the accepted sense, being energy transfer by electromagnetic waves. All bodies emit radiation at various wavelengths, including those of visible light and heat. 'Radiant heat' intensity increases with the temperature of the source – although 'radiant heat' is actually something of a misnomer since it is radiation *energy*, and not heat, that travels: the heat is only induced when the radiation strikes other matter, as when the radiation from the stove converts to heat in the house around it.

Climate influences on design

Obviously, transmission through the skin will vary with the temperature difference across it, so the first determinant is climate. Early design decisions can reduce climate impact, even before consideration of the insulation of the skin itself.

The influence of site location

The site location is the starting point of the small house. In the extremely unlikely event of there being a free choice, and assuming the climate is temperate so that cold stresses in winter count more than hot stresses in summer, the site (2.6) has advantages. It avoids the valley floor, where cool dense air tends to collect and hold the temperature several degrees below the prevailing average. Similarly, it avoids the wind-prone hill crest – convection losses increase sharply with the speed of the surrounding air-stream. There could be a 30 per cent, or more, heat-loss different between the small

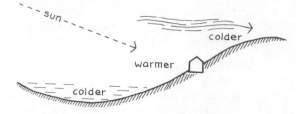

2.6 Site affects thermal performance

house in exposed and sheltered locations. Half-way up the sun-facing slope would be a lucky site!

Conversely, in hot climates, the criteria may reverse, with buildings sited specifically for shade, or to catch whatever cooling breezes are going.

The influence of climate on building form

A building's heat loss or gain must increase with the area of surface it exposes to the air outside and so there is a thermal argument for keeping this to a minimum for the volume enclosed. This means a hemispherical building theoretically, but practically something approaching a cube. The unkinder the climate, the stronger the argument becomes. Nature adapts form to climate and so do traditions in building (2.7). In the cold latitudes, the conifer with its short spines is a compact, 'introverted', sort of tree, closing itself away from the air around it, just at the log cabin is compact. In temperate climates, vegetation can merge more freely with its surroundings, and so can buildings there. Freer planning still involves some thermal penalty, especially in winter, but it is less severe and, hence, is acceptable if other planning advantages result. Hot humid climates see similar expansiveness, with buildings shaded against sun but open-walled for ventilation. In hot dry climates, the cactus and desert buildings are emphatically back in compact form.

But, even in temperate regions, a two-storey, i.e. cubic version of the small house, could be around 20 per cent cheaper to heat in winter than the flatter single-storey version with the same floor area. Thermally speaking, the house should, ideally, have no projections on plan, no small extensions. Just as a heating radiator has fins to increase heat transmission, so the building skin should be plain to minimise it.

The influence of solar radiation on optimum plan shape and orientation

But there is another thermal influence on shape which, in temperate climates especially, tends to offset the compactness argument. It would obviously be a good thing if a building could be shaped to collect as much sun as possible in winter, and yet avoid collecting too much in summer – and, interestingly, it is possible. (2.8)

Take the winter situation in the northern hemisphere. Most of the sun's heating effect occurs in the middle of the day for, in the morning and afternoon, the sun is low on the horizon and its effect is obscured and weak. So, if the building is elongated on the east-west axis, giving it a relatively longer southern wall, it will be exposing a larger collecting surface to the available radiation. So far so good: but what is, at first, surprising is that this plan shape is also the one best suited for *avoiding* excessive summer gain. The long south wall is not so vulnerable then, simply because the summer sun is so much higher in the sky, meaning that the radiation on the wall is very oblique and, hence, diluted. In other words, from the sun's viewpoint, less

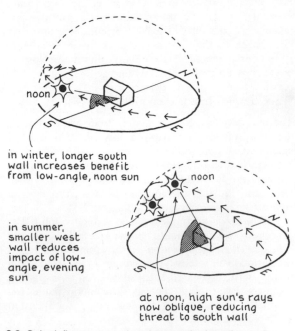

in winter, longer south wall increases benefit from low-angle, noon sun

in summer, smaller west wall reduces impact of low-angle, evening sun

at noon, high sun's rays now oblique, reducing threat to south wall

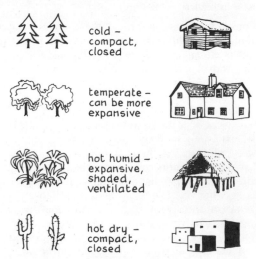

cold – compact, closed

temperate – can be more expansive

hot humid – expansive, shaded, ventilated

hot dry – compact, closed

2.7 In building, as in nature, form adapts to climate

2.8 Solar influences on plan shape and orientation

area of south wall is 'visible' from the higher elevation. The roof is more vulnerable, but the plan shape does not directly affect the roof area. In summer, the vulnerable times of day are fairly early morning and late afternoon, when the sun is lower and the radiation arrives at an angle more normal to the walls and windows, and this is exactly why the elongated east-west plan scores again, because it presents its shorter east and west elevations to the sun at those times.

In fact, as a further refinement, the ideal plan would turn, so that its main 'south' aspect faced slightly east of south. The reason for this is twofold. The sun's path through the sky is such that the daily solar radiation in winter peaks in value when the sun is slightly east of due south and, in summer, when it is to the west of due south. Add the criteria of winter collection and summer avoidance, and the eastward adjustment is logical. Also, in the morning, the air is still cold from the night before and an earlier radiation gain will help offset this, increasing morning insolation when the temperature build-up is most useful. Again the optimum adjustment varies with latitude but it is about 15° east in a temperate climate.

The air temperature/solar compromise
The air-temperature compactness argument and the solar elongation argument pull in opposite directions, and the importance of one in relation to the other varies with latitude. In zones of extreme cold or heat, the air-temperature influence dominates, compressing the building towards the cube. In more temperate zones, this requirement reduces, allowing the rival solar influence to elongate the plan. It would be unrealistic to give optimum dimensions because the calculation is complicated by other variables. For example, a building's site exposure and the enclosure's relative insulation efficiency both increase the air-temperature influence. Window sizes and allocation to different elevations affect the solar influence. But, for interest, one is talking of an axis ratio in the order of 1/1.1 in cold latitudes, 1/1.5 or more in temperate regions and returning to around 1/1.3 in hot dry regions.

The effect of windows
The balancing act between reducing heat transmission and yet capturing solar radiation applies equally to window sizing on the respective elevations, save that the overriding influence is more urgently between providing adequate daylighting while satisfying thermal needs as a whole. Even double glazing has less than 40 per cent of the insulating value of a block/brick cavity wall, and is at least 20 times more admissive to radiation. So, thermal questions arise sharply. The extent to which daylighting and thermal needs align or conflict depends on climate. In the hot, dry climate they align, the very bright, hot conditions favouring small windows in both cases. In moderately warm climates, windows can be

larger, and the southerly window may usefully add solar gain in the underheated winter condition – as explained, it will be less affected by the high altitude of the summer sun, but there is a case for the smaller westerly window to avoid the summer afternoon sun pervading and delaying the welcome cool of evening. In the temperate/cool climate, such as in most of Northern Europe and the northern USA, daylighting and thermal needs tend to conflict. Basically, windows there should be as small as daylighting needs allow, save that a larger southerly window does at least have the merit of allowing solar gain in winter. Of course, larger southern windows increase *conductive* losses to the outside air, which may persist even when the *radiation* gain is occurring, and so are prime candidates for double glazing. That said, this is all very general – latitude, climate, internal planning and other design factors all interact.

Solar shading
Good shading can reduce solar radiation impact by 80 per cent or more. In small buildings, this is useful in temperate climates and very important in hot ones – in large buildings, shading is very important almost everywhere, for reasons which will be explained later.

At domestic scale, trees are a first line of defence – deciduous varieties, especially, offering an opaque screen in summer but being conveniently bare in winter when the sun's rays are welcome. Architecturally, roof overhangs and wall projections offer partial screening. In view of the sun's path through the sky, it follows that overhangs are more effective on southerly walls, while vertical fin-like projections from the walls are more effective to the east and vulnerable west. One is reminded of the boldly projecting roof profiles of the California and Prairie schools' houses much earlier this century, notably by the Greene brothers and by Frank Lloyd Wright. There was a sharp environmental awareness there, although it has to be admitted that the highly fenestrated walls, which were another characteristic of these houses, brought thermal penalties, both in summer radiant gain and, especially unfortunate today, in winter heat loss.

In the ordinary way, window vulnerability to solar gain is answered by blinds or shutters and 2.9 shows the importance of their placement in the inside-to-outside enclosure sequence. A Venetian blind inside may be valuable in cutting out visual glare (See Chapter 5 Lighting) but is a poor defence against solar heat gain. Even a fairly light and shiny blind will only reflect about 40 per cent of the incoming radiation back through the glass, absorbing the rest, heating up and, itself, becoming a radiator *within* the room. A blind inside a double-glazed cavity is a common modern solution and is rather more efficient. It, and the cavity, similarly heat up, but the arrangement transmits more of this heat back outwards. However, the best shade position is

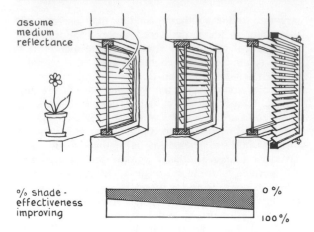

assume
medium
reflectance

% shade-
effectiveness
improving

0 %

100%

2.9 Shading's effectiveness varies with its placement in the enclosure sequence

outside the glass, such as the louvred shutters and screens that we associate, in any case, with hot climates. They intercept the radiation early, and the absorbed portion is then harmlessly re-radiated and air-convected away. Shutters still allow a healthy flow of natural ventilation and, since they can be opened or closed at will, they give us a first idea of the 'tunable' building, alterable to suit the prevailing conditions.

Summary

So, in the cool/temperate climate, from the strictly thermal point of view:

- The ideal site is a sheltered southerly aspect.
- The climate, though generally kind, is still under-heated in winter, so that a fairly compact, two-storey form is desirable.
- The ratio of the plan axes reflects the air temperature/solar compromise – about 1/1.5.
- The optimum orientation is slightly east of due south.
- Window sizes are as small as is compatible with daylighting need and views out.
- Trees offer useful summer shading. Other shading devices are as described although comprehensive shading at domestic scale is really only necessary in hot climates.

Insulation

When the enclosure insulates us from the temperatures outside, there are two quite distinct processes at work, *resistance insulation* and *capacity insulation*. Resistance insulation is generally the most important and, in fact, is the one people usually mean by 'insulation', the enclosure acting as nearly as possible as a thermal barrier, slowing down the rate at which heat is lost, or gained, through it. Clothes and blankets are resistance insulators. Capacity insulation, sometimes called ther-

mal capacity, is where the enclosure acts as a heat reservoir. The material in the enclosure must, itself, take time to heat or cool and this creates a thermal damping effect, reducing and delaying the inside's response to outside peak temperature stresses. Barrier and damper are the key words.

Resistance insulation

You may have seen houses photographed on film sensitive to infra-red (heat) radiation showing temperature differences across the enclosure as a colour-coded, heat-loss picture. Warmer parts are obviously those losing heat faster and, as such, are a symptom of poor insulation behind – not to mention areas leaking out draughts. The insulation or heating-products catalogue is similarly fond of showing the 'typical' heat-loss picture for a house – 30 per cent through the roof, 10 per cent through the windows, 20 per cent owing to draughts and so on. But these values can only be unhelpfully approximate, depending as they do on so many variables – the relative areas of walls, windows and other elements, the particular construction, the building siting. We must consider the factors more generally.

Draught exclusion. The first thing is for the enclosure to be reasonably airtight. There will be unavoidable losses through necessary ventilation, but wrapping a building in an otherwise resistance-insulating skin that is pierced by draughts convecting heat away through ill-fitting doors and windows is about as sensible as running an electric fire on the doorstep outside. The colder and windier the climate, the more critical does draught-sealing become. In new buildings, the main thing is to have well-fitting windows and doors with adequate overlapping checks between openable and fixed parts of the frame – see 2.18 and 2.22. In an existing house, adding compressible draught-stripping around closures is certainly a measure to take before installing expensive double-glazing.

Reducing conduction. The resistance insulation of the enclosure itself depends on the material composition and thickness of its elements. Dense materials, like stone and, especially, metals, in which the component particles are tightly packed, conduct heat more easily and are poor insulators in consequence. Heat passes rapidly up a metal soldering iron but not through its wooden handle. Brick, being lighter and more porous, is a fairly good heat insulator, and wood and aerated building block are better still. Porosity is a positive factor in insulation because the myriad of air pockets contained breaks the material's continuity; air, like all gases, is a poor conductor, its matter being so widely dispersed and, moreover, when pocketed in a material, it cannot flow and convect heat either. It follows though, that porous insulating materials must remain uncompressed and dry to be effective – we become cold when our clothes are wet simply because water has displaced the insulating air.

So, typical insulating materials are light, and porous or cellular – cork tiles, spun-glass quilting, aerated concrete blocks. Expanded polystyrene is aerated plastic and is such a good insulator that it feels warm when we touch it, though really we are only feeling the heat build-up from our own hand. The air gap in a cavity wall, and between the panes in double glazing, helps insulate. (Interestingly, the optimum gap is around 20 mm only: with larger gaps, efficiency actually reduces, owing to convection currents developing between the inner wall faces.)

Surface resistance. The surfaces of materials offer a small added resistance to heat transmission – another reason why double glazing and cavity walls, which present extra surfaces to the heat flow, are good insulators. Convection losses are greatest through textured or corrugated surfaces – they are not 'minimum' surfaces. Also, the radiation through a surface, emitting or receiving, is greater when it is dark and matt-textured and less when it is light-coloured and reflective. Survival blankets, polar exploration shelters, thermos flasks, are all reflective-lined to reduce the radiation component and, in everyday construction, the building skin can usefully incorporate a layer of reflective foil, perhaps as a backing to the plasterboard or insulation quilting, or placed in the wall cavity if there is one. (Admittedly though, another purpose of foil is as a vapour check, reducing condensation, details of which are given later). Conversely, reflectivity is also a protection against incoming solar radiation, hence, the advantage of white clothes in summer and of light coloured, whitewashed, building surfaces in hot climates.

Conductance and U value. The *conductance* of a particular building material or construction – cavity wall, double-glazed window or other composite – is the amount of heat that will conduct through unit area in unit time for unit temperature difference between the faces. But, in practice, the *U value* scale is a more useful insulation measurement, since it takes account of the surface resistances of the material as well. Imperial measurement uses British thermal units/per square foot/per hour/per °F; metric units describe the power flowing at any instant, i.e. watts/per square metre/per °C. The *lower* the U value, the *better* the insulation. Most developed countries now have minimum legal requirements on insulation, their severity tending to reflect the severity of the prevailing winter conditions. 2.10 shows U values for some common constructions.

The enclosure elements. If the lowest floor is an oversite slab, it is not necessarily insulated – not in reasonably temperate regions anyway – on the argument that the mass of the slab and ground underneath act as a 'heat sink' dissipating the heat more slowly

construction		U value
floor		
	carpeting; underlay; 20 mm timber deck on joists; cross-ventilated floor void under	1.90
	wood-block floor on bedding compound; 105 mm concrete slab; (membrane on hardcore - see 'Water exclusion' section); ground	0.80
wall		
	105 mm outer brick leaf; unventilated cavity; 100 mm lightweight block inner leaf; plaster	0.95
	... and with 50 mm insulating slab in cavity	0.70
	outer weatherboarding; building-paper backing; timber frame; incorporated 50 mm glass-fibre quilt; foil-backed plasterboard	0.60
	105 mm brick outer leaf; unventilated cavity; incorporated 50 mm quilt; polythene vapour check; plasterboard	0.50
	window; single-glazed, wood frame	5.60
	... and double-glazed with 20 mm airgap	2.50
roof		
	tiles; battens; felt; rafters; 50 mm incorporated quilt; foil-backed plasterboard. Construction ventilated - see 'Condensation'	0.60
	20 mm asphalt / felt; 25 mm minimum screed (i.e. at bottom of drainage fall); 50 mm wood-wool deck; joists; vapour check; plasterboard. (Construction ventilated)	0.80

2.10 Resistance insulation. U values ($W/Ø^2\,°C$) assuming 'normal' weather-exposure. Low value means high performance.

than does the rest of the enclosure above the ground. Sometimes one sees slab insulation restricted to the perimeter band near the external walls where the heat loss is more significant. However, recent research is suggesting that the 'heat sink' effect may be less helpful than had been thought and slab insulation is likely to become more common practice. The insulation can occur as a permanent, rigid shuttering underneath the slab or there are various inherently insulating lightweight slab mixes and topping screeds. The traditional timber suspended floor has the disadvantage of requiring fresh-air ventilation through the floor void underneath. This is to remove moisture build-up from ground damp and from vapour which has permeated from the house interior. Tight-fitting floor boards or (better) sheet-decking, and underlay and carpet reduce the loss.

The traditional 225 mm (9 in) brick wall is unacceptable as an insulator today: instead, there will usually be some form of composite, such as an inner leaf of blockwork with outer leaf of timber construction, externally-rendered block or facing brick. The all-masonry cavity wall can have an insulating fill, for example, polystyrene sheeting. This locks in the air, again, reducing heat convection to the outer leaf. There are also modern techniques for filling existing cavity walls with insulating foam.

The unfilled block/brick cavity wall will normally only meet present UK requirements if it is unventilated – and here a point. Obviously, the air in a cavity wall, like that in the underfloor void, cannot insulate if it is ventilating to the outside, and yet building practice has often argued that cavities be ventilated to dry out any residual dampness resulting from rain penetration or from vapour permeating from the building interior. Perforated air bricks are often inserted in the outer leaf. But, in truth, slight damp in the masonry wall cavity is not a serious defect – that is, provided any collected water at the foot can be drained to the outside. So the ventilation intended for 'good measure' is merely robbing the wall of its insulating air-gap by continually convecting the heat to the outside. At worst, the cavity temperature might be the same as that outside so that, thermally, the outer leaf might as well not be there at all! Recent thinking is that masonry cavities should be sealed, except for the occasional drainage holes at the foot. On the other hand, any dampness in the timber-frame wall would be very damaging. Timber frame construction is proven and reliable, *provided* (as past failures show) that it is knowledgeably designed and constructed – we will come on to this.

Windows obviously benefit from being double-glazed, owing to the air-gap and additional surface resistances that result. But double-glazing is expensive and the reduced conduction will not always, in itself, be sufficient argument for having it. The point is that there are other advantages. The cold pane of the single-glazed window, quite apart from its extra heat loss, tends to cause uncomfortable downdraughts and induces uncomfortable radiation losses from people

nearby (see Chapter 3 *Climate services*). It is much less effective as a noise barrier (see Chapter 6 *Acoustics*). And double-glazing units, added to existing windows, may cut draughts.

Pitched roofs will be insulated at the level of the ceiling joists or, if the attic space is to be used – as it jolly well should be – then the insulation can be placed between the sloping rafters. Insulants between ceiling joists are glass-fibre quilt, which comes in rolls, and loose particle fill, which comes in bags. The insulation should not be carried through under the water tank there in case the water freezes: instead, this should be boxed in, e.g. by polystyrene sheeting. Insulation between rafters can again be by polystyrene sheet cut to size, or by quilting if there is internal tongue-and-groove boarding or other lining to support it. There are condensation implications in all this which we shall come on to.

But to sum up, even in today's energy context there must be limits to how much insulation is sensible – limits that obviously vary with the length and severity of the heating season. Insulation is only one factor, albeit very significant, in good thermal performance, and the budget allocated to it has to be weighed in terms of the whole thermal strategy of the building. We will be elaborating on this. Also proper insulation strategy in 'total energy costing' has to consider the total energy cost in producing insulation, e.g. at the mine and factory and in transport. As with all thermal measures a common yardstick is to balance the initial capital cost against the *pay-back* time, the time it takes the resulting fuel-saving to repay the first investment. But here there is a qualification the other way. One has to allow that fuel costs will almost certainly go on rising in absolute terms, potentially distorting the original equation. At least it is better to tend towards 'too much' insulation rather than have 'just enough'.

Capacity insulation

A church on a hot day seems relatively cool inside. Its heavy walls are protecting the interior but, here, it is owing to *capacity insulation* as well as to resistance insulation. Matter requires energy to raise its temperature and heavy walls will require more energy than would light ones. They will take longer to heat up and longer to cool down. Whereas resistance insulation in the church walls reduces the rate at which heat flows through them in response to temperature differences either side, capacity insulation, sometimes called 'thermal capacity', reduces, and delays by several hours, the effect inside of temperature fluctuations outside (2.11). The walls and, of course, the floor and, to a lesser extent, the roof, pews and so on, bring a thermal inertia or damping so that *the temperature inside remains closer to the 24-hour average, i.e. the average for that time of year, than in a building with a lighter construction*. Think of the difference between a heavy boat and

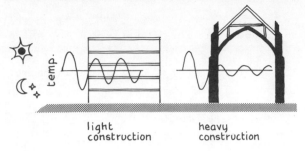

2.11 Capacity insulation reduces and delays the effect felt inside of temperature changes outside

2.12 Resistance insulation placed towards the inside isolates the high-capacity wall and so speeds response to the heating system

a light one on a choppy sea. The temperature inside the great Egyptian pyramids must hardly vary ever.

Resistance or capacity insulation?

With capacity insulation increasing with density, and resistance insulation with air-entrainment, it follows that the two properties tend to be mutually exclusive in any one material. Usually, resistance is more important than capacity, but not always, and the importance of one relative to the other varies with climate.

Capacity is very important in *hot dry* climates, where days are hot and nights, with no insulating cloud cover, are very cold. There, heavy walls can 'store the chill of night', so to speak, and use it to temper the impact of the peak tempratures by day and vice versa. The heavy-construction, small-windowed house benefits from this kind of thermal trade-off. In *temperate* climates, in central Europe and part of the USA for example, capacity insulation still has summer value but, with the peak impact and day-night variation less, the need is only moderate. Most of the time, temperatures are below comfort levels, so the need is primarily for resistance insulation. Insulated cavity walls, or composite block and timber walls, give the best of both worlds, but the lighter all-timber construction is at no serious thermal disadvantage there.

In *prevailing cold* climates, the need for capacity insulation virtually disappears. It would only serve to stabilise temperatures in the discomfort zone. There has to be heating and good resistance insulation. It was lucky that timber, such a good resistance insulator, was widely available to foster the timber-building traditions in North America and Scandinavia.

The effect of resistance and capacity insulation on heating response speed

Often, capacity and resistance elements will combine in one composite wall, for example the masonry wall with an insulation layer, and the order in which they are placed will affect the interior's speed of response to the heating or cooling system. If the resistance insulation is outside, the thermal inertia of the heavy wall, exposed towards the inside, will slow the response. If

the building is continuously used, like an ordinary house, this is undesirable rather than serious. But, if it is intermittently used, like a weekend cottage, then it is not much use if the occupants are leaving before the walls finally warm and the heating starts to tell. There, resistance insulation on the *inside* would speed the response by isolating the heavy wall from the heating-up process (2.12).

Summary

- Resistance insulation is a thermal barrier – factors are material conductance and surface resistance.
- U value is measure of a construction's resistance – metric units are $W/m^2 \, °C$.
- Heat-saving strategy includes draught prevention and reducing enclosure element U values, mainly by use of air-entraining materials, double glazing and air cavities.
- Appropriate insulation levels are assessed in context of total energy costing, and pay-back time versus capital cost.
- Capacity insulation is a thermal damper, an inherent property of heavier construction: helpfully reduces amplitude of internal response to outside temperature change – most important where daily variation is extreme.
- Resistance and capacity insulation placement affects speed of interior response to heating (or cooling) system: intermittent heating asks for resistance insulation towards inside.

Condensation

Condensation attaches logically to thermal performance, since the temperature reduction inside to outside, through the enclosure, is a principal factor in its occurrence. *Surface condensation* is a familiar nuisance – undesirable when humid air hits cold kitchen and bathroom windows and potentially damaging when, in extreme form, it leads to residual dampness in plaster and other finishes, possibly causing mould growth in time. But the most serious type and one which has caused notorious problems with modern lightweight enclosures, is *interstitial condensation*, condensation occurring within the constructional thickness of the enclosure itself.

The air in a house readily absorbs water vapour from people, washing, cooking and so on, and the amount of vapour it can absorb, its water-carrying capacity, increases with temperature. Thus, a given volume of air, only normally humid at room temperature, becomes progressively more humid if it is cooled, although the actual amount of water present remains the same. Eventually, it reaches saturation temperature and condenses into water droplets. The temperature at which saturation and condensation occurs is appropriately enough called the *dew point*.

The relative 'wetness' of a given volume of air at a given temperature is described as its _relative humidity_. This is simply the amount of water vapour present in the air relative to the amount of water vapour the air *could* carry at that temperature, expressed as a percentage. If the relative humidity is 50 per cent, half the air's water carrying capacity is being used. Reducing that air's temperature reduces its potential capacity, eventually raising the relative humidity to 100 per cent or dew point.

In understanding the principles (2.13), it is not important to be precise about the temperatures and relative humidities either side of the enclosure. Suffice it to say that 20 °C and 50 per cent RH inside is a fairly normal sort of domestic atmosphere, and 0 °C and 95 per cent RH outside represent cold winter conditions in which condensation is most likely to occur. Note that the air outside has higher RH than the air inside only because it is colder – *it actually contains less water vapour*.

In the wall section (a), the solid line shows the actual temperature gradient towards the outside, and the dotted line shows the dew-point temperature gradient, the temperature at which air of that particular vapour content would start to condense. The reason for the actual temperature gradient is obvious, but a word on the dew-point gradient. Masonry is fairly permeable to water vapour, which wants to spread from places of high concentration to places of low concentration almost in the way an ink drop wants to spread on the surface of water. In fact, we talk of *vapour pressure*. The result is a vapour-content gradient and, hence, dew-point gradient through the wall, and the inside and outside dew-point values give us a likely line for this. Assuming the wall is of regular consistency, it follows that the temperature and dew-point gradients will also be regular. By geometry, if condensation fails to occur on the inside face, i.e. the dew point there being safely below the actual temperature, it is unlikely to occur interstitially in the wall either.

The trouble starts where the nature of the enclosure construction is such as to distort one or both of the gradients so that the lines cross. In the window section (2.13b), it is a simple case of warm air hitting cold glass, the humidity outside being irrelevant, since glass is

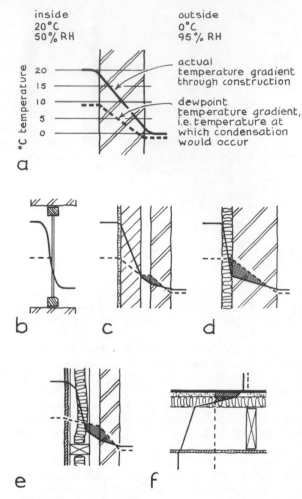

2.13 Causes of condensation

impermeable to vapour. The temperature gradient falls, but the vapour pressure and, hence, dew-point temperature gradient does not. The lines cross and condensation occurs. And similar physics can occur with more damaging consequences at any point where the enclosure is bridged by an element with poor resistance insulation, for example, a concrete window lintel. This so-called *'cold bridge'* is avoided by insulation wrapping the exposed outer face or the interior face – we will return to this later.

Distortion can also occur interstitially in cavity construction (2.13c) although 'trouble' might there be putting it too strongly, since some condensation in a cavity is normal and can drain away. But there could certainly be trouble in 2.13d where the added insulation layer sharply distorts the temperature gradient, inviting condensation on the resultingly cold inner brick face immediately beyond it. In fact condensation could occur anywhere within the line overlap and apart from the inevitable damage in time, the lowered insulation value of the dampened construction could all but defeat the point of adding the insulation in the first place! The masonry-wrapped, timber-frame wall (2.13e) would invite similar problems if no other precautions were taken.

The timber flat roof (2.13f) has the added problem that the dew-point gradient is distorted. The continuous membrane keeping the rain out *has the unfortunate, related property of trapping the vapour in*, preventing its escape to the drier, outside air. Assuming, for the sake of argument, there is no ventilation connection with the outside, the vapour pressure and, hence, dew-point level right through the construction will build up to, and stabilise at, the higher indoor level. Actually, roof constructions are all the more at risk in that their external surfaces are liable to suffer lower temperatures than the other parts of the enclosure. Their very open exposure to the elements can be one reason for this, but another factor is *sky radiation*, the roof's further *radiant* heat loss to the sky – in effect, to space. On a clear winter's night, this added component can drop roof temperatures to several degrees below the ambient air temperature. This and, in the recent past, the popular adoption of flat membrane roofs, coupled with the rise in required insulation levels, all these led to failures wholesale until the ways of preventing interstitial condensation came to be properly understood – if they are properly understood even now.

A final point in passing, emerging from all the above, is that the extent to which a room's internal surfaces, or its enclosure construction, are prone to condensation, further depend on its purpose. The living room is inherently safer than the unventilated, and seldom used, spare bedroom on an exposed flank wall, or under the roof.

Preventing condensation
Heating and ventilation. There are two immediate steps in avoiding visible condensation on windows and inside wall surfaces. One is to improve the heating and, thus, raise the actual temperature gradient line, the other is to improve ventilation to the outside and, thus, reduce the vapour content and lower the dew-point line. Of course, ventilation is limited by the need to conserve heat, but direct extracts from kitchens and bathrooms are sensible in attacking the vapour problem at source. However, raising the room temperature will do little to reduce interstitial condensation occurring in the outer colder parts of the enclosure construction and, although improved ventilation helps, it cannot be relied on alone. More positive constructional methods are as follows.

Placing the insulation towards the outside. 2.14a has the insulation outside the solid wall, so that the temperature gradient distortion is a safer one. This is not always a practical solution though. Most insulation

products are vulnerable to damp so there are risks in moving them towards the weather side. Also, the insulation, being outside, must slow the house's response to its heating system.

Vapour checking. The masonry-wrapped, timber-frame wall (2.14b) distorts the dew-point gradient in the safe direction by introducing a positive *vapour check*, typically thin plastic sheeting or metal foil, on the *warm* side of the insulation. Foil-backed plasterboard is a common type and, as said, reduces the radiant heat loss into the bargain. If the check is effective, the vapour pressure and, hence, dew point in the enclosure is at the lower outside value.

Vapour checking and construction ventilation. But, in the most testing case, the flat roof with outer membrane (2.14c), a vapour check alone is not enough. Vapour checks cannot be relied on as 100 per cent effective. Inevitably, there will always be some outward seepage through the membrane joints, staple fixings, nail holes and so on – for which reason the term vapour *check* is to be preferred to vapour *barrier*. In the previous example (2.14b), the amount of any vapour seeping through the check would be too small to raise significantly the dew-point line and would permeate harmlessly through the brick to the outside. But the roof's further check ('barrier', in fact) on the outside will cause any seepage to build up in time within the cavity and, depending on the relative vapour-trapping properties of the inner and outer membranes, the dew-point line could again stabilise closer to the indoor value. The answer is to give the cavity some alternative communication with the outside air. A gradual through-ventilation via occasional air-bricks

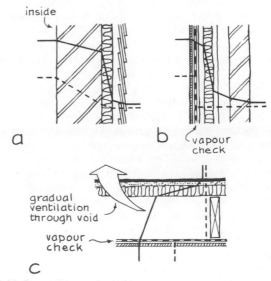

2.14 Preventing condensation

(perforated bricks), or other slots at opposite roof edges, will carry any vapour safely away. A vent area of 0.3 per cent of overall plan area is the rule-of-thumb for flat roofs.

The tendency for the vapour checking and ventilation precautions to be seen as mutually exclusive is curious. Those in favour of vapour checks are fond of pointing out that construction ventilation convects away heat as well as vapour and so reduces insulation. This is true. The ventilation lobby, with almost tiresome predictability, point out that 'it takes just one nail to puncture the check . . .' and this is also true. In the end, fortunately, most architects do both, though some may be hedging their bets rather than realising that this is the sensible compromise? Vapour checking and *gentle* ventilation avoid condensation without significant added heat loss.

Summary

- At its simplest, condensation is caused by warm, moist air reaching cold places. Its preventions are:
- Improving heating and ventilation generally.
- Possibly, putting the insulation towards the outside.
- Vapour checking on the *warm* side of the insulation.
- Ventilating the construction to the fresh air if there happens to be a vapour-trapping membrane on the outside.

Some further problems and innovatory preventions are described under *Larger buildings*, pp 69–70.

WATER EXCLUSION

Rain in the living room is rather obviously undesirable but, in fact, there is a whole variety of water threats to buildings – rain, collected surface water on paths and so on, damp subsoil and the water table – and, when the enclosure resists them, it is not only keeping the interior dry, it is keeping itself dry. In practice, costly damage to the enclosure itself is usually the first result of water penetration, and uncontrolled damp will eventually destroy most constructions. Untreated timber is liable to fungal attack. Continued dampness in a warm, unventilated space, like a roof cavity, encourages the attack rather misleadingly called 'dry rot'. Alternate soaking and drying is more likely to cause 'wet rot'. There are timber preservatives but of course they are not a front-line substitute for keeping things dry – externally-applied timber treatments will, at best, reduce the rate of deterioration, and the more effective pressure-impregnated timber from the factory is expensive. Plaster deteriorates on wetting. Many insulating materials perish. Saturated masonry can deteriorate if

conditions turn cold enough to freeze the absorbed water, for the resulting expansion can progressively crack off the surface particles and break the mortar joints. As a worst case, there is the unprotected wall head. If the roof or coping fails in a derelict building, destruction can proceed apace as the alternate soaking-freezing regime successively erodes the courses down the wall.

Impermeable and porous materials
Impermeable building materials are generally those which have cooled from the molten state at some stage in their formation, such as naturally occurring igneous rocks like slate, marble and granite, and metals, bitumen, plastics and glass. The porous ones contain air voids. Timber, for example, is cellular, which leaves pores when the sap dries out. Water-mixed materials acquire pores on drying – bricks and tiles when the water is driven from the clay in the kiln, and concrete, mortar and plaster when they dry naturally. Sedimentary rocks, like limestone and sandstone, are porous and so, obviously, are strawboard, wood-wool decking, clinker building blocks and the many other insulation products so vulnerable to damp.

Capillarity
The problem of porosity in building materials, where it *is* a problem, is vastly accentuated by capillary action. Water permeates a brick or concrete wall not only owing to gravity or external water pressure but, also, because the capillary effect draws the water inwards or upwards, sponge-like, through the pores. The common demonstration of this is to stand a narrow hollow tube in water – the water rises a few inches until the capillary forces acting upward at the surface are balanced by the weight of the water column below. (The operation of the forces is complex but, essentially, they result from intermolecular attractions within the water, and between the water and tube walls).

Rain on the roof

Clearly, the roof is the hardest tested surface in wet weather. Before the end of the last century, roofs were nearly always pitched, simply because there was no cheap sheet material available for making flat areas waterproof. There were flat timber roofs in the dry climates, topped with cement or clay, but, elsewhere, the pitched roof was the mainstay, its slope shedding the water quickly and allowing a waterproof covering to be achieved with rows of overlapping single elements – the straws in thatching, flat wood shingles, clay tiles and naturally-occurring slates. Modern materials, such as 'corrugated iron' (usually galvanised steel), corrugated asbestos cement, and plastic sheeting or tiles, have extended the range of pitched coverings but, more than that, modern bitumen-impregnated fabrics and plastic membranes have made the flat roof a workable alternative. Pitched construction is traditionally more associated with the domestic scale, where it has a 'rightness' aesthetically and creates the useful or upper floor space inside. Flat roofs are more part of the larger public scale, where pitches would create internal spaces more obviously redundant, and where a large complicated ground plan would mean awkward peaks and valleys. Also, in multi-storey buildings, it makes constructional and economic sense to repeat the first-floor construction method at roof level.

Pitched roof

The roof (2.15) would normally have a 35°–40° minimum pitch and, ideally, more in high rainfall regions. This ensures a quick run-off and helps the overlapping cover to block driving rain which, in hard conditions, may strike horizontally or even upwards. It also makes for worthwhile attic space.

The horizontal rows of tiles are staggered and sufficiently overlapped down the roof slope to ensure that no joint is located over a joint in the layer below. There is never a direct water route through the barrier. The clay tiles shown are made with a slight curve, so that the interface within the overlap progressively widens, breaking the possible capillary paths into the roof. Other coverings range from thatch and traditional slate to modern lightweight tiles in plastic and other synthetics. There are proprietary clay and concrete tiles with profiled edges interlocking along the slope, improving the water-proofing and reducing the amount of lapping, and roof pitch, required. Here, tiles are shown nailed to wooden battens underlaid by a continuous damp-proof membrane (DPM) of bituminous felt or plastic sheeting, to exclude wind, improve insulation and create a second line of defence against seepage. An arguable improvement is to have a second layer of counterbattens running down the slope under the lateral battens, raising them slightly clear of the DPM and allowing a clear run for seepage down to the gutter. Often, nail fixing is through the DPM to overall timber sheet decking or boarding. The continuous DPM brings the need for ventilation under. In practice, the roof will never be airtight around its edges and the pitch will helpfully induce convection from the outside. But still, some purpose-made air opening is wise. The detail abutting the raised gable end is shown for interest though, in truth, it would more likely apply to the party wall in terraced housing. The small-house roof would probably oversail the gable end, in which case the roof edge can be weathered in various ways, including fascia boarding, cement grouting and turned-down verge tiling. But this is to digress. There are, of course, endless pitched roof constructions and our purpose here is to illustrate only some basic water-proofing principles.

Flat roof

The modern flat roof has had its teething problems, to put it mildly, which is perhaps not surprising considering the functional demands placed upon it. The continuous outer covering must:

- Be waterproof.
- Withstand the attacks of ultra-violet in sunlight and acids in rain.

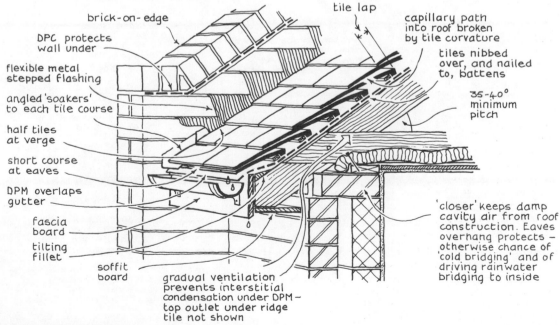

2.15 Some waterproofing principles for the pitched roof

- Withstand the physical stresses of wind uplift and foot traffic.
- Be flexible enough to accommodate movement relative to the sub-decking, owing to thermal expansion and contraction.
- Be fire retardant.

The pitched roof and its materials cope with all this most naturally. In the flat roof, the need is for an impervious, chemically inert, tough, flexible membrane, and new products are appearing all the time. However, bituminous felt is still the common membrane for small roofs. It comes in rolls and is usually bonded down in two layers, staggered to break joints, and with overlapping edges sealed with hot asphalt or bitumen or, as is becoming more common, synthetic adhesive. 2.16 is a simple application at domestic scale. Stone chippings or lightweight tiles can protect the membrane from foot traffic, direct sunlight and fire. If they are light-coloured, they will also act as heat reflectors, helpful to the building in summer and, by reducing thermal movements in the underlying roof structure, protecting the membrane further. And remember that with the continuous flat membrane, the need for gradual ventilation underneath is *critical*.

Of course a 'flat' roof is not flat! It sheds water by breaking towards edge gutters or dishing into central outlets. Falls are achieved by tapered or progressively smaller wooden firrings in timber construction (or, on larger scale decking, by carefully laid concrete screeding) and they have to be steep enough to avoid puddles remaining on the roof – standing water can lead to high, dissolved acid concentrations as it dries out and, also,

can cause damaging differential movements as the sun's heat plays on alternately wet and dry patches. Theoretically, the minimum fall is around 1:80, increasing to 1:60 for rough-textured surfaces. But this has to be the *achieved* fall. The designed fall may have to be more, to allow for the inevitable construction inaccuracies and the offsetting effect of a roof deflecting under its own weight.

As well as the membrane, the whole structure has to be tied down against wind stresses – uplift under the eaves and suction over the top surface. The joists can be tied down to the wall head, as shown, or the wall plate tied down and the joists skew-nailed to that.

Gutters and downpipes

In both types of roof, the collected rainwater has to be got away. Obviously, if it were just left to be shed from the edges, it would soak and damage the wall, and turn the ground immediately around the house into a swamp. Rainwater drainage is included in Chapter 4 *Utility services* but, basically, pitched roofs will have eaves gutters leading to the downpipes, and flat roofs either recessed gutters as shown or, in dished surfaces, outlets connecting to the downpipes direct. The downpipes themselves, carrying the water to the main sewer outlets, are normally run outside the wall. If they have to be inside, there will have to be encasement for noise reduction and yet access for repair or unblocking.

Rain on walls

The simple house wall is shown in three kinds (2.17). The solid wall (a) is porous. Capillary action draws water

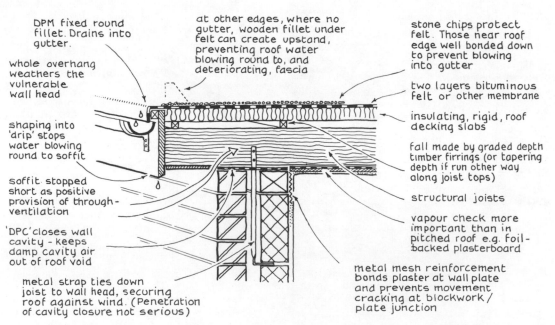

DPM fixed round fillet. Drains into gutter.

whole overhang weathers the vulnerable wall head

shaping into 'drip' stops water blowing round to soffit

soffit stopped short as positive provision of through-ventilation

'DPC' closes wall cavity - keeps damp cavity air out of roof void

metal strap ties down joist to wall head, securing roof against wind. (Penetration of cavity closure not serious)

at other edges, where no gutter, wooden fillet under felt can create upstand, preventing roof water blowing round to, and deteriorating, fascia

stone chips protect felt. Those near roof edge well bonded down to prevent blowing into gutter

two layers bituminous felt or other membrane

insulating, rigid, roof decking slabs

fall made by graded depth timber firrings (or tapering depth if run other way along joist tops)

structural joists

vapour check more important than in pitched roof e.g. foil-backed plasterboard

metal mesh reinforcement bonds plaster at wall plate and prevents movement cracking at blockwork / plate junction

2.16 Some waterproofing principles for the flat roof

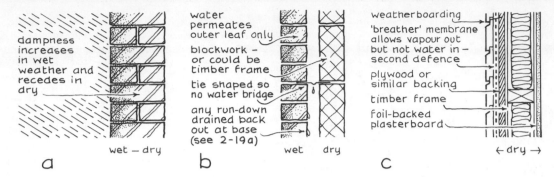

2.17 Waterproofing by solid, cavity and impervious-faced walls

gradually into the stone or brick and, especially, through the mortar joints, and the defence is simply to have the wall thick enough for fine weather to arrive and dry it out again before damp can ever reach the inside face. But, save in very kind climates, solid walls are now uneconomic, since adequate water-proofing and insulation demand a masonry thickness greater than that needed for structure alone.

The modern cavity wall is more efficient, shown here (b) with a facing brick for weather-proofing and appearance, and with an internal leaf of lightweight concrete or e.g. cinder block, cheaper to buy, faster and hence cheaper to lay, and giving better insulation. (Or the inner leaf could be timber-framed). There would be open weep-holes at the foot of the cavity, just above ground, to allow drainage out – this is because rain-water could penetrate to the cavity in severe conditions or there could be condensation moisture there, as vapour migrated from inside the building. It is also argued that these holes allow a small amount of cavity ventilation, especially if a perforated air-brick is inserted towards the top of the wall. The ventilation must help drying out and, moreover, by ensuring that the air pressure in the cavity is the same as that of the air outside, may prevent excess wind pressure from increasing the water penetration of the brick. On the other hand, air exchange in the cavity means insulation loss and, anyway, the cavity fills described earlier obviously cut off ventilation, apparently with no ill effects. But this raises a further question. Does the cavity fill, although itself impervious, not increase the chance of water crossing to the inner leaf? Current thinking is that it may, but only in very exposed situations.

Of course, avoiding bridging the cavity is a preoccupation in standard cavity detailing. It affects the design of wall openings and footings, as we shall see, and note now the wall-ties are shaped to shed any drips crossing.

Thirdly, there is the *impervious wall*. This can still be masonry, but externally water-proofed with a paint or dense cement rendering. These should prevent water

penetration, but the catch is that, if they are not well maintained and cracks develop, water washing down the wall face can enter and, because of them, be unable to dry out again. Walls can end up wetter than if they had had no coating at all. This is not to discount these applications – renderings are pretty much standard where there is a lack of local materials for making good quality facing brick.

But 'impervious faced' more commonly means where the water-proofing is by some form of separate front layer or screen, such as tiles or weatherboarding on a backing block wall or timber frame, operating similarly to the overlapping covering on the pitched roof (c). Alternatively, especially in the larger buildings we shall be discussing later, impervious sheet or panels can form a water-proof cladding which, as we might anticipate from the flat roof membrane, can pose its own problems in jointing and condensation.

Rain on wall openings

Window and door openings break the wall's continuity and are, therefore, the more vulnerable. The simple timber window (2.18) with opening casement and top-light shows most of the weather-exclusion principles common to windows (and doors), whatever their type. From the drawing, note the following points:

- It is important to reduce the amount of water streaming down the window face. Ideally therefore, the window is set back within the wall depth. This helps protect the upper part of the frame from direct rain, it 'weathers' the window and also sheds clear the wash of water from the wall above. (The head mouldings in classical architecture are elegant but they are there for weathering too). Similar projections, timber weather mouldings, protect the vulnerable points lower down the window. The top surfaces of mouldings and sills are sloped to shed water clear and avoid its collecting where it can be blown back and lie against the glass/frame joint.

- Joints that have to open, i.e. where the window opens, are protected by their shaping, and by overlapping the abutting frame parts – joint with a

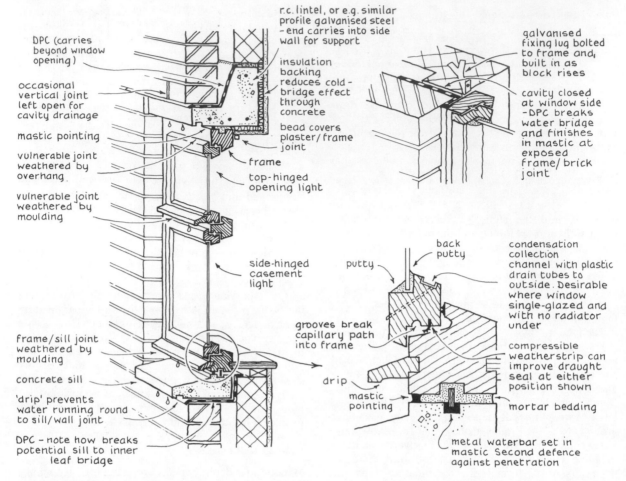

2.18 Waterproofing the openings in walls – a 'typical' timber window

straight travel through the frame would never keep the rain out.

- All fixed joints, e.g. frame to surrounding wall, need to be protected from water penetration.
- In all joints, openable or fixed, the effects of capillary action must be foiled – possible capillary paths are broken by grooves or by widening the joint interface.
- Capillarity can be assisted by wind pressure if there is a draught through the frame – draught penetration is minimal if the openable joint is well fitted, but weather-stripping creates a more positive barrier to draughts and wind-driven water, let alone the associated thermal benefits.
- Even fixed joints should, ideally, include some positive stop or direction change along their travel. The concrete sill to timber joint incorporates a metal water-bar as an additional defence should the mortar fail.
- Mastic sealants can further improve water-resist-

ance at joint faces. There is a cautionary point here though and it applies to mastics on the enclosure as a whole. Modern sealants, the development of which can be traced back to the needs of early curtain walling in the 1940s, are undeniably durable and efficient. But their efficiency has sometimes been their drawback, in the sense that they are too often seen as the universal water-proofing panacea – 'It'll be all right, just gun in some mastic.' The mastic seal needs a proper rebate to sit in physically, needs to be physically protected and needs to be gunned in with care that can be guaranteed. It is not, and never will be, a substitute for good detailing.

Structurally, of course, the window has to be secure. The frame can be bolted to the masonry or fixed by all manner of lugs and ties – an example is shown in the detail at the top right. Another method is to close the cavity in such a way that one of the leaves projects sideways slightly more than the other, i.e. making an offset before or behind the window against which the frame can physically butt.

Damp from below ground

The wall base and lowest floor must exclude ground damp. Porous and water-vulnerable building materials must sit on the muddy site and remain dry. For illustration 2.19, we will take the cavity wall, the weatherboarded timber frame and the masonry-faced timber frame and, respectively – and quite at random – combine them with the suspended timber floor, the concrete slab and the suspended ground floor with basement construction below. In each, the position and detailing of the damp-proof membrane is the pivotal point in the design.

As with most masonry, the brick and block in 2.19a can act like a wick, drawing up moisture from the damp ground by capillary action in the material pores. Unchecked, 'rising damp' can lead to wet, damaged interior plasterwork and eventual rot in the timbers. And damp masonry is less insulating and, moreover, its resultingly colder inner face may locally cause condensation, meaning more damp. Hence, the DPC. The traditional damp-proof course really was a 'course' of slates or impervious dense brick but slate, in particular, being brittle, was prone to failure as a building settled over the years. Lead and copper strip were often used. The modern DPC is a thin flexible strip of bituminous felt or plastic. In the drawing, note especially the need to keep the DPC a sensible, safe distance above ground level, and to have the oversite concrete inside the wall at least at ground level – if not, the head of ground water in wet conditions could cause seepage into, and even flooding of, the underfloor void! The sleeper brick course is added protection, distancing the timber wall

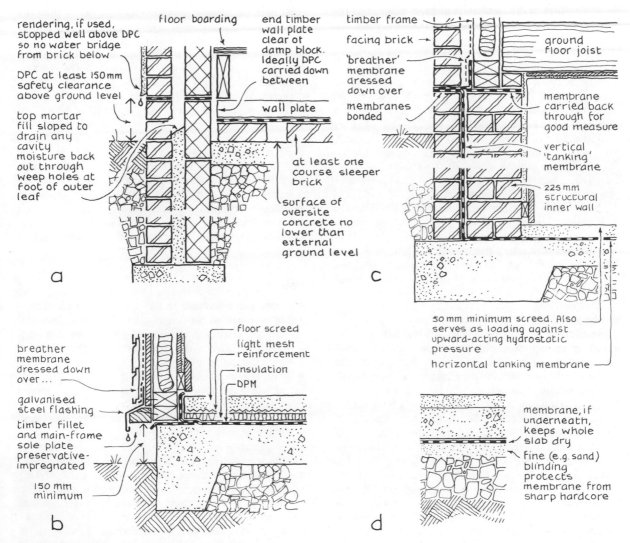

2.19 Waterproofing the wall base and lowest floor, including basement

plate from the potentially damp concrete and improving the route through for the underfloor ventilation.

2.19b turns the corner between the timber frame wall and the solid concrete floor. For interest, an under-screed insulation layer is shown (with top mesh to stabilise the screed and prevent its cracking). Insulation here is important, where the screed is electrically wired or piped with hot water for underfloor heating (see Chapter 3 *Climate services*) but is in any case, thermally more comfortable underfoot, and more common as insulation standards rise.

2.19c combines the masonry-faced timber wall and basement wall. The problem for any basement wall or floor slab is that it may be subject not only to damp, but to damp under hydrostatic pressure, if the water table rises around the construction. Just as a swimming pool has to be sealed to keep the water in, so a basement in wet ground has to be *tanked* to keep it out. The traditional tanking membrane is asphalt, but various synthetic sheet and applied liquid membranes are now available. As shown, the tanking membrane wraps around the basement and mates with the DPC above ground level. Moreover, it has to be physically re-strained, since hydrostatic pressure can readily force a membrane inwards off a wall or up off a slab – so, apart from their other functions, the 215 mm inner leaf and the 50 mm screed are *loading coats*. Certainly, unloaded, liquid applied water-proofings are used in less testing circumstances – more on this under *Larger buildings*. But loaded tanking is a fail-safe answer anywhere the water table may rise.

2.19d shows the membrane *underneath* the slab. This is becoming more common and has the advantage that the whole slab is dry and therefore more insulating but, at the perimeter, the membrane has to wrap up around the slab edge and mate with the wall DPC, perhaps with an external brick or other facing to protect it.

Damp-proofing existing buildings

Remedial damp-proofing is worth a few words. It is a common enough need but, for all that, there seems to be a variety of misconceptions on the subject. Old DPCs can have failed or there may never have been a DPC, historic demands of dryness, perhaps, being rather less than today's. A house-buyer's loan is often conditional on damp-proofing being carried out.

One system is to insert physically a new DPC. Usually this is done in stages progressively around the wall base – making a horizontal saw-cut, inserting the DPC length and re-grouting – so that the wall's overall load-bearing function is never impaired. The method is sure but expensive. A simpler system reduces the masonry's porosity by saturating the lower courses in water-proofing fluid, often silicone. You may have seen this – a row of inverted bottles sticking into holes drilled at intervals along the wall, gradually dripping-

in the fluid. It has been called 'the drip and hope method', correctly implying that the saturation is too hit-and-miss to guarantee complete impermeability. But does this matter? Again, there is the tendency to think in absolutes for, in practice, as long as the treatment slows the rise of damp to the point where the building's natural drying-out regime can cope, then it has worked. It is the method now the most widely used; and there are now pressure impregnation methods on the same lines. A third method inserts copper electrodes into the wall, on the basis that electric potential across a porous material has been observed to affect the moisture flow within it. But opinions differ on why or, indeed, how far, it works. A final method inserts a row of small, hollow pipes along the wall, ventilating it to improve the drying. Complete remedy involves replacing any damaged plaster – and not only because it is unsightly. It may have salts leached into it from the wet masonry, making it hygroscopic and ever after prone to attract damp from the ordinary humidity in the house (clothes wetted in sea water have to be washed before they will dry out properly). The replacement plaster can be made additionally water-retardent.

Actually, lack of data makes it hard to appraise damp-proofing treatments. Any unoccupied house will get damp, in time. New owners carry out damp-proofing and are pleased to find in a couple of months that the treatment has 'worked'. But it is often forgotten that the house has since been ventilated and warmed, perhaps with central heating installed. In other words, it would be hard to know whether it was the damp-proofing that cured the problem or, indeed, whether the damp-proofing was necessary in the first place. Often the loan office/ surveyor/ contractor involvement is cautious to the point of absurdity.

Summary

- First water threat is to the enclosure fabric itself.
- Impermeable materials generally have 'molten' origins; porous materials contain air voids. Capillarity through voids and joints accentuates the penetration problem.
- The common domestic roof is pitched at around 35° minimum – reliant on overlapped/staggered coverings with e.g. felt underlay.
- The 'flat' roof is reliant on an impermeable membrane.
- Wall categories are principally solid (but now virtually obsolete), cavity or waterproof-faced.
- Windows and doors are normally set back – frame water-proofing includes mastic sealing, with openable joint faces checked and with e.g. capillary grooving and draught-stripping.
- Wall DPC and slab DPM exclude ground damp. Suspended floor void is ventilated. Basements are normally tanked and with loading coat.

DAYLIGHTING, VIEWS, VENTILATION – WINDOWS

Windows – 'wind eyes', historically – are very much a barrier and filter combined, even in the small house responding to a complex interaction of needs (2.20).

Daylighting

Daylighting is properly covered in Chapter 5 *Lighting* but naturally affects window design. Window *area* needed relates to daylight needed, a fact that is admittedly rather easy to grasp. Room geometry and internal reflectance, and the anticipated tasks within, play a part – but always remembering the thermal penalties that big windows incur. Window *shape* is also influenced. High, narrow windows tend to admit more useful daylight than low, wide ones, principally because they offer more chance of the unobstructed sky source being 'seen' by the back of the room, where they improve lighting levels in consequence.

Views out

The desire to look out also influences shape, as well as suggesting where windows should be located. A low, horizontal window tends to be claustrophobic if it cuts off the sky view, and offers a less representative view of what is outside than one with a vertical emphasis. Sills too high can block views out when people are seated; transoms can be awkwardly placed.

Of course, a good view out can mean a good view in, and the need for greater privacy in rooms like bedrooms can dictate a higher sill.

Ventilation

Air for breathing may be vital but it is not, in fact, the first consideration in ventilating. The needs for cooling, and removing humidity, odours and bacteria, and supporting efficient combustion in fires and boilers, are generally more pressing. If anything, a vertical window is more comfortable. Summer ventilation may ask for as large an openable area as possible and, preferably, some of it should be at low worktop level, where cooling draughts are most felt. Conversely, the winter need is for a small opening and, preferably, at high level, where draughts will pass unnoticed. And of course any ventilation needs to be readily adjustable.

Sensible structure

Any opening will interrupt the structural continuity of a wall, requiring head lintels to bridge the vertical loads. Here, again, the narrow window makes more sense, a wider window having greater structural penalties, particularly with heavier wall constructions. In fact, taking all the above points together, it emerges that on strictly functional counts, the modern, fashionably wide 'picture' window – the familiar Corbusian aesthetic – scores virtually nil.

Cleaning

Windows have to allow external cleaning. In casements, for example, the hinge geometry can be arranged to create an access gap at the hinged side when the window is open. Sash windows can be detached and pivoted windows reversed.

Window types

The simple casement (2.21a) is shown side-hung. It can also be top- or bottom-hung, swinging outwards if top-hung and inwards if bottom-hung, i.e. always shedding the rain to the outside. Casements offer 100 per cent of the window area for open ventilation, favouring their use in hot climates. Size is limited by the need for stiffness in the open position and, also, if they project too far on the lower storey, they could obstruct pathways around the building.

The vertically sliding sash (2.21b) achieves only 50 per cent ventilation area but, since this can be selected at high or low level, it gives good summer and winter control. The sashes can be rather larger than single casements because they are always structurally restrained by the fixed surrounding frames in which they slide, and because they do not project from the wall. To ease opening and closing, the weight of the moving lights is offset by springs or cord-hung counterweights within

- daylighting may argue larger size and higher, oblong shape
- thermal performance may argue smaller size and double glazing
- view out may argue higher, oblong shape
- privacy asks e.g. high sill level
- ventilation asks for adjustable opening and slightly prefers higher oblong shape.
- structure prefers narrow shape

2.20 Main influences on window design

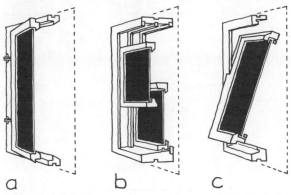

2.21 Common window types, (a) Casement (b) Sash (c) Pivot—shown in timber the fixed side frames. Horizontally sliding sashes are located in metal or plastic runners at top and bottom.

The pivot window (2.21c) can be horizontally pivoted, as shown, or vertically pivoted like a revolving door. Sometimes, the pivot points are offset from the window centre line in order to reduce the dimension of the inward opening part, preventing its fouling blinds and curtains.

The types are often combined, perhaps having a fixed light, a casement for adequate summer ventilation and a top light for winter ventilation. Top lights are common in many combination types, although a more recent trend is to provide winter ventilation through openable holes or slots in the solid head of the frame itself.

Materials

Timber. This is the traditional material, readily available and easily shaped into the required frame sections shown earlier. Softwood is common, rot-protected by preservative treatments and painting, but hardwoods can be used for the more exposed weather-mouldings and sills.

Metal. The modern steel window developed at the end of the last century, its strength allowing large lights with minimum frame obstruction (2.22a). The sections are mainly produced by 'hot-rolling' molten ingots, or 'cold-pressing' mild steel sheet, processes which gear the industry mainly to the mass production of simple standard sections. Steel is normally rust-proofed by hot-dip galvanising, coating it with a thin zinc layer. Modern aluminium frames are lighter than steel and usually stiffened by integral timber sub-frames (2.22b).

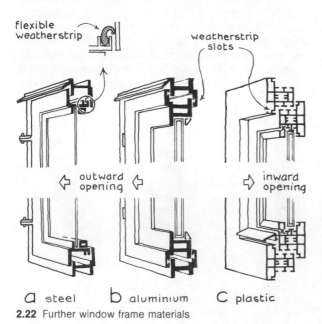

2.22 Further window frame materials

In manufacture, the molten aluminium is extruded toothpaste-like through specially shaped dies and, since the dies are easy to change, a far greater variety of sections is practicable. Aluminium is corrosion-protected by anodising, coating in oxide film. Bronze and stainless steel frames are durable, and elegant in context, but almost certainly too expensive for the small house.

Plastic. Plastic frame sections are also extruded. The casement sections (2.22c) are, in fact, similar in outline to the timber equivalent: the apparent complexity results only from the added internal stiffening fins – extrusions cannot just be thickened up at points for added stiffness, because the thicker parts would cool unevenly from the die and become distorted.

Composites. Materials can be combined. A wood 'sub-frame' is often set between a metal window and the surrounding wall opening. Plastic outer sections can be stiffened internally with steel, i.e. the weather-resistant properties and sectional variety of plastic combining with the strength of the steel. Conversely, metal outer sections can have timber or expanded plastic cores, combining stiffness with improved thermal insulation – cold bridging leading to condensation is a problem with unprotected metal frames.

The glass

Glass is mainly the product of fusing silica with basic oxides such as phosphorus and lead. The terms 'sheet' and 'float' glass describe the methods of manufacture. Sheet glass is drawn from the tank in a continuous, molten ribbon, and is flat enough for ordinary window purposes. In the float glass technique, developed by Pilkington Brothers in the late 1950s, the molten ribbon is floated on the surface of molten metal, making large glass areas that are flat and optically excellent. Transparent plastics and tough plastic/glass laminates are another development, but tend to be expensive as yet.

Securing the glass. 2.23 shows some glazing methods. Traditional putty is attractively simple, if slow to apply and slow to set. Beads give a faster and more positive fixing, and are easily removable when the glass has to be replaced. Fixing is simpler if the beads are on the inside of the frame. Timber beads can be used on timber window sections, but the illustration shows two metal variants. The synthetic PVC or rubber gasket, familiar to the automobile industry, is also used, but is mainly applicable to larger buildings.

ACCESS – DOORS

Obviously, a domestic door operates similarly to a casement window and it will probably be more helpful

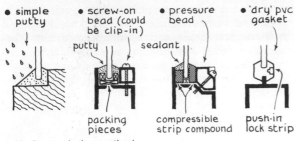

• simple putty • screw-on bead (could be clip-in) • pressure bead • 'dry' pvc gasket

putty sealant

packing pieces compressible strip compound push-in lock strip

2.23 Some glazing methods

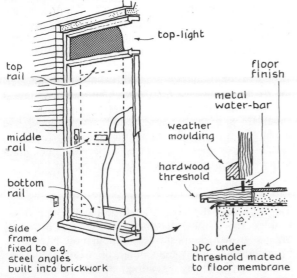

top-light

top rail

middle rail

bottom rail

side frame fixed to e.g. steel angles built into brickwork

floor finish

metal water-bar

weather moulding

hardwood threshold

DPC under threshold mated to floor membrane

2.24 Door construction – frame detailing is similar to casement window

to go and look at one than to read a description here. That is not intended facetiously. It is only too easy to divorce building theory from practice, to spend a tedious time over a pile of text books solving, say, a metal window detail, when a trip to the window by the light of which one is working will serve the same purpose.

Doors, like windows, are filters, maintaining the weather barrier, but allowing access – access to friends that is; locks are people filters? Dimensional constraints include the need for adequate head room and clear width for the passage of furniture, prams and such and, important, disabled persons' wheelchairs. In housing for the elderly or disabled, there must be no steep steps or awkward threshold details impeding easy, unaided access.

The traditional door has main frames with timber or other panelling set in. 2.24 is 'flush-faced'. Framing tends to be heavier than for windows, since doors get rougher treatment and, being larger, are more prone to cumulative distortion and consequent loss of weather-tightness. Doors have to open in all weathers and provision of draught lobby and inner door behind is good practice.

SECURITY

Historically, in our earlier fortified tower, defence against enemies was part of survival, and the simple house written about in those times would have had security as a major heading. Today, measures excluding intruders from *domestic* buildings are mainly confined to the design and locking of doors and windows.

Fire danger is another security influence on the enclosure. In a house, the escape need is normally satisfied by the ordinary doors and windows. If the small house has close neighbours, the need to restrict fire spread could control the area and position of all openings, and the combustibility of wall materials (see Chapter 7 *Fire safety*).

NOISE (AND POLLUTION) EXCLUSION

The only real way to exclude noise or atmospheric pollution is to have a sealed outer skin and, since this is incompatible with ordinary ventilation, the proper answer is that houses should never be sited where such needs may arise. In the ordinary way, street noise can be a problem but the simple house can achieve a fair degree of suppression owing to the mass of its walls, and if its windows are double-glazed – but see Chapter 6 *Acoustics*

COST

Effect of scale

In many countries, detailed costing is a job for the specialist: for example, in the UK, it is for the quantity surveyor. In the US, it falls more to the architect. But, whatever the system, costing must influence design decisions from the start. Obviously, the main determinant is size but, in terms of *cost per unit floor area*, there can be economies of scale. Expensive items, like roof perimeters, window surrounds and foundations, are more felt as scale decreases – the principle that prompted Henry Ford's comment, 'small cars, small profits'. The small house will be more expensive than the equivalent unit in a terrace, which has less external skin per house unit enclosed and enjoys economies in constructional and component repetition.

Effect of shape

Shape is an influence. The enclosure is expensive, and cost consideration wants to reduce surface area just as did heat conservation. In fact, the cheapest house shape is a square box. This is not to advocate that houses be boxes of course, only to say that if the plan, for example, is articulated, as it may be for good reason, there must be an associated cost penalty, since projections will increase the wall area and complicate the construction of the roof, floors and foundations.

It is often asked whether a single-storey house is cheaper than one which has the same total floor area but two storeys. Certainly, the single-storey saves in that it needs no stairs, which saves money, space and, hence, more money. But the single-storey's advantages are partly eroded by the greater enclosure area (slightly less wall but much more slab and roof) and this, taken with its less economic land use (and less efficient thermal shape, incidentally), normally tips the scales towards the two-storey version.

Capital outlay balanced against running costs

Other things being equal, one can expect to get the standard of enclosure one pays for. Money spent can have a bearing on the finish achieved, both visually and in durability – durability reducing the later cost of maintenance. This link between initial capital outlay and later running costs is a fundamental theme in cost planning. As already said, it is nowhere more crucial than in energy conservation – put simply, spending money on insulation today to save fuel tomorrow.

Detailed costing

Intelligent costing operates in all the ensuing, more detailed, aspects of the design. Take windows, for example, which are up to three times as expensive as the equivalent area of solid brick wall. Costing might show two medium-sized windows to be the cheapest way of meeting the glazing requirements in the small-house elevation, where three smaller windows would require a relatively greater amount of expensive window surround and framing for the same glass area, and one large window would incur a cost penalty owing to the larger lintel span and heavier glass needed. At times, such costing exercises seem a tedious, even cruel hindrance to the designer but, in the widest sense, they are a reliable barometer of what is sensible and practicable to do.

DURABILITY AND MAINTENANCE

The simple house will visibly age over the years. There will be mechanical erosion, chemical action and accretion of dirt.

Mechanical erosion

Here, an obvious example is the wetting-freezing regime on porous masonry, and the denser types of brick and cement renderings would be the wisest masonry choice where such conditions prevailed. Wind causes mechanical erosion. As an extreme case, if the simple house were on an exposed coastline, it would need tough finishes to survive the repeated sand blastings.

Chemical action

Rain and, to some extent sun, attack many materials. Rain is weak carbonic acid, owing to the carbon dioxide it dissolves from the atmosphere, and it is this acidity that oxidises materials. In industrial areas, there may be additional dissolved impurities. Of course, the agreeable green attained by a copper roof over a couple of years is a perfectly acceptable design intention, but the corrosion of aluminium and rusting of ferrous metals are not. Water washing down a building may reveal chemical incompatibilities between adjacent materials, though most of the unhappy bedfellows are known and avoidable, for example, aluminium and damp concrete; certain hardwoods and metals, especially zinc; limestone and sandstone; and rubber and copper.

Dirt accretion

This is perhaps the most difficult problem from a visual standpoint. Porous masonry will collect dirt and organic growths in its surface, although the effect may be mellowing rather than defacing, with the natural colours and textures tending to camouflage later change. If anything, the more impervious skins present a greater visual problem. Their surface absorbs less water and the flow down the face is consequently greater, tending to take preferential paths as it spills off roof parapets and round balconies, or responds to local wind patterns. In time, this can cause unsightly, uneven weathering. Concrete walls in urban areas are particularly at risk because they show the results so readily.

Typical precautions include the avoidance, as far as possible, of horizontal projections, with their dam-like effect, and contouring or texturing the surface to help regularise the flow. To this end, you will have seen concrete surfaces with vertical grooves cast in by the shuttering profile or roughened, after the shuttering has been removed, by mechanical hammering or sand-blasting.

All types of surface are prone to the effects of differential washing, the visible result of rain getting a better chance to *clean* the surfaces which are most exposed to it. It is not always a problem. Since both the prevailing rain direction and daylighting are from above, the washing often emphasises the existing daylight shading rhythms on an elevation. The weathering of cornices and mouldings on classical elevations is a striking example. Weathering effects can be controlled by the use of materials appropriate to the given circumstances – building location, exposure, scale – and by good detailing. *In general, any change that mellows the small house or enhances its designer's original intentions on the drawing board is acceptable, but change that deteriorates or distorts is not.*

APPEARANCE

By association, we come last, though hardly least, to appearance – that is 'architectural' appearance as controlled by the enclosure design. Of course, aesthetics are incapable of technical definition, here or anywhere else, if indeed they are capable of definition at all. Certainly, good appearance depends on more than the visible surface of the external skin. It depends on form and mass, on how the outside expresses the functioning plan inside, on many things. Essentially, we see a building and know if we like it. But, more tangibly, the appearance of the enclosure itself must depend on modelling, shading, texture and colour, both of the whole and of the component parts. The nature of the surface and the 'rightness' of its detail, the close-up view, will be likely to count more at the small-house scale than in a larger building, and it is no misfortune, perhaps, that the small compass of the domestic enclosure facilitates the use of traditional materials, like brick and timber, with natural textures and long-familiar look. Also, for most people, the smaller building has the happy attribute of being closer to the human scale and, arguably, 'small' is indeed 'beautiful'. 'Big' can also be beautiful, but curtain-wall claddings and sweeps of glass and concrete are more innovatory and controversial, and pose challenges to the designer of a wholly different kind.

LARGER BUILDINGS

Obviously, there is no magic dividing line between enclosure construction for small buildings and for large buildings. There are large buildings with brick walls and small buildings with structural frame and lightweight cladding, and large buildings with pitched roofs and small buildings with flat ones – but they are exceptions and, in general, the enclosure construction does change once away from the domestic scale. So for convenience, we will in fact draw a straight dividing line between 'small' and 'big' buildings, allowing that in practice, it would be a rather wavy one. Again, we will take the functional headings in turn, only moving more rapidly now the basic principles have been covered.

STRUCTURAL PERFORMANCE

As before, we will work up to constructional details, but the outline (2.25) will give us something to start with. It may seem perfunctory to have so small a sample standing for the world of larger enclosure types – Chapter 1 *Structure* indicated some of the wide variety – but, remember, the intention here can only be to illustrate some of the more important principles, reasonably to select some illustrative themes. Bear with it!

Roof – flat and pitched

For the flat-roof type (2.25a) we take the monolithic option, the reinforced concrete slab, but clearly there are all manner of joist and decking panel systems in concrete and steel. The pitched-roof type (2.25b) we show as the portal frame and panel cladding. Generally, the flat roof in multi-storey buildings will repeat the construction of the floors below. The pitched roof, at larger scale, will be mostly confined to industrial buildings.

Walls – cladding

The masonry wall option is, of course, widely applicable at larger scale, including multi-storey work up to 15 storeys and more. And the monolithic reinforced concrete wall has its civic and public functions. In either case, there may be an external cladding. However, the *enclosure performance* principles of these solid-wall types are not particularly different from those at small scale and, here, it will be more useful perhaps to confine ourselves to the cladding wall (2.25c,d,e). Chapter 1 *Structure* told how increased scale generally prefers the structural frame with a cladding wrapping that has no direct supporting role, with scale increase upwards eventually making this generality a rule.

The brick spandrel wall (c), built up off the slab edge or perimeter beam, is still a masonry choice but now non-load-bearing. It needs no crane but is quite slow to erect, if anything suggesting the smaller end of the larger building scale. However, there are faster systems, using whole-bay brick panels, preformed on site or in the factory. Some bond the bricks with special 'sticky' mortar, others use a consolidating backing of insulating concrete. Any masonry cladding will be relatively heavy, restricting its use to low- and medium-rise.

Panel claddings are a more direct and, nowadays,

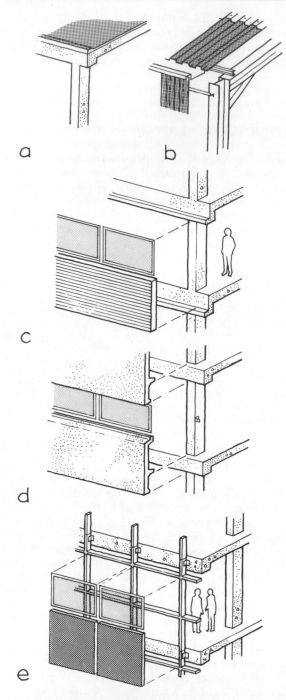

a

b

c

d

e

2.25 Sample larger enclosure constructions – see also expanded details (2.32, 2.33)

with a concrete frame. Its weight requires a crane and, again, restricts the use to low- and medium-rise. The panels are usually precast in the factory but, in a very large project, it might be worth setting up a special panel-casting plant on site to reduce lorry transport and handling costs.

The curtain wall (e) is one of the lightweight skins – panel systems chiefly in glass, plastics or metal sheet – which are fast to erect and appropriate at all possible scales and heights. Note how their lightness allows the column grid to pull back as explained in Chapter 1 *Structure*. We will see, though, how lightweight skins, so versatile in many ways, can pose their own problems in heat loss and gain.

Fixing, dimensional tolerance, strength and size

Constructionally, the panels can run storey-height, spanning between successive floor edges. Or there can be separate spandrel wall and window bands spanning horizontally between the columns or, as shown for the lightweight curtain, between columns and intermediate, vertical mullions. Lugs or hooks on panel backs can be located on the supporting frame, the connections being adjustable (e.g. bolts with variable spacers or through holes with oblong slots), thereby allowing adjustment during erection to cope with the inevitable slight discrepancies in panel size or main frame alignment – tight tolerances so easy to specify on the drawing board are harder to achieve in the reality of putting up a building. Concrete panels are normally profiled to be located on the floor edge and/or column face. Glass or plastic cladding panels are usually glazed, window-like, i.e. with bead or gasket, into the secondary metal framing. Alternatively, whole-storey preformed units can attach to the main frame directly. Enlarged details are shown in the later sectional drawing (2.32, 2.33).

Panels must cope with their own self-weight, wind loads and a reasonable amount of hardship in the truck and on the building site. Sizing is ultimately limited by what is structurally sensible and by the erection method intended, i.e. whether craned or finally manhandled into place. Beyond this, it is a matter of individual design, reflecting the dimensions of the secondary support system or structural main frame and, in fact, the modular co-ordination of the whole project.

The lowest floor

The lowest floor will generally be a monolithic concrete slab, usually sitting directly on the ground. But, occasionally, a combination of poor soil bearing capacity and high floor loads may demand a scaled-up version of the suspended floor of the simple house, a concrete slab spanning between intermediate beams set on piles. A factory on reconstituted ground is a frequent example. If there is a basement, the vertical walls below ground will usually be concrete as well – the potential

familiar consequence of the structural frame, wrapping round high-rise offices, schools, factories, hotels, in fact, all kinds of modern public and commercial buildings. The concrete cladding (d) would be used typically

soil and water-table loads are correspondingly greater in large-scale construction and, whereas the masonry retaining wall of the simple house relied mainly on its mass to withstand the overturning forces, the monolithic concrete wall is additionally stiffened by its cantilevering up from the ground slab or by its being integral with the main frame or walls behind – see the buoyancy raft in Chapter 1, p 24.

Movement joints

Again, the movement risks are structural, thermal and through moisture change. Joints to separate the structure, such as those designed to accommodate differential settlement between low- and high-rise blocks, are described in Chapter 1. Thermal movement may be automatically catered for where such joints occur but, otherwise, will need their own provision – expansion and contraction of the enclosure fabric, negligible in the small house, could build up to spectacular failure if unrelieved at the large scale. Large areas of roof slab and long runs of masonry or concrete wall need expansion joints at around 30 m intervals (2.26a,b), and cladding panels and their support members (c) come together with some sort of sliding overlap rather than abutting directly. Large window panes must be glazed

into their frames with enough clearance margin or gasket flexibility to allow for differential movement between the brittle glass and its surrounds. Where materials of different expansion coefficient adjoin, for example, a large timber window frame set in a concrete or steel main frame, the junction design must allow each to move relative to the other by having a sliding overlap or a gap filled with flexible sealant.

Actually, sky radiation again merits a mention. The solar component on a hot day can raise the external surface temperature of buildings scorchingly above the prevailing air temperature and, conversely, the component to the cold night sky, to space, can reduce temperatures below the air temperature. The range and, therefore, movement is greater than recorded weather temperatures might lead the designer to expect. Interestingly, this has been a contributory cause of recent constructional failures in the hot, dry tropics, where western design practice has failed to allow for the marked radiant effects such climates' intense sun and clear nights can bring – there, a surface temperature range from 70 °C to -10 °C, or wider, is possible.

As to moisture movements, a common problem is the shrinkage of large concrete elements on drying out. The oversite slab in a large basement or industrial floor incorporates construction joints (2.26d), or grooves cast in the top surface so that any shrinkage cracks occur along clean lines and can be filled in.

Summary

- Flat roofs generally repeat the floor construction below. Pitched roofs most familiar in industrial-type buildings.
- Wall types with frame include built-up spandrel bands or, more commonly, panel cladding or complete 'curtain'.
- Cladding is designed for strength in transport and use, and for easy fixing and jointing, especially with regard to construction tolerance.
- Joints are frequently needed to safely accommodate movement, e.g. structural movement joints between low- and high-rise, thermal expansion joints at intervals along building, contraction joints in large lowest floors slab.

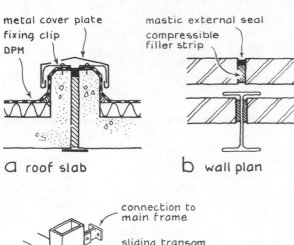

a roof slab   b wall plan

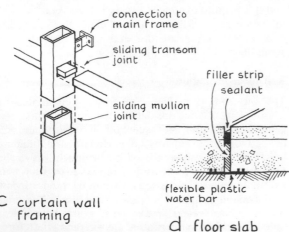

c curtain wall framing   d floor slab

2.26 Movement joints

THERMAL PERFORMANCE

Scale effect

In the small house, the main thing was to reduce heat loss, with overheating really only a problem in hot climates. Scale effect, already explained in Chapter 1 *Structure*, has an important bearing on thermal performance, too. A two-fold linear increase in size brings

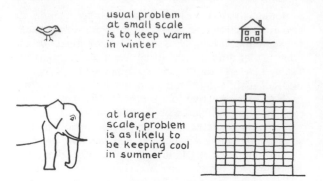

usual problem
at small scale
is to keep warm
in winter

at larger
scale, problem
is as likely to
be keeping cool
in summer

2.27 Scale increase brings a thermal shift

a four-fold surface area increase and an eight-fold volume increase. Volume increases faster than surface. In animals, the rate of heat loss is related to skin area, whereas the amount of metabolic body heat produced is related to body volume. Larger species have a thermal advantage in winter. Insects die, and so may small birds unless they can eat often, but livestock generally survive. Hannibal's elephants at least had scale to help them in the Alps. But conversely, large animals may have a cooling problem in hot weather, and there are thermal as well as structural limits to size in the animal kingdom (2.77). The brontosaurus must have been thermally as well as structurally relieved by wallowing in water.

Buildings, similarly, have metabolic or 'free' heat, produced in proportion to their volume and inside activities. Lights, electrical machinery, plant and, of course, people, they all produce heat and, by the scale effect argument, it follows that large buildings can go a long way towards keeping themselves warm in winter, needing less heat input than a scaled-up increase in the needs of the small house would seem to indicate. Size brings a thermal shift, automatically moving the large building a few degrees up the temperature scale and, potentially, this is a big bonus. The snag, of course, is that the shift commensurately increases the risk of overheating in warm weather, and cooling, like heating, consumes energy. This is an avoidable snag, but many large, modern buildings have not avoided it. The potential advantages of scale increase have often been eroded by the lightweight, unshaded, highly-glazed cladding skin. Pre-war the Bauhaus ethic had long heralded the widespread adoption of the industrial type glazed curtain wall, and with post-war classics like the Seagram Building it became the common, international practice for the high rise enclosure. But the associated thermal problems have only relatively recently come to be properly recognised and responded to. Unprotected glass walls can bring winter penalties owing to their poor resistance insulation. This is especially felt around the plan perimeter. Heat loss is increased directly and

there are uncomfortable downdraughts off the cold glass panes. But the warm weather penalties may be more significant. All the free heat is still being produced: for example, even in broad daylight, most of the lighting will be on in a deep-plan building and, added to this, are the glass walls, acting like a greenhouse, admitting the incoming short-wave solar radiation but less transparent to the much longer wave radiation from the warming interior. The result of this sorry state of affairs can be a quite massive demand for cooling, with the building services making up for the skin's deficiencies at a high energy price. Even in a temperate climate, such a building's annual cooling bill could exceed that for heating.

But, as said, the lightweight skin on a structural frame is a natural consequence of scale increase. The point is that there are ways of reducing the heat loss and overheating problems it brings. Again, we have the heading sequence of the small house.

Shape and orientation

The thermal shift reduces, but does not eliminate, the importance of cutting heat loss. It increases the importance of cutting solar heat gain. So minimising the building surface area is still helpful. And, similarly to the small house, it is some help to have an oblong plan with the longer axis east-west, although this has now more to do with solar-gain avoidance than collection. Of course, the large-building plan shape is more than ever overridden by the hard influences of available site shape and internal organisation. In fact, earlier this century it was also influenced in as much as plan depth was limited by the needs for cross-ventilation and artificial lighting, needs which generated elongated, cruciform, re-entrant and central well plans irrespective of the thermal penalties these could incur. Modern servicing and lighting capabilities have of course changed all that, but it is safe to say that where a high rise building like a hotel or office does have an elongated plan – and this is quite common, having two banks of accommodation on each floor, with windows to the outside and central access corridor between – then there is some thermal benefit if the axis is east-west.

Resistance insulation

The shift similarly reduces, but does not eliminate, the need for resistance insulation. Obviously, this is a generalisation and the reduction in importance will depend on other factors, like climate and building function. The insulation need is that much less in the temperate climate, i.e. away from cold, or very hot, external temperature extremes. From the point of view of the cooler climate, the need is also less in buildings like warehouses and sports halls, which have lower indoor working temperatures anyway and, therefore, less temperature drop through the enclosure, and less also in those factories which happen to enjoy abundant free heat from their industrial processes.

But, of course, there has still been a case for improved standards, even in larger buildings, and double glazing for example, which was often the first thing to be dropped when contract costs started to overrun, is now more established practice. Remember, single glazing has a U value around 5 W/m² °C, whereas double glazing reduces this to 3 or less. Further, reducing the window areas must help again – the solid enclosure wall should achieve a U value well under 1.0, in fact, normally around 0.6 or less. Spandrel walls and solid cladding bands under windows, typically, have thermal block backings and/or other insulation linings. Cladding panels can have an integral insulation layer.

Insulation linings and deckings (again, see 2.32, 2.33) are an important part of roof construction. In fact, roof insulation standards often need to be higher than for walls, not only in low-rise buildings, where the roof area may be a relatively large proportion of the enclosure but, also, in multi-storey buildings where, from the point of view of the top storey, heat loss through the roof may have more immediate effect than the thermal advantage of the building bulk below. Industrial roof-insulation standards are rising and the solar gain penalties through excessive roof-light areas are more recognised (see Chapter 5 *Lighting*).

Capacity insulation

The thermal shift tends to *increase* the importance of capacity insulation, mainly owing to its ability to reduce the impact of peak day-time temperatures. Fortunately, the greater capacity of large buildings automatically increases their thermal capacity as well. A big building is thermally more stable than a small one simply because there is more of it and, to some extent, this offsets the inherently lower capacity of the lightweight construction and cladding. But, in the large low-rise building or in the high-rise 'slab' block with its narrow plan, this capacity advantage reduces and the thermal penalty of lightweight construction increases – and the warmer the climate, and the greater the day-night temperature variation, the harder that penalty is to justify. Again, one has to talk in generalities. But, in a medium-rise building, consider the choice between a fairly heavy construction – say, a reinforced concrete or masonry main structure with a masonry or precast enclosure – and a lighter steel construction with glazed curtain wall. In a temperate climate, the lightweight alternative's possible advantage in structural economy would at best have to be assessed against the undoubted thermal disadvantages it would incur. If you were building in the middle east, or the hot southern states of the USA, there would be no question – the lightweight choice would be architectural nonsense. In both resistance and capacity insulation, the compact,

2.28 The heavier construction, small windowed building is better defended thermally

heavy and small windowed building is better defended than its lightweight counterpart (2.28).

Reducing solar gain

Significantly, the small windows in the heavy building also minimise solar gain. As already said, proper plan orientation will reduce the gain and it will help to have the solid parts of the skin light-coloured and reflective – light-coloured cladding panels, and white chips or pavings on flat roofs. But, again, window effect is crucial, for the unprotected ordinary glass, already poor

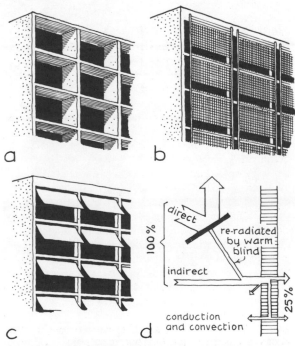

2.29 Solar shading by structure, screen and blind

as an insulation, is virtually useless as a solar defence. Elevations incorporating windows in solid cladding or alternating windows with solid spandrel bands are one move towards the sensible thermal-daylighting compromise.

Solar shading

Shading devices are an obvious protection. In a modest way, this can mean a reflective Venetian blind dropping within a double-glazing cavity. But the frontline defence is external shading, masking off the radiation before it ever reaches the enclosure proper. There are many types. So-called 'structural shading' includes window bays overhung by running balconies or flanked by projecting cross-walls or fins, incidentally often doubling as structural wind bracing in the direction of their plane. Assuming the northern hemisphere, horizontal overhangs will be most appropriate towards the south, where the sun angle is higher. Vertical projections flanking the windows will be appropriate towards the east and vulnerable west where the sun is lower – they can be angled off on plan better to screen windows from radiation arriving near normal to the elevation plane. Combination, 'eggcrate' shading is particularly associated with hot climates (2.29a).

An alternative is louvres and mesh screens (b), with

fins so angled that they admit daylight and view but mask off the sun. They can hang clear of the elevation, as did the simple house shutter, with cooling convection flowing around them and across the shaded building face, and with very little of the heat they acquire conducting back to the building (d).

There are shading devices which automatically adjust. These can be pre-programmed or connected to outside sensors, responding to temperature, radiation intensity, and to time or season, and acknowledging the sun's changing position, its daily path in altitude and azimuth (bearing) through the sky. Common types here include roller Venetian blinds at each window head which drop down externally, tent-like awnings (c) which break outwards over windows, or rotatable louvres and fins. This is as close as one comes to the *theoretical* ideal, the building which is thermally tunable to the changing weather. But, of course, such systems are expensive to install – and they can be a headache to maintain.

Special glasses

The glass areas themselves can be treated to reduce solar penetration (2.30). In fact, solar-control glazing has been one of the really major enclosure developments in recent years and has gone a very long way towards remedying the glazed curtain wall's former, chronic

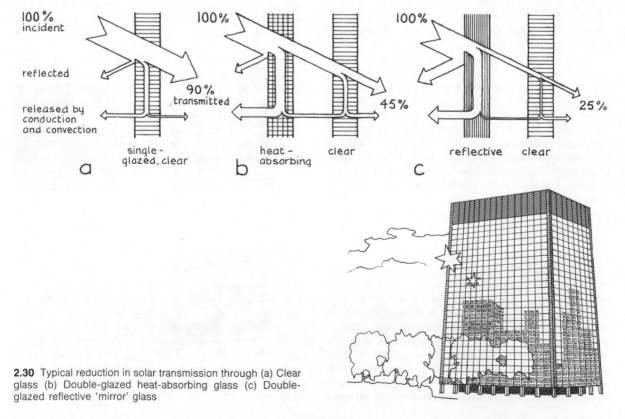

2.30 Typical reduction in solar transmission through (a) Clear glass (b) Double-glazed heat-absorbing glass (c) Double-glazed reflective 'mirror' glass

vulnerability. There is colour-tinted, heat-absorbing glass. Single-glazed (a) its main use is to reduce visual glare – in intercepting radiation, it heats up and itself becomes a radiator into the building. But, used double-glazed, it becomes an effective solar check (b). Then, there are reflective glasses incorporating fine metallic coatings. Among these, the really high performance 'mirror' glasses, so striking visually, can cut solar admission by 75 per cent or more (c). To some extent, the reflectivity will also help cut heat loss in cold weather by reflecting back the outgoing radiant component, improving the U value for ordinary double-glazing to 1.9 or better. But still, it will be thermally better if there are solid spandrel areas incorporating insulation backing.

Incidentally, it is often argued that radiation intercepting and reflecting glasses must also dull the view of the outdoors and cut useful daylighting – by nearly 50 per cent in fact. But, in practice, the eye's ready adaptation makes the change in exterior brightness scarcely noticeable, and the daylight reduction may not signify so much in the deep-plan building relying, in any case, on supplementary artificial lighting (see Chapter 5 *Lighting*).

Summary

- Scale effect brings a thermal shift, moving the large-building interior a few degrees up the temperature scale. In the cool or temperate climate, this is a thermal bonus, except that scale increase also argues having a lightweight enclosure.
- The significance of heat loss reduces but is not eliminated.
- The significance of heat gain from 'free' heat increases. So does that of the 'greenhouse effect', trapping insolation – especially in respect of the thermally transparent lightweight wall.
- The slab block axis is ideally orientated east-west.
- Resistance insulation is still beneficial against heat loss and, sometimes, gain, although appropriate U values further depend on climate severity and building function.
- Capacity insulation is valuable against the daily peak external temperatures, hot or cold.
- External shading and intercepting glazing are effective against insolation.

Condensation

There are factors both increasing and reducing the condensation risk in large buildings, as compared with the small house. On the one hand, offices, public buildings, factories, tend to have drier atmospheres than do houses. There are obvious exceptions of course – swimming pools, sports changing rooms, commercial laundries. On the other hand, the large building is more likely to have a vapour-trapping impervious outer skin, e.g. the flat roof membrane or continuous wall cladding.

In brief, the normal design approach would be this. The likely humidity for the occupation inside would be established, and so would the 'typical' low temperature for the particular winter climate outside. Knowing these, and the proposed enclosure construction, it can be fairly simply calculated 1 whether the temperature of the inside face of e.g. the spandrel walling, or of the window glass or its frame, will fall below the particular dew-point and cause surface condensation, and 2 whether the dew-point line will be crossed by the temperature gradient within the construction, causing interstitial condensation. This kind of procedure would be standard practice for curtain-wall manufacturers.

Avoiding surface condensation

In fact, conditions would have to be fairly extreme for there to be condensation on the inner surfaces of walls and ceilings. But windows will often need condensation channels. Metal window frames, or cladding secondary framing, are a potential cold bridge and so will often be filled with, or made thermally discontinuous by, insulation packing. Structural elements, like concrete columns or floor slab edges, may be a potential trouble source if they happen to break through the cladding to the outside, and will need an insulation wrapping, usually round the outside face.

Interstitial condensation in roofs

Again, there normally needs to be a vapour check on the warm side of the insulation and gradual through-ventilation in the construction above. True, pitched industrial roofs are often built without vapour checks and, even if there is insulation incorporated, as there usually should be these days, the relatively dry atmosphere in the building and the outside air ventilation of the roof void avert the condensation risk. But large roof areas make ventilation through the construction, between vents around the perimeter, hard to achieve effectively. This is the more true if the roof is flat, for example if it is a slab, for there is then no rising pitch to allow natural convection to get the ventilation flowing. Stagnant air pockets may develop. There needs to be a vapour check under the construction and, also, intermediate ventilators can be introduced, poking through flat roofs at intervals (2.31a). Mind you, they are prone to get tripped over by maintenance staff and, like any membrane interruption, must be that much more leak-prone in time. It is sometimes possible and, perhaps preferable, to vent intermediately at the other inevitable breaks, e.g. at expansion joints.

The studded plastic 'isolating mat' shown is providing the lateral ventilation route. But note that such mats have the further function of isolating the membrane from differential thermal movements in the roof construction as a whole.

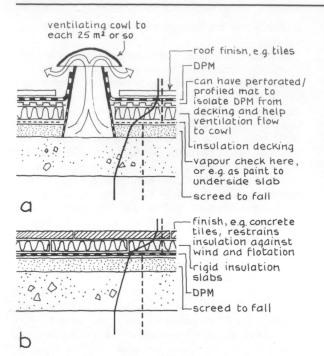

ventilating cowl to
each 25 m² or so

roof finish, e.g. tiles
DPM
can have perforated/
profiled mat to
isolate DPM from
decking and help
ventilation flow
to cowl
insulation decking
vapour check here,
or e.g. as paint to
underside slab
screed to fall

a

finish, e.g. concrete
tiles, restrains
insulation against
wind and flotation
rigid insulation
slabs
DPM
screed to fall

b

2.31 Avoiding interstitial condensation in large-area flat roofs

Avoiding trapped construction water. Roof-void ventilation has another important function. The slab and screed (and, indeed, much of the building below) are of wet construction and, even in dry weather, it is impossible to ensure complete drying-out before the top membrane is laid. This fact, and the presence of the vapour check beneath, could make for a permanently damp sandwich were there no way for the construction water to dry out.

Other solutions, including the inverted roof. Dry screeds, like lightweight aggregate in bitumen, are increasingly used. They avoid the construction water problem and allow the top membrane to be laid immediately. Then there is the development of *breather membranes*, top membranes, which are water-proof but, nevertheless, allow water vapour to permeate up through them, and out.

Most intriguing is the *inverted roof*, 'the upside-down roof', which actually places the insulation layer *above* the rain membrane (2.31b). Obviously, the insulation has to be non-absorptive and non-rottable, and the most common type is slabs of extruded (not expanded) polystyrene. With the membrane on the *warm* side of the insulation, the need for vapour barrier and ventilation is eliminated and, as a further bonus, the roof construction is thermally more stable, reducing expansion and contraction. A chipping or paved topping protects the insulation from foot traffic and, most important, holds it down against wind suction and flotation in heavy rain.

The inverted roof may be the best answer to the condensation problems with which the modern, highly-insulated flat roof has been so notoriously plagued.

Condensation in cladding walls. Condensation problems are much less frequent in cladding walls than in flat roofs, because their construction is seldom so vapour-trapping. Masonry and precast concrete are, to an extent, vapour-permeable but, more important, many cladding systems can be ventilated slightly through their interlocking joints. Nonetheless, having a vapour check is good practice and there will be ventilation to the outside or, and this is quite common, some positive drainage to the outside from small condensation collecting channels at the bottom of each panel. Obviously, the need is most marked in the plastic or glass curtain wall with sealed joints. These points are illustrated in 2.32, 2.33.

Summary

- Basic condensation prevention in roof construction is vapour check and through-ventilation, large roof areas requiring surface ventilators at intervals. But also:
- 'Upside-down' roof places insulation on outside and
- 'Breather' membranes are waterproof but allow vapour out.
- Trapped construction water is avoided by complete drying before applying top membrane, or by use of dry screeds.
- Similar provisions apply for imperviously skinned walls: claddings can be drained.
- 'Cold bridges' are avoided, or external projections are insulation-wrapped.

WATER EXCLUSION

Adding water exclusion to the structural and thermal functions pulls the threads together, as it were. We can now consider in more detail, and illustrate, how the main performance needs reflect in the larger enclosure construction above and below ground.

Roofs

Pitched

In the industrial-type pitched roof (2.33b), the pitch will generally be less than at domestic scale, simply following the shallow profile of the trusses or portals below. The 'tiles' can take the form of corrugated metal or plastic sheets, and the points of overlap down the slope can have a mastic-type sealant as added protection against driving rain. The construction shown is assuming the need for a good standard of insulation and vapour migration control.

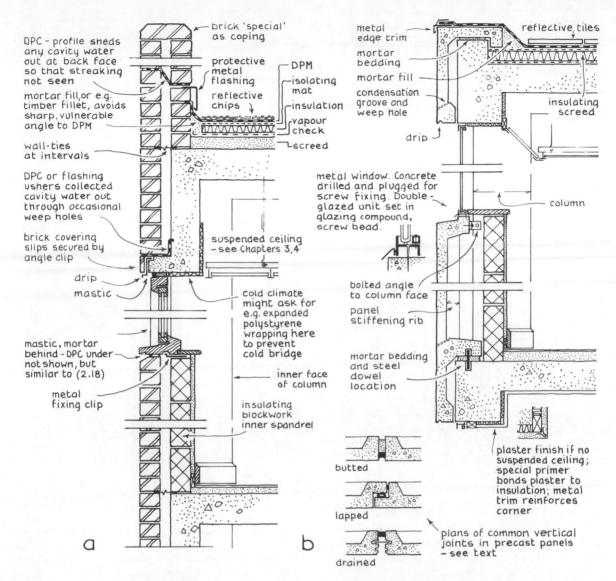

DPC - profile sheds any cavity water out at back face so that streaking not seen

mortar fill, or e.g. timber fillet, avoids sharp, vulnerable angle to DPM

wall-ties at intervals

DPC or flashing ushers collected cavity water out through occasional weep holes

brick covering slips secured by angle clip

drip

mastic

mastic, mortar behind - DPC under not shown, but similar to (2.18)

metal fixing clip

brick 'special' as coping

protective metal flashing

reflective chips

DPM

isolating mat

insulation

vapour check

screed

suspended ceiling - see Chapters 3,4

cold climate might ask for e.g. expanded polystyrene wrapping here to prevent cold bridge

inner face of column

insulating blockwork inner spandrel

a

metal edge trim

mortar bedding

mortar fill

condensation groove and weep hole

drip

reflective tiles

insulating screed

metal window. Concrete drilled and plugged for screw fixing. Double-glazed unit set in glazing compound, screw bead.

column

bolted angle to column face

panel stiffening rib

mortar bedding and steel dowel location

butted

lapped

drained

plaster finish if no suspended ceiling; special primer bonds plaster to insulation; metal trim reinforces corner

plans of common vertical joints in precast panels - see text

b

2.32 The cladding enclosure, (a) Brick infill (b) Precast concrete

Flat

Actually, we have already illustrated a couple of built-up flat roof constructions, but it will help to note the example wall-to-roof junctions in the constructions (2.32, 2.33). As to the water-proofing, specifically, note that while the fall can theoretically be achieved by tilting the main beams or slab, or tilting the decking on top if there is one, it is usually more practicable to have a level structure and decking, and then achieve the fall by screeding. The screed can break towards collecting gutters and outlets around the roof perimeter but, for simplicity, the wall and roof junctions illustrated assume screeds dishing towards central outlets. The

bituminous felt shown for the small-house roof usually becomes liquid asphalt or, in the USA especially, one of the many recent multi-coat synthetic applications. In the case of asphalt, application is normally in two coats onto a sheathing-felt underlay. The felt acts as underlay to the new asphalt when it is wet (especially appropriate when asphalting onto insulation decking slabs direct rather than onto screed) and, thereafter, as an *isolating layer*, reducing the chance of cracking in the set asphalt owing to thermal and structural movements in the main construction below. As shown previously, there can be a separate mat as isolator/ventilator combined.

Walls

The constructional and thermal themes in 2.32, 2.33

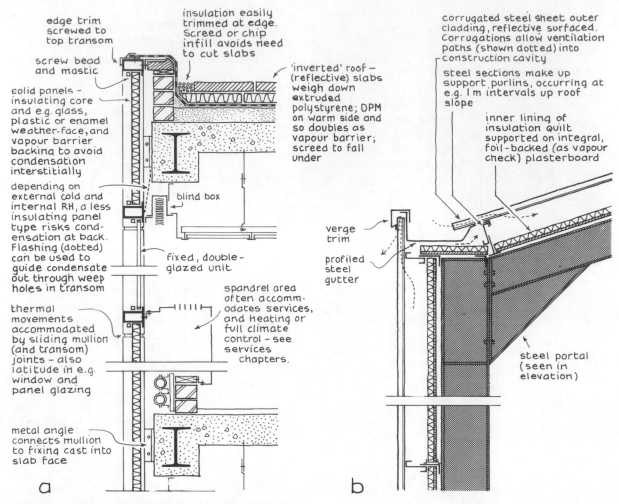

edge trim screwed to top transom

screw bead and mastic

solid panels – insulating core and e.g. glass, plastic or enamel weather-face, and vapour barrier backing to avoid condensation interstitially

depending on external cold and internal RH, a less insulating panel type risks cond- ensation at back. Flashing (dotted) can be used to guide condensate out through weep holes in transom

thermal movements accommodated by sliding mullion (and transom) joints – also latitude in e.g. window and panel glazing

metal angle connects mullion to fixing cast into slab face

insulation easily trimmed at edge. Screed or chip infill avoids need to cut slabs

'inverted' roof – (reflective) slabs weigh down extruded polystyrene; DPM on warm side and so doubles as vapour barrier; screed to fall under

blind box

fixed, double- glazed unit

spandrel area often accomm- odates services, and heating or full climate control – see services chapters.

corrugated steel sheet outer cladding, reflective surfaced. Corrugations allow ventilation paths (shown dotted) into construction cavity

steel sections make up support purlins, occurring at e.g. 1 m intervals up roof slope

inner lining of insulation quilt supported on integral, foil-backed (as vapour check) plasterboard

verge trim

profiled steel gutter

steel portal (seen in elevation)

a

b

2.33 The cladding enclosure, (a) Curtain walling (b) Indus-trial-type sheeting

have been enough explained by now (it is to be hoped) and can be taken happily in our stride. But some points on water exclusion are relevant. The brick infill span-drel wall (2.32a) is, of course, semi-permeable and therefore has a cavity behind. The method is more or less as for the small house. Cladding panel systems, on the other hand, are generally waterproof, the testing part being the joints between the panels. Thus, in the precast concrete cladding (2.32b), the vertical joints could be butted, lapped or drained, as shown. Follow-ing these joints through from the weather side, the but-ted joint has external sealant with mortar backing; the lapped joint offers more dimensional and movement tolerance, achieving its seal by compressible front strip and gasket behind; and the open, drained joint is ingenious in locating its seal well back from the force of the weather, having drainage slots, baffle and (important) a backing seal to stop wind-driven penetration.

The curtain wall (2.33a) is shown with solid insulating panels alternating with the window bands. Of course, there are all sorts of other panel types, or the glazing can be full storey-height. Panel and window units are secured into the transoms and mullions by beads, gas-kets or sealants, as already described. It might seem a tall order to achieve seals that can withstand for decades movement, possible chemical attack, extreme exposure and gale-driven rain but, assuming they are properly detailed, modern synthetic rubbers and seal-ants show every sign of doing so.

The industrial-type enclosure (2.33b) would, in fact, meet relatively high criteria as regards thermal insu-lation and vapour control – many a warehouse will do with less. Again, there are other claddings, and masonry or precast wall wrappings and clerestory lighting, which are familiar in the industrial town-scape. Incidentally, the purlin centring in the system shown would need to be close enough to ensure the inner lining kept stable despite changing temperature and humidity, and the construction sequence would need to advance inner and outer linings apace to keep the insulation dry. Modern proprietary systems add simplicity, for example, sandwiching insulation core between inner/outer sheet-metal facings.

Lowest floor – damp from below ground

We have already looked at water-proofing the lowest floor slab. Here, we will take the case of the all-concrete basement, a common construction under the multi-storey building. And, as before, we will assume that the good sense measures have been taken to reduce the problem at source – drained paving and, possibly, sub-surface field drains around the building.

Integral water proofing – no membrane

One often hears talk of 'water-proof' concrete but, actually, the description is misleading. The material is not inherently waterproof and – to air another popular misconception – there is no 100 per cent effective, magical additive to make it so. But techniques have improved and, given a reasonable constructional thickness, properly designed and placed concrete can be made, to all intents and purposes, *water tight*. One is minded of the distinction between the ordinarily 'water-proof' and the undersea water-resistant watch.

The concrete is made dense, with as few water-permeable voids as possible, by keeping the initial mix very dry – and, since dry means less runny, vibration by mechanical poker and, possibly, use of admixtures restoring some of the lost workability, are needed to ensure proper compaction and distribution within the shuttering. Then there is the problem of seepage through shrinkage cracks which, in the ordinary way, will occur at intervals of 5 m or so in the setting concrete. One remedy used is to cast small areas at a time and protect the resulting construction joints between consecutive pours by 'water bars', plastic or rubber strips either covering the joints externally (as shown under the movement joint, 2.26d) or straddling them inside the concrete thickness. But many engineers now argue that the bars – especially internal bars – just further disrupt the concrete continuity, aggravating the very problem they are intended to solve. An alternative is to use strip projections on the shutter to cast grooves at intervals in the concrete surface, stimulating the cracking along controlled weaker lines, where it can be sealed over after the shutter is struck.

But there is still a point. Achieving *water*-resistance is not the same as achieving water-*vapour*-resistance. It takes some 300 mm thickness of concrete to control vapour penetration inwards from surrounding wet soil. There is a thickness analogy here with thermal cold bridging.

The decision on whether or not to go for integral water-proofing, with its attendant simplicity, must be influenced by the height of the water table outside, the wall thickness intended and the sensitivity to damp of the particular basement use. It will do very well for a car park in dry ground but, even at today's state of the art, would be risky for living accommodation with only modestly thick walls in wet ground. There is a grey decision area between these extremes and, frankly, different engineers will say different things.

Introducing a membrane

Membranes have the enormous potential advantage of being 100 per cent water- and vapour-excluding. Their snag is that they will not be so if they are badly designed or installed, apart from their bringing inevitable penalties in cost and constructional complication.

It is best we take the most testing case for the basement membrane, where it has to act as full tanking, excluding water under hydrostatic pressure from the surrounding water table. Molten mastic asphalt is the traditional tanking material. Others are bituminous and synthetic coatings, and heavy plastic sheetings.

Tanking membranes are generally categorised as either 'internal' or 'external'. There are completely internal membranes with enough adherence – even to damp concrete – to withstand *limited* hydrostatic pressure without a loading-coat backing. This may seem to fly in the face of the conventional wisdom, but their application is a specialist matter and mainly confined to remedial water-proofing of existing construction. 2.34a is a pretty standard, internal membrane, loaded but internal to the main structural part of the enclosure. Note that the structural continuity of the reinforced concrete, between floor, wall and structural frame above, is maintained. However, where the basement happens to have intermediate floors, or walls or columns, connecting with the main structural enclosure, the respective needs for structural and membrane continuity will tend to conflict.

External membranes (b) have the immediate advantage of being forced onto rather than off the construction by the water pressure. Also, having the main structural part of the enclosure entirely inside, automatically removes the chance of membrane/structural continuity conflict. The completely external membrane, i.e. outside the whole construction rather than sandwiched as shown, is on the face of it a very neat solution. But there are points. Clearly, for the basement floor part, the small-house option of laying plastic sheeting under the slab is less practicable on the large, rugged site and, obviously, quite impracticable for liquid-applied tanking. There will need to be, at least, a protective layer of site concrete laid first. Access to install the wall part can call for a wider excavation around the basement box and there will need to be some form of rigid-sheeting protection to stop the membrane from being punctured by large stones when the excavation is backfilled. The approach illustrated

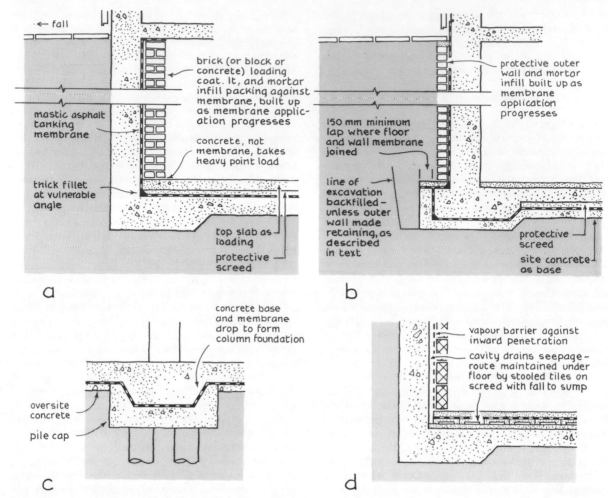

2.34 Concrete basement waterproofing, (a) Tanking internal to structure, (b) Tanking external to structure (c) Column foundation (d) Drained cavity

is, in fact, the most common external membrane, 'external' in as much as the membrane is outside the *main* structure. At the wall, we show a single brick skin as the outer, protective part of the sandwich. (A variant is to have heavier masonry or reinforced concrete here, built first off the slab and forming a shoring to the ground behind. The so-called 'external membrane' can then be applied *internally* to this – the membrane application and inner main walling then progressing up together to maintain the loading-coat function.) (c) details a typical column foundation solution. (d) is a solution whereby any water ingress is drained to a central collecting sump.

DAYLIGHTING – WINDOWS

The small-house conflict on window size, where daylight needs vied with those of thermal performance,

would occur strongly in large buildings were it not for the common reliance on supplementary artificial lighting. The situation would be worse, in fact, with the need to reduce both solar gain and heat loss wanting to reduce window size, and deep floor plans more than ever wanting to increase window size. But, in practice, the lights are on and the windows are as often for psychological needs as for desk-top illumination (see Chapter 5 *Lighting*).

Structurally, the windows can be regarded as cladding panels, subject to the same wind loads, thermal movements, tolerance requirements and need for dimensional co-ordination within the elevation module. Most likely, the frames will be of metal or plastic and the glazing by the faster clip-beading or gasketing techniques. Occasionally, the windows will be openable for ventilation and, may be, reversible for cleaning – see *Durability and maintenance*, p. 76. In either case, their design must allow for the powerful wind-buffeting that can result from the disturbed air stream around large buildings. There again, the large building may be air-conditioned, with the windows fixed closed and cleaned from outside cradles.

ACCESS – DOORS

The entrance area has considerable visual importance in a large building – which is to say, it helps if you can find it. The change in elevational rhythm on the ground floor, and attributes such as approach steps, podium and canopy, all help it read as a focal point, implying 'entrance' to the person arriving.

Naturally, the sizes and number of access doors are dictated by the kind of occupancy. The regular flow of people through a public foyer asks for doors that are fast and easy to operate, but still draught-sealing – the familiar revolving door ('always closed' was their inventor van Kennell's early slogan), double swing doors, or any of a variety of automatically operated swings or horizontal sliders. Normal exits are often the final exits of emergency escape routes, too, and this can affect their location and design (see Chapter 7 *Fire safety*).

Access needs, similarly, influence the entrance design of industrial buildings – an extreme example is the factory or warehouse floor deliberately raised to form a loading bay, level with lorry tail-gates. Doors large-scaled for vehicle or goods access pose problems in achieving adequate stiffness and wind stability. One is reminded of the older type of aircraft hangar or shipyard door, sliding quite clear of the building elevation and top-stabilised by trussed outrigger arms. Today, the many sophisticated types of industrial door, usually electrically operated, include those sectioned vertically or horizontally and which concertina upwards or sideways, clear of the opening, or roll up and over like a blind. Thought has to be given to the insulation of such large doors, and to the associated problems of cold bridging and condensation.

SECURITY

Providing security in large buildings is now very much a growth industry, protecting against theft, industrial espionage, fire outbreak and even sabotage. Generally, the architect will be advised by security specialists, probably in collaboration with the building insurers.

Physical protection in industrial buildings can extend to having a security fence, flood-lit, fitted with alarms and, possibly, patrolled. In buildings generally, security needs influence the enclosure design. Heavy, solid construction is the traditional and still most common approach to enclosing buildings sensitive to security.

But, in fact, the well-lit, transparent glass box is an alternative security philosophy – banks have used it. Also although lighter, solid-clad enclosures, say to industrial-type buildings, are on the face of it more easily penetrated, there are effective protections. These include strong bolting of the cladding to the framing, heavy mesh interlayers in the enclosure construction, and use of toughened glass and sensitive, vibration-triggered alarms.

As might be expected, much of the recent development has been electronic, in alarms and control monitoring. Alarms include ultrasonic and microwave scans detecting movement, infra-red body-heat detectors, inertia vibration detectors, beams, pressure pads, and door and window contact alarms. Closed circuit TV monitoring, linked to a central control room, is now common, scanning the designated entrances and vulnerable points in the enclosure, such as the back basement premises. Movement of people through the enclosure control points (and, indeed, within the building) can be subject to electronic locks and recorders activated by identity cards, digital combination 'keys', and innovations like voice and palm prints. Even in quite ordinary applications, electronic systems can show a fast pay-back against the cost of traditional security patrols.

NOISE (AND POLLUTION) CONTROL

The importance of excluding noise and/or pollution obviously depends on both the nature or the activities inside the building and the nature of the immediate surroundings outside. Ideally, you do not want hospitals or hotels in noisy places, nor non-industrial buildings in dirty places but, generally, the large building will be well defended if, as is quite likely, it is air-conditioned, with its windows sealed. Occasionally, there is a requirement to keep process noise *in*, or to keep pollution *in*, say, where a laboratory works with dangerous cultures or chemicals. (See Chapter 3 *Climate services* and Chapter 6 *Acoustics*, respectively).

COST

The costs in large buildings, indeed, in any building, cannot be compared on the basis of enclosure alone. All the elements play a part obviously, height, shape, specification, apart from which the picture is complicated by the varied circumstances in which buildings get built – different sites and subsoils, different labour and material costs in different places. But there are some general pointers.

Some factors in building high

The small-house discussion compared one- and two-storey housing, but consider some of the cost factors in building high. In housing, it used to be held that going high saved cost per unit of floor space and, incidentally, land space. These assumptions are now rightly questioned. We have shown that single-storey is normally more expensive than the equivalent two- or three-storey but, over this, there probably needs to be a lift and that is expensive. And not only will the structural frame with cladding option become increasingly attractive with height but, also, once the shuttering system and/or cladding run is established, there come further economies in repetition. In other words, once there are lifts and structural frame, cost-wise there may be an argument for going up a few storeys more. But there comes a point where increasing height starts to attract structural penalties, both on the frame and on the foundations, which may require to be expensive piles. So, very broadly speaking, cost decisions occur around four storeys and then, somewhere upwards of twelve, allowing that the second point especially is loose and dependent on a host of other factors.

Cladding costs

As with the small house, the enclosure will be cheaper if the elevations are uncomplicated and flat. Claddings are often thought to be cheaper than masonry but this is seldom true of the cladding itself. Even the simpler types of plastic composite panelling are likely to be dearer than the equivalent brick infill. Precast concrete claddings might be double the brick cost, and a double-glazed curtain with aluminium framing and reflective glass outer leaf might well be treble. Claddings are really chosen for their lightness, erection speed and ease of maintenance. And, to quote one reputable cost consultancy, there are also those 'frankly architectural' reasons!

Cost philosophy

The percentage of the total budget allocated to the enclosure must acknowledge the need to reduce maintenance and, need it be said, energy consumption. A typical cost breakdown for say a twenty-storey office with basement would be in the order of walls 20 per cent, roof 2 per cent, work below ground 15 per cent, structural frame and floors 20 per cent, services 30 per cent, and other things making up the total. Naturally, a school complex with low height and large plan area would have much reduced wall costs and increased roof costs, and this kind of variation in the relative significance of elements affects budgeting philosophy. For example, small savings per m² of roof would have far less cost impact on the office than the school. Wall costs would have great impact on the office and less on the school. But, although this has understandable influence on cost allocation, there are tempering arguments. Even though saving or over-spending on a small element may be apparently insignificant in a total budget, it is still money to the client. Equally, economy in large elements might be false economy. In the office, the walls are critical to cost but, also, critical to thermal performance and, from that point of view, are the place where money ought to be spent, not saved. There is no simple picture. A final point is that, whereas traditional builders, when they thought about it, would have regarded their buildings as fairly permanent, modern budgeting tends to take some account of a building's estimated life-span. There is an effort to tailor durability, as it is reflected in cost, to time. There is a certain realism in this but there also has to be caution, in as much as no one can really know how long a building will last. 'Temporary' war-time prefabricated homes are still in use today.

DURABILITY AND MAINTENANCE

Cladding and traditional walling compared

In theory, the cladding wall of a large building should need less maintenance and last as long as the small-house masonry wall. Generally, this holds true but with qualifications. Durability can suffer where less is known about the long-term performance of synthetic facing materials, gaskets and so on. Here again, innovation has sometimes caused a troublesome information gap for the architect. There can be a tendency to accept reduced durability and, resultingly, more expensive maintenance in the long term, as the price for the immediate savings in initial capital outlay. The *materials cost* to *later durability* equation is similar to the insulation cost to later energy cost equation, in as much as capital savings which make initial sense to the developer may be economic nonsense in the long term. Lack of 'durability' can even lead to a building's being closed down for repairs.

Visually, impervious skins are more prone to weather-streaking than traditional masonry ones. As explained, modern claddings can either be boldly profiled to accentuate the weather patterns, or else kept flat to smooth the rainwater flow so that whatever dirt is not washed away is, at least, deposited evenly.

Window cleaning

From inside
Outside window cleaning in large buildings poses the simple question of how to get there. Openable windows

can be made fully reversible or, at least, designed so that all outside glass faces can be reached from inside the building (2.35). This would be sensible for a private occupancy like a block of apartments, where windows are small and openable, in any case, for ventilation. But internal access is often inappropriate. In many building types, maintenance staff tramping around would be disruptive. There are ergonomic limits in reaching out to cover larger window surfaces. And often there will be air-conditioning, with windows fixed closed.

From outside

External access has the immediate advantage of allowing coverage of the whole exterior for maintenance, not just the windows, and it lends itself to more rapid systems. The simple lean-to ladder is, of course, limited to a few storeys and would be uneconomically slow where the elevation is wide. Vertical ladders clipping onto the building are more secure and the more convenient in placing the operator a constant distance from the work. There are travelling ladder systems that slide along runners on the elevation. Fixed walkway access tends to be uneconomic for maintenance alone and visually inappropriate to most building types. But, of course, there may be running balconies anyway, and horizontal shading projections can be designed to double as access routes.

Cradles are the accepted way of getting to the large, high-rise elevation. Typically, they hang from the overhanging gantry arms of a moving trolley, fixed on rails

2.35 The large façade has to be reached for cleaning and maintenance

around the roof perimeter. Long drops call for additional cradle restraint to guide rails on the elevation face, for example, to the curtain-wall transoms. But the sophistication of the system again depends on building size, ranging from the simple hand-operated pulley of the bosun's chair, to the electrically powered sorts, positionable by the cradle operator. And there are fully automatic cradle systems, programmed to cover the elevation like a car-wash.

APPEARANCE

It has been said that the best thing a modern building can do is disappear, an unkind view but one more probably provoked by large buildings than small. Certainly, large buildings can have enormous impact visually and this partly explains occasional reaction against concrete tower blocks and 'those awful glass boxes'. But, where there is adverse reaction, can it not be argued that many of the things disliked – the visual repetitiveness of the cladding, the hard modern materials, the size – result as much from the expanded scale of the way we live, the expanded purposes we have for architecture, as from anything the enclosure design is doing in itself? Of course, there are bad designs, dishonest, illogical structural grids confusing the enclosure rhythm, new claddings used in wrong ways in a rapidly expanding technology. And, undeniably, increased scale has often sprung from fashion, not need – thankfully, the appropriateness of high-rise housing and ultra-high-rise commercial building is questioned more realistically now than in the 1950s and 1960s. But the point remains that modern building criticism is the more valid when it acknowledges the whole social and economic context in which buildings are built.

It has also been said to the point of cliché that, in good design, form ought to follow function and that if a building or anything else is designed expressly to fit its purpose, then it will automatically look well. 'Form from function' was virtually a slogan in architectural schools a few years ago. Essentially, most people would agree with the principle. Although aesthetics can never be wholly subordinate to any set of rules and although functionalism as a single principle would be hard put to explain the design of many good buildings, for example the Sydney Opera House, the form of which is scarcely connected with the needs of structure, internal planning or even acoustics, yet, for most buildings – and most enclosures – function must be a prime determinant of good appearance. The modern, large building enclosure has indeed suffered its various technical and visual doldrums in its development this century, multi-storey cladding notoriously so. But the uses are coming to be understood and, properly executed, the large enclosure can have clarity, simplicity and beauty.

3

CLIMATE SERVICES

ENGINEERING FOR COMFORT

The heating, ventilating and, possibly, air-conditioning that ensure a comfortable indoor climate are all part of what is now usually known as environmental engineering. This is an umbrella term that also includes utilities like water, electrical and drainage services, artificial lighting, lifts and other installations generally. Our concern in this chapter is with the services for physical comfort, so we had better start by saying what 'comfort' means.

What affects comfort

The food we eat fuels our metabolism which, in turn, produces heat which we have to lose at a certain rate to be comfortable, in fact, at a rate keeping our body temperature around 37 °C – it is a matter of balance. Physical exertion increases the metabolic heat to be lost, inactivity reduces it. An active task asks for a room temperature around 16 °C or less to allow the appropriate heat loss without discomfort. A sedentary task asks for a higher temperature, around 21 °C. But this is a generalisation. For one thing, 'optimum' comfort is subjective, varying slightly between one person and the next and, for that matter, it depends on what they are wearing – a curious recent statistic quoted the number of million dollars in energy costs the US could expect to save if everyone wore an extra waistcoat to work. This is perfectly true. Also, and this is the main thing we have to deal with here, while air temperature is a principal factor determining whether or not we feel thermally comfortable, it is not the only one. Air movement, the temperature of surrounding surfaces and relative humidity play a part as well.

We lose heat by convection, radiation and by the latent heat of evaporation absorbed by perspiration. Assume the room (3.1) has an air temperature that feels just warm enough, say 20 °C. Our bodies are warmer than that, and heat convects gently away at a rate which comfortably balances the metabolic heat we produce – we do not feel cold. But bring in other factors. If the air starts to move around us, if we are in a slight draught, the convection loss will increase and we may start to feel cold *even though the air is at the same temperature*. (Shake your hand and it feels colder.) The room temperature would have to be raised two or three degrees to compensate.

Or, suppose there is a large cold window in one wall of the room. In nature, warm matter will radiate energy to cooler matter nearby. Our bodies radiate to cooler surfaces around us, and windows can be particularly cold. In fact, we may need a slightly higher air temperature to compensate. So double glazing, with its warmer inner pane, makes comfort sense as well as cutting heat loss directly.

Relative humidity, described in the last chapter, is an important comfort parameter. In a cold room, dampness is an adverse factor in that it reduces the insulation of our clothes, albeit the effect is slight. More significantly, our ability to perspire relies on there being a vapour pressure drop between our skin surface and the surrounding air. At 21 °C, 50 per cent RH would be comfortable but 80 per cent would feel clammy.

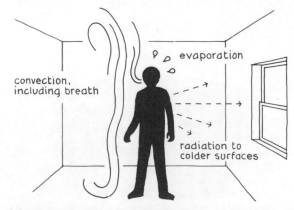

3.1 We lose heat by convection, latent heat of evaporation of sweat, and radiation

There are corollaries in the overheated condition. More blood flows close to the skin surface for cooling then, and our sweat glands produce moisture which assists the cooling by taking up latent heat from the skin when it evaporates. (Shake your hand and it feels even colder if it is wet). If a room becomes too warm for us, air movement can compensate, up to a point, by increasing the rate of heat convection from our skin and assisting perspiration evaporation. 'Punkah' ceiling fans helped cool people before air-conditioning, even though they did nothing to reduce temperature – their motors and turbulence must have increased it slightly in fact!

Clearly, relative humidity and our related ability to perspire and cool are highly significant in the over-heated condition – hot humid climates are much more uncomfortable than hot dry ones.

So, though we can adapt, acclimatise, physiologically as well as by altering the amount of our clothing, the range of *comfortable* adaptation is really rather small. And uniformity of temperature throughout a room is generally helpful, since even slight temperature contrasts can be noticeable and bring a sense of discomfort. The comfort factors are all interlinked. A deficiency in one can be compensated for by another – but only up to a point.

Summary

- Air temperature
  A principal factor – affects the rate at which convection removes body heat.
- Air movement
  Increases convection, reducing comfort in an underheated room and increasing comfort in an overheated room.
- Surrounding surface temperatures
  Affect radiant losses from, or gains to, the body: cold windows reduce comfort in a room otherwise warm enough. Radiant heaters are especially pleasant in a room otherwise too cold.
- Relative humidity
  Comfort band as wide as 30–65 per cent – high levels increase discomfort in overheated room.
- Uniformity
  Helps adaptation.

Scales other than temperature

So the simple thermometer is only a loose indicator of comfort and, although there is no single instrument capable of measuring the effect of all the factors at once, there are instruments and scales which take more than temperature into account. For example, the 'equivalent temperature scale' combines all the factors except humidity, and the 'effective temperature scale', mostly applied to air-conditioning, combines all but radiation.

'Enclosure + Services input = Comfort'

Picking up from Chapter 2 *Enclosure:*

$$enclosure + services\ input = comfort$$

was the loose equation applied to the 'small' house. Enclosure design is crucial but, even with the best of enclosures, the services are left with a balance to make up between conditions outside and those desired inside. Again, it may help to start by thinking in terms of the 'small' house.

THE 'SMALL' HOUSE

HEATING

Assume the small house is in a relatively temperate climate, away from the hot or cold extremes. Heating is then the main input for comfort. Cooling equipment will be unnecessary, provided the enclosure is properly designed and shaded – actually, even in the tropics, it would be an expensive luxury. Summer cooling will be mainly by natural ventilation through windows, as described in Chapter 2 *Enclosure*, with some mechanical assistance from local extracts to remove at source odours and surplus heat from kitchens and bathrooms.

The typical heating process has a fuel, converts it in a furnace or boiler into heat, and either emits this heat to the house directly or, more usually now, transfers it via a carrying medium like water or air to room emitters (3.2). A control tailors the output to the varying need.

Fuels

Essentially, the four choices for fuel for the small house are coal, oil, gas and electricity – devices like solar panels and heat pumps, deriving heat from natural sources, are described later. Cost, convenience and availability are the basic considerations, and all three are, to some extent, influenced by locality. Cost is particularly hard to assess, since trends can vary unexpectedly – the forecast in the 1960s was for cheap electricity 'as a result of nuclear power and a coming oil glut'. Instead, all fuel costs have spiralled.

Coal

Coal and its various smokeless and other solid derivatives declined as a domestic fuel until the 70s. Since then, for economic reasons, there has been a revival. It is relatively cheap but inconvenient and rather dirty. Its needs for dry storage, stoking and ash removal are a nuisance

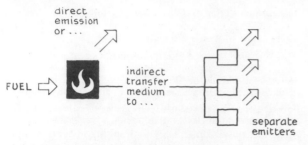

FUEL

direct
emission
or ...

indirect
transfer
medium
to ...

separate
emitters

3.2 Typical heating process converts fuel to heat in a burner and either emits the heat directly or transfers it via a carrying medium to emitters

and, like most fuels, it needs a chimney, a flue. There is a control disadvantage – the burning rate can be slowed but not switched off at the flick of an automatic thermostat, like the other fuels.

Oil

Oil costs continue to rise, particularly owing to the political instability of supply. There has to be a storage tank and, in large buildings at least, this has to be enclosed in such a way as to contain the possible spillage of the whole contents and, if inside, to compartment against fire spread (see Chapter 7 *Fire safety*). The furnace or boiler needs a flue and needs particularly careful servicing for high efficiency.

Gas

This can, in fact, take the form of liquefied petroleum gas, propane and other varieties needing a storage tank, but they are expensive. The real contender is piped gas, derived from coal or oil, and supplied on the city grid. It has most of the points of oil, including the need for a flue, but is the more convenient for requiring no storage.

Electricity

This is energy rather than a fuel but, of course, it comes from the conversion of a fuel at the power station. A cable delivers nice, clean, convenient power: there is no store, little servicing and no flue. It would be the natural choice, were it not so expensive. The expense is primarily because fuel conversion to energy at the power-station is only about 40 per cent efficient. Some countries have pricing schemes encouraging 'off-peak' use, exploiting the spare capacity of the power-stations at night, but even off-peak tends to be expensive and the thermal storage systems it requires have their own disadvantages as we shall see.

The main point for the future, though, will be the increase in primary electricity, i.e. electricity as the distributor from nuclear sources (despite their contentiousness) and from other natural sources – hydro, geothermal, wind and solar.

Summary

- Very broadly, and in *today's* circumstances, if the small house were in a town, then the tendency is towards gas but, if off the gas grid, then oil, or less frequently but increasingly, coal.

Direct heating

Open fires and stoves

The traditional open fire (3.3a) was only about 20 per cent efficient owing to the heat and unburnt fuel that was convected up the chimney. But we like fires – they look good, feel good (lots of radiant heat) and help ventilate the room, and modern fires are achieving efficiencies up to 40 per cent. The traditional fire entrained primary air to feed the fire-bed and secondary air to assist the fire draught. Most modern fires do the same, but improve the burning efficiency by controlling the air-flow patterns through specially shaped grates and flue throats, with adjustable venting. There are many types. In 3.3b, the primary air is drawn through a duct from outside the house: it can be fan-boosted, in fact. There are no cold intake draughts through the room, instead, fresh-air ventilation is achieved by having only part of the primary flow going to feed the fire-bed – the rest circulates around the hot fire-box and is then convected into the room directly, supplementing the radiant-heat output. Stoves (3.3c) similarly make the best use of the combustion process within the design of their casing and, of course, the fact of their being free-standing adds to their radiant and convective output. Efficiencies over 60 per cent are possible. Many are multi-fuel, capable of burning wood, smokeless fuel and so on. Modern fires and stoves can still incorporate the traditional back-boiler arrangement, – in fact, there is as much as 10 per cent added efficiency if they do. Here, the firebox and flue gases heat a water jacket behind, from which pipes run to two, maybe three, additional radiators and/or the hotwater supply cylinder. Flues are preferably incorporated in an internal wall, the better to capitalise on their added warmth. If in the external wall, then at least the outside leaf must be weather-resistant and insulating – if flue gases cool too much, the fire draught can be upset and, also, there can be troublesome condensation.

Oil stoves would hardly be chosen as part of a designed heating system. They can be useful where there is no other heating but have to be filled, can be smelly and older types can be dangerous if knocked over. Gas fires are popular, convenient and around 60 per cent efficient. They need a flue, but the flue requirements are simpler than for solid fuel, requiring only the removal of the products of combustion and not needing to induce a draught. There is a slight ventilation bonus. Electric heaters off the domestic circuit are clean, convenient but expensive.

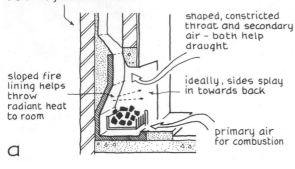

if wall external then
insulated to reduce
heat loss and prevent
flue gases cooling to
condensation, e.g. could
be cavity construction

shaped, constricted
throat and secondary
air – both help
draught

sloped fire
lining helps
throw
radiant heat
to room

ideally, sides splay
in towards back

primary air
for combustion

a

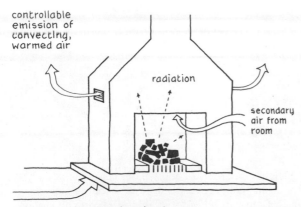

controllable
emission of
convecting,
warmed air

radiation

secondary
air from
room

underfloor pipe supplies air direct
from outside – part is primary air
for combustion, part is diverted around
hot firebox then emitted at grills

b

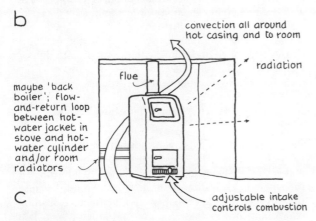

convection all around
hot casing and to room

radiation

flue

maybe 'back
boiler'; flow-
and-return loop
between hot-
water jacket in
stove and hot-
water cylinder
and/or room
radiators

adjustable intake
controls combustion

c

3.3 (a) Traditional open fire (b) Improved open-fire design
increasing efficiency (c) Stove making best use of combustion
process within the design of its casing

Both gas and electric heaters can take the form of
radiators or *convectors*. The terms are misleading,
implying that there is either radiation or convection

exclusively when, in fact, both do both. With 'radiators', the flame or hot element is visible (except in the
electric oil-filled type which looks like a central-heating
radiator) and the radiant component feels warm. With
'convectors', the radiant component is negligible – air
flowing past the internal heat source, i.e. inside the
protective outer casing, convects heat to the room.
Fan-assisted electric convectors are now popular, in as
much as they give a quick air heat-up and are convenient. But they are very expensive to run (also, being
electric, there is the added cost of the fan power), and
the induced air movement outside the warm stream and
the lack of radiant component both tend to reduce
comfort.

But, although some of these heaters have been
described as 'convenient', they are only relatively so.
Switching any heater on and off when you feel cold or
hot is inconvenient and, if you are cold or hot, you are
already uncomfortable, so the control is already too
late, inefficient – and wasteful. True, the air in a room
can be warmed quickly owing to its low thermal capacity but, by the time that has happened, when you get
up in your bedroom, it is time to go down to the cold
kitchen. The heavier building fabric takes hours to
warm so, even though the air is warm, you still feel
uncomfortable radiation losses to your surroundings.
Stop-start heating on a room-by-room basis tends to
give least comfort at most cost.

Whole house heating

The idea of having overall heating systems, or 'indirect'
systems as they are sometimes known, is hardly new.
The Greeks had a go at it. The Roman 'hypocaust' system passed hot gases from a furnace through the
underfloor space and up ducts in walls. Modern central
heating has its origins in the late eighteenth century,
when direct heaters were found inadequate for the
large industrial and public buildings then starting to
appear. Piped circuit systems were developed to distribute heat more widely, using steam or hot water as
the heat-carrying medium. Today, with our high expectations of comfort and technical ability to meet these
expectations, it would be rare to find even a small
house being built without an overall, self-regulating,
heating system. Basically, we are talking of either heating on an electrical circuit or, more likely, 'central heating' where a central boiler or furnace transfers heat to
room emitters by piped hot water or ducted hot air.
The ideal system is:

- Simple to install.
- Convenient to run.
- Cheap to run.
- Fast-responding for both comfort and economy, its
  output quickly controllable to suit the immediate
  need.
- Flexible, allowing separate room-by-room control.

- Capable of delivering its heat in a comfortable way.

Electric 'central' heating

To be pedantic, electric heating systems are not true 'central' heating since there is no central boiler, unless one calls the power-station the boiler. But it is with central-heating systems that they have to compete. Having no boiler or flue, and no pipes or ducts to thread around the house, all tends to simplify the installation. The very simplest thing would be to have thermostatically controlled heaters in every room, run off the ordinary power circuit. This would satisfy all the above criteria save one – it would be horribly expensive.

Electric storage heaters. Storage heaters are designed to exploit the cheaper (but not cheap) off-peak rates at night. They have heavy blocks or other mass inside their metal casings to give them high thermal capacity. Each has an internal heating element, run off the special off-peak meter and circuit. Heat is pushed into the mass at the cheapest time but released to the house continuously over the 24-hour period. There are snags though. The heaters naturally tend to run down towards evening, probably when they are needed most, and may need an on-peak, afternoon boost. Also, their output is rather uncontrollable.

The fan convector types are an improvement (3.4a). They have an insulated outer casing, so that the output is mainly by convection when the fan operates in response to a thermostat or timer. But the intermittent noise annoys some people and, as with all storage systems, one is left with the problem that the night's intake is the following day's output, irrespective of what that day's outdoor temperature turns out to be. There can be an outdoor thermostat influencing the night intake. But things are still imprecise and storage systems are mostly seen as background heating, topped up by a secondary, fast-response system, controlled as required.

Electric underfloor heating. Underfloor heating (approximately 50 per cent radiant and 50 per cent convective) exploits the inherent storage capacity of the heavy concrete floor by heating cables laid in the screed (3.4b). This is again a background system, particularly since the limit of unobstructed floor area, and the limit in the temperature that is comfortable to people's feet (about 24 °C), limit the output. The floor construction is slightly complicated by the thicker screed required (for adequate storage mass as well as cable housing) and by the insulation layer and, although there have been improvements here – including withdrawable cables – failed cables can be troublesome to get at and repair.

Electric ceiling heating. Here, electric heating cables are used to turn whole ceiling areas into heating

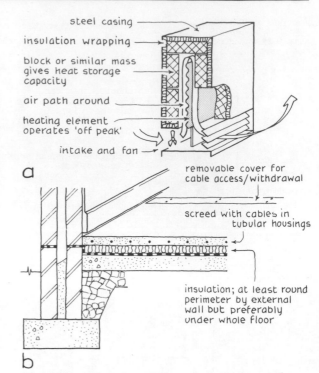

3.4 (a) Fan convector storage heater (b) Electrically-heated floor screed – this can also exploit the storage principle

panels. The output is about 90 per cent radiant and 10 per cent convective, downward convection being against the natural order of things. Having no storage involved makes the system more responsive but more expensive. Higher temperatures are possible than with underfloor, about 40 °C, but the effect is oppressive with ceiling heights much under 3 m and, over 5 m, the system starts to get inefficient.

One of the commoner *electric* systems for new houses is to have underfloor cables in the ground floor as background and a faster response top-up, possibly including ceiling heating in the bedrooms above.

Introducing indirect systems – hot-water central heating

In today's conditions, it is a more efficient use of fuel to convert it directly into heat in the house itself. It makes further sense to have one central fuel burner for the whole house rather than many small ones, because there is then a scale economy, and the fuelling, flueing, controlling and maintaining are that much easier. But the heat has to be distributed around the house and, as you will know, the principal domestic systems are piped hot water and ducted hot air. We will compare their relative merits later but suffice it to say that hot water is favoured in most of Europe and, while hot air has enjoyed considerable popularity in the USA, hot water now seems to be on the increase there also.

In hot water systems, the central *boiler* converts fuel to heat, the *piped water circuit* carries the heat to the rooms, and the *emitters* give out the heat from their surfaces by radiation and convection. The cooled water returns to the boiler. Simply, one is heating water, piping it around and widening the pipe to increase the emission area where heat is needed.

The boiler

Boiler choices are essentially divisible into cast iron, the traditional material, and steel. The latter is cheaper albeit less durable, tending to corrode inside if sulphur-bearing fuels like oil are burned at an inefficiently low temperature (causing condensation and the formation of sulphuric acid). The fuel is normally burned in a central combustion chamber with the water to be heated circulating in a 'jacket' surrounding the chamber or passing through metal tubes in the chamber itself (3.5a).

Location. The boiler must be sensibly sited for maintenance and, if coal-burning, for fuelling. There needs to be ventilation for good combustion. The flue need is often the dominant location factor, especially where central heating is being installed in an existing house.

Flues. All combustion boilers need a flue, preferably vertical and with the boiler close under, i.e. avoiding horizontal runs that would upset the natural draught – this is particularly so with solid fuel. An alternative with gas boilers and, sometimes, oil ones, is the *balanced flue* (3.5b), where the concentric intake and exhausts are directly through the external wall onto which the boiler backs. The flue is 'balanced' in that wind pressure fluctuations outside affect intake and exhaust equally and, therefore, do not upset the draught balance inside the boiler itself.

Size. Boiler sizes are most usefully described in terms of their heat output, Imperial measurement talking about quantity of heat, BTU per hour, and metric measurement quoting in kilowatts the power flowing at any instant – it comes to the same thing. A 15 kilowatt boiler is capable of producing that amount of heat energy from the fuel it burns and, incidentally, the *thermal efficiency* of a boiler is the actual output achieved, expressed as a percentage of the calorific value of the fuel put in – 70 per cent, 75 per cent and 80 per cent are the sorts of efficiency modern domestic boilers get with coal, oil and gas, respectively.

Obviously, the anticipated heat loss from the building is the basis for sizing the heating system, plus an added margin for acceleration, the system's ability to heat up the building reasonably quickly from cold. Briefly, the starting point is the temperature difference between the internal design temperature, which varies with the activity in a building but is, typically, 20–22 °C for a house's living areas and perhaps 16 °C for the bedrooms, and the external design temperature. This latter is the typical low temperature the building is assumed likely to suffer, not the lowest possible temperature – sizing the system to cope with the rare cold extreme would be rather redundant (except in a building the temperature of which absolutely must not fall below a stated minimum, an art gallery perhaps). The lowest monthly average, the 'mean monthly minimum', is one measure. In parts of the UK, it might be taken as −1 °C. Under a similar system, New York might arrive at −18 °C. Whatever the accepted datum, there can be slight increments to allow for cases where the building construction is relatively heavy and, therefore,

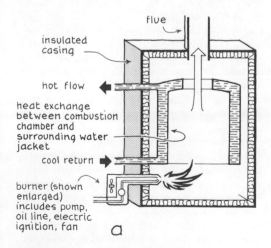

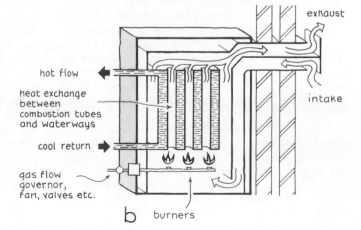

3.5 Domestic boiler. (a) Conventional flue boiler – shown as oil-fired but could be gas (b) Balanced-flue boiler, usually gas.

There are also compact, wall-mounted versions but the principles are similar with extra thermal capacity to cushion it against short-term cold extremes, or where the central heating system is to be of a type having some built-in overload capacity for extremes.

As explained in the last chapter, the U value or thermal transmittance (they are synonymous) of each roof, wall and window element in the enclosure (W/m² °C) can be found from standard tables. Thus, if the inside/outside temperatures were 22° and –1 °C, i.e. 23 °C difference, and the U value of a particular wall element were 0.8, then the heat loss in watts contributed through that element would be 0.8 × its area in square metres × 23. Finding the watts heat loss per hour through the whole enclosure is a matter of repeating this simple sum for all the elements and totalling the results. Added to this is the heat loss through ventilation, i.e. the volume of air inside × the estimated number of air changes per hour (perhaps from around 1 to 3 in a house, depending on its shape, construction and exposure to wind) × density × specific heat × 23 °C. A further small allowance can be made for hot washing water losses through drains, and an allowance for acceleration, as said. Well that is the regular theory, though with practised experience, the calculation for a small building is rather more rule-of-thumb, as one might expect.

The hot-water circuit

It is rather lucky, when you think about it, that water, so readily available, happens to be physically well suited as a heat-carrying medium. It stays chemically stable no matter how many times it is heated and cooled, and has a particularly high specific heat, i.e. heat-carrying capacity for a given volume. Its slight disadvantage is the relatively narrow band between freezing and boiling, 0° to 100 °C, the former obviously rendering the system inoperative and, perhaps, damaged, the latter causing steam and dangerous expansion. Where temperatures could drop low enough to freeze water, e.g. in greenhouses or refrigeration plant, brine, light oils or water with other chemical additives can be used. At the other end of the scale, we shall see how higher heat-transfer rates can be achieved with temperatures well over 100 °C, yet without boiling, provided the system is sealed and pressurised.

The immediate connections – the expansion tank and hot-water cylinder. We will compare the various heating circuits in a moment – first, a look at the immediate connections around the boiler, i.e. its links with the expansion tank, cold-water supply, if desired the domestic hot-water cylinder and, of course, the central-heating circuit itself (3.6). The main point to notice is that the boiler system water, the pipes shown in black, is kept separate from the ordinary tap-supply water. This is because the boiler water, remaining unchanged save for rare top-ups to replace evaporation losses from

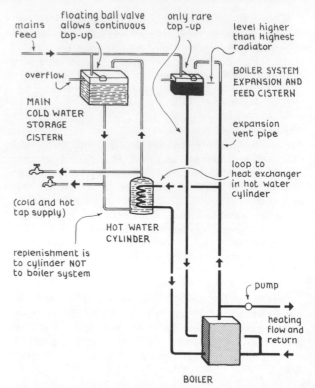

3.6 Immediate connections around the boiler. Boiler primary circuit to hot water cylinder and central-heating system water is shown dark. (Domestic water-supply details are more fully described in Ch. 4. 'Utility services'.)

the heating system, would tend to contaminate the tap supply if it mixed with it. So the main cold-water storage tank supplies the taps and the smaller tank provides the rare top-ups of the heating system. The boiler-system water's main function is to supply the central-heating circuit, but its supplementary role of heating the hot-water supply cylinder is achieved by a circuit linking with a heat-exchange coil in the cylinder. The device, indirect cylinder and exchange coil, is also called a *calorifier*. Note that both the hot-water cylinder and the boiler-system water require some means of pressure release, in case either malfunctions and overheats – hence, the traditional fail-safe, the expansion vent pipes, discharging back into their respective feed tanks as logic would suggest.

The gravity heating circuit. The original and one of the simplest types of heating circuit is powered by gravity, that is to say, there is no pump to push the water around, reliance being placed on natural convection, on the slight positive buoyancy that hot water in the initial flow-pipe rising from the boiler has over the cooler denser water in the return-pipe dropping back. But the flow is rather weak and the response sluggish, meaning that the pipes have to be fairly large to get enough heat around the building. The boiler, obviously, has to be at the lowest point in the circuit,

and horizontal parts of the circuit, i.e. where there was no contribution to the convective effect, have to be limited in length.

Electric pump systems. Things have been revolutionised by the electric pump, the electric heart enormously improving the modern circuit's performance. The reliable, higher pressure speeds the response, frees the planning layout, allows smaller pipes and reduces the chance of air locks. The smaller pipes, 15–25 mm or so diameter, have given rise to the term 'small-bore'. The pipes themselves are still usually copper. Stainless steel is rather expensive and mild steel tends to corrode. Plastic piping, already common for domestic cold-water supply and waste plumbing, had traditionally found it harder to cope with the higher temperatures in hot-water supply and, especially, heating circuitry, but now 'hot-side' plastics are also appearing on the market.

Single and two-pipe circuits. 3.7 shows single and two-pipe, small-bore circuits. The traditional single-pipe circuit has one flow pipe routed round the house, with simple flow and return connections with the radiators it serves. The two-pipe circuit has a hot-water flow and a *separate*, cooler water return. Obviously, the single-pipe circuit is simpler, so why the two-pipe alternative? In fact, the single-pipe has drawbacks. Obviously, the water flow giving off heat to successive radiators gets cooler and, strictly speaking, the radiators would need to be progressively larger or more numerous to maintain the same output. The flow through the radiator loops is by gravity and not very positive (although to be fair there are improvements where the radiator pipe to main pipe connection is so shaped that the main pumped flow induces a better

radiator flow by Venturi effect). The radiators are hard to control independently: turning one up or down alters the heat reaching the others along the circuit.

The two-pipe system is now preferred because it is easier to tailor to the needs around the house and easier to keep balanced in use. The hot-water flow supplies all radiators at virtually the *same* temperature, because the cooler water leaving a radiator goes *straight to the return pipe and back to the boiler, not on to the next radiator*. Usefully, there is a pressure reduction between the flow and return, since the pump is on one or the other, not both, and this means the water is positively pumped through the radiators. Altering one radiator setting does not significantly affect the others. However, a point that can be appreciated from the diagram, is that each radiator needs a valve to ration its supply – if not, the radiators nearest the boiler would starve the later ones, at worst the first radiator simply short-circuiting the whole flow to the return and, hence, allowing the others no heat at all!

Ultra small-bore systems (microbore and minibore). Combining the inherent efficiency of the two-pipe circuit with more powerful pumps has led, rationally enough, to even smaller pipes. Bores as little as 6 mm are available, about the width of the average pencil. 'Microbore' and 'minibore' are two of the better known systems in this developing market. The pipes, copper and, occasionally, stainless steel or nylon, are extremely neat and their flexibility makes for easy installation. High flow speeds around 1.8 m/sec. further speed the response time. The typical layout (3.8a) has a flow and return loop between boiler and central manifold under the floor, from which secondary loops radiate to a maximum of nine or so radiators. The radiators often come as pre-packaged units, each with

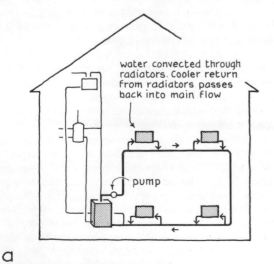

water convected through radiators. Cooler return from radiators passes back into main flow

pump

a

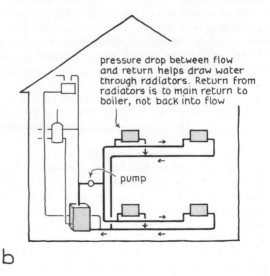

pressure drop between flow and return helps draw water through radiators. Return from radiators is to main return to boiler, not back into flow

pump

b

3.7 Domestic 'single-pipe' and 'two-pipe' central-heating circuits

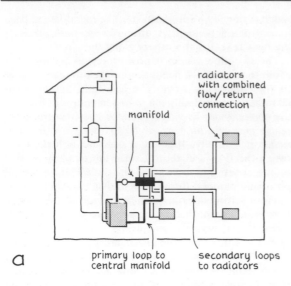

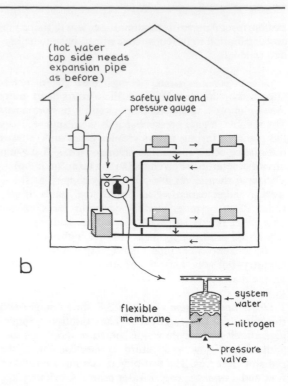

3.8 (a) Domestic microbore system, comprising central manifold with circuit loops off (b) Sealed system with expansion chamber

a combined flow-and-return connection at one end, there being a pipe along inside the bottom of the radiator to ensure distributed hot-water flow – all very neat and easy to fit.

Sealed systems – possibly for pressurisation. We have assumed the systems so far are low-pressure, i.e. open to the atmosphere via the expansion pipe. And note, they are said to be 'low-pressure' even though they are pumped. But another way of increasing the heat flow is to have the water above the normal 100 °C boiling point, by sealing the whole system and running it at higher than atmospheric pressure. This can be done with small- and ultra-small-bore systems, and can achieve a heat transfer many times greater than with the old gravity flow. In one of the earlier ideas, pressure was achieved from the expansion of steam at the top of the boiler – the vent pipe and expansion tank was replaced by pressure release valves on a principle not unlike that of the pressure cooker. But it is now more common to incorporate an expansion chamber in the circuit, usually near the boiler (3.8b). This has a flexible diaphragm across the middle, with the system's water flow on one side, and air, or now preferably nitrogen which is less soluble in water, at a predetermined pressure on the other. There are further back-up safety valves.

Let it be said right away that pressurised systems are rare in domestic heating. The boiler, valves, in fact the whole installation, has to be tougher, which is more expensive, and the higher temperatures mean that pipes and radiators have to be encased for people's safety. Of course, sealing a system does not have to mean pressurisation and high temperature as such and, in the small house, the reason for a sealed system would as likely be to avoid having an expansion tank as for any real need to deliver larger amounts of heat.

The room emitters

Central-heating emitters, like direct heaters, are generally described as radiant or convective. But, remember, both types radiate and convect, it is just that, whereas the radiant component in a radiator is small but significant (about 20 per cent), in a convector it is very small and insignificant.

Radiators. The modern steel radiator (3.9a) is ribbed front and back, adding strength, increasing the emission surface, and creating internal water columns to improve the internal convective flow. The valve at one side is a regulator, and the valve at the other is an isolator, allowing radiator removal without draining the whole circuit. The blend of radiant and convective output is useful. Both warm the room generally but convection warms the air, particularly, and radiation has its special relevance to comfort. The latter is especially so where radiators are intelligently sited. A radiator next to a cold external wall or under a cold window is immediately helpful to a person nearby, partly because its convection counteracts cold downdraughts, and also because the perceptible radiant gain is tending to neutralise that person's radiant losses to the cold adjacent surface. A small radiator under a window could well do more for comfort than a larger one on an internal wall. Radiant emission, generally, can lead to environmental temperatures (i.e. 'comfort' temperatures, as earlier explained) as much as 4 °C higher than the real air temperature – in other words, the room can be cooler and yet more comfortable, and energy is saved.

Building surfaces used as radiators. The hot-water circuit can be threaded through the floor and, possibly, ceiling, turning them into large radiator panels. This is similar to electric underfloor and ceiling heating except that a pipe from the boiler is replacing the electric cable. The same surface temperature and response-speed limitations apply, so it is not a common heating type. One application is to have pipes heating the ground floor slab-where it was desired to avoid ground floor radiators perhaps – and ordinary radiators in the bedroom above. Since the underfloor is restricted to lower temperature, there requires to be a mixing valve at the point where the flow from the boiler diverges to underfloor and ordinary radiators.

Convectors. In the convector (3.9b), fins on the circuit pipe are increasing the heat transmission and this, and the casing shape, induce the convective through-flow and output. Convectors have little radiant value – a significant disadvantage – but tend to have a higher overall output than radiators and can be more compact. Units can be built into the inner wall construction below windows, and skirting convectors (3.9c) can be run unobtrusively round rooms, especially where windows are too low to have ordinary units under. The response is rather faster than with radiators, mainly because of the quicker air heat-up, though the lower water/metal mass can also help by lowering the units' thermal inertia, i.e. reducing the time they take to heat themselves

up. Also, it is important to note, they are the proper emitters with pressurised systems, where naked radiator surfaces would be dangerously hot.

Output and response improve further when the convection is fan-assisted. Having a fan also allows sharp on-off control, possibly by an air temperature thermostat, although there will then need to be an overriding water-temperature thermostat to cut out the fan and prevent its blowing cold air around a cold room when the hot boiler supply is switched off. Fan convectors can 'throw' the warm air, freeing their positioning and usefully helping the air mixing in deeper rooms. Units on external walls can be designed to double as mechanical ventilators in summer (just possibly, with a cooling coil for refrigeration in hot climates).

The high output and quick response of convectors are advantages in heating any building but more so in large buildings than small. With the small house, special needs for ventilation in hot climates or for fast heat-up owing to intermittent occupancy might ask for convectors but, otherwise, the choice is rather more open and one is left with that radiant lack – fan-convected hot air can lead to environmental temperatures as much as 4 °C *below* real air temperature.

Summary

Boilers

- Cast iron or steel.
- Sizing factors – fabric losses, ventilation losses, acceleration, hot water supply.
- Location factors – fuelling, flueing, ventilation, noise.

Circuits

- Gravity less usual – modern circuits usually pumped.
- Single pipe possible but two-pipe likely.

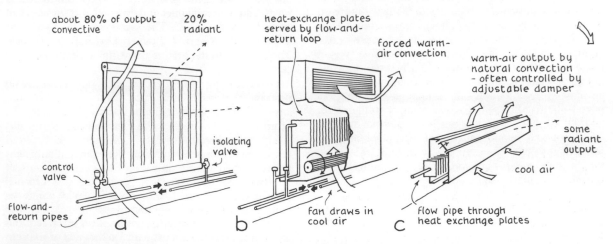

3.9 Room emitters. (a) Common 'radiator' (b) Forced convector (c) Skirting convector

- Expansion to, and top-up from, own tank.
- Loop to hot-water cylinder calorifier.
- Microbore systems on increase.
- Sealed systems need no expansion tank. Pressurisation for high-temperature output is unlikely at domestic scale.

Emitters

(a) radiators
- Common – useful mix of radiant and convected heat.

(b) convectors
- No radiant component.
- Faster response.

(c) fan convectors
- Even better response and output.
- Sharper air thermostat control.
- Useful throw of air.
- Possible use for ventilating and maybe cooling.

Hot-air central heating

Warm air or hot water?

There are debits and credits in using air as the heat-transfer medium. Debit one is that air distribution ducts are much larger than water-circuit pipes – air, being a gas, has a minute thermal capacity compared with water and, therefore, it takes a much greater volume to carry the equivalent heat. Finding space for duct runs is easy enough when designing new houses but often harder or impracticable in existing ones. Debit two, hot-air output is all convection, so there is no comfortable radiant component. Debit three, some find the fan noise a disadvantage, although it is pretty well muted in a properly designed installation. Debit four, air systems are harder to control on a room-by-room basis.

Credit one is that hot-air systems can be cheaper in a new house designed to accommodate them – more on this in a moment. Credit two, hot air gives a rapid heat-up and response: the chill can be off the house in five minutes and the fans have an immediate on-off response to the control system. Credit three, blowing hot air around the house can incorporate positive fresh air intake for ventilation – this is imperative if windows have to be fixed closed against airport or traffic noise, and in all houses is marginally helpful in summer or in hot and humid climates generally, in bringing some slight convective relief. Credit four, related to this, if the climate is hot and the services budget high, refrigeration plant can be added.

So factors suggesting warm-air choice
- Cost.
- Windows fixed closed (no option here).
- Intermittent occupancy requiring fast heat-up.

- Warm climate requiring forced ventilation and maybe cooling.
- Local supply conditions favouring it.
- You happen to like it.

The warm air source – the furnace

The central warm-air unit (3.10a) is a furnace rather than a boiler – there being nothing to boil. Its flue needs are the same as for a boiler but it will tend to be taller, maybe full room height. The fan powering the system draws air into the unit through a filter which then passes around the combustion chamber (or, less usually, electric element). Most furnace combustion chambers are direct heating, which is to say that the metal walls pass heat to the air-stream directly. With indirect heating, there is an intervening water jacket around the chamber, inevitably slowing the heat-up time, but preventing fumes from being blown around the house if the combustion chamber ever failed. The hot water in the jacket can be conveniently circuited with a calorifier in the domestic hot-water cylinder – direct heating chambers have to have a hot-water heating unit incorporated if the furnace is to heat the hot-water supply. The unit is best sited near the centre of the house, minimising duct runs.

The warm air 'circuit'

The air ducts are usually of galvanised steel or aluminium sheet but, as we might expect, there are alternatives in glass fibre and other synthetics. Ducts have a much greater surface area than hot-water pipes and have to be insulated where they travel under suspended floors or in cold attic voids. The size, shrinking in logical hierarchy, is obviously dictated by the need to deliver enough warm air to heat each particular room, 'enough' meaning a sufficient volume, at a given temperature, in a given period of time. In fact, the air temperature is limited to around 30–40 °C, otherwise, the flow from the grills gets obtrusive and also tends to smell, and the metal duct and exposed grill surfaces can cause burns if touched. The speed is limited to around 3 m/sec in ducts and 1½–2 m/sec at grills, to limit noise. The cross-sectional area is the variable that tailors the capacity.

Unlike the pipes in a hot-water system, there is no flow-and-return circuit as such – not in a small house anyway (3.10b). Ducts carry the warm air to an inlet grill, or grills, in each room, while the cooler air permeates freely back through outlet grills and so to the unit intake. Gaps under doors are often thought sufficient return to the corridor, but it only takes the addition of a thick carpet to block them, and a stifled return will pressurise a room and stifle the flow. 'Why aren't we getting warm?' people wonder.

In a compact house, the duct layout is often reduced to a single vertical run, probably in the area of the central stairwell, with very short, horizontal 'stub ducts' on

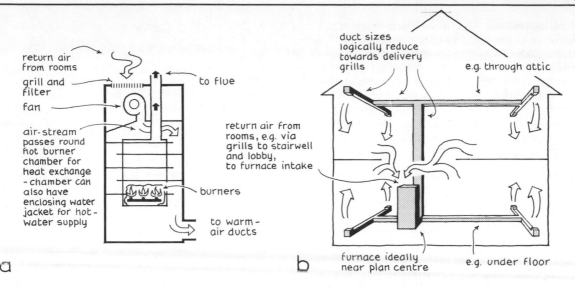

return air
from rooms

grill and
filter

fan

air-stream
passes round
hot burner
chamber for
heat exchange
– chamber can
also have
enclosing water
jacket for hot –
water supply

to flue

return air from
rooms, e.g. via
grills to stairwell
and lobby,
to furnace intake

burners

to warm –
air ducts

duct sizes
logically reduce
towards delivery
grills

e.g. through attic

furnace ideally
near plan centre

e.g. under floor

a

b

3.10 (a) Domestic warm-air furnace (b) The warm-air circuit

each floor, punching through to each room. This arrangement saves cost, but care has to be taken that the inlet and outlet grills in rooms are sufficiently separated to prevent the air from simply short-circuiting between them.

Many systems include a direct intake from outside, and a stale air exhaust, so that fresh air mixes with recirculating air at the unit. The proportion of fresh to recirculated air varies. In winter, most of the heating air would be recirculated to conserve the heat, with only a minimum intake for ventilation. In summer, there could be 100 per cent intake for ventilation, that is, assuming the air outside remains at least cooler and less humid than that within (if not, then the ventilation would be off and reliance would be placed on the house's thermal capacity for day-time comfort).

But there is a further humidity point. Whereas, in summer, the ventilation can contribute to comfort, in winter, it may actually make things too dry, giving people itchy throats and maybe making timber start to crack. The problem is that heating the cold outside air all the way up to the indoor temperature reduces its RH in consequence, possibly below the 30 per cent comfort minimum. This is the reverse of the condensation mechanics in the last chapter. Systems in places prone to cold, dry weather should include humidification which, essentially, consists of a water spray across the airflow in the unit – but more under air-conditioning later. A curious, related phenomenon with over-dry air, often found in large centrally-heated buildings, is that people tend to get electric shocks off door handles and other metal objects. Managements will rightly point out that it happens because people pick up an electric charge as their shoes scuff along the synthetic carpets but add that it only seems to happen in cold

weather. But how does the carpet know what the weather is like outside? What is happening, of course, is that the over-low RH of the air in cold weather abnormally lowers the air's electrical conductivity. The gathering charge cannot dissipate from people's bodies in the ordinary way, and so builds up, pending contact with a conductor. Anti-static carpet sprays are tried but it seems with only limited success. The potential benefit of air humidification is seldom realised.

Care has to be taken that blowing air around the house does not blow cooking smells, and worse, around too. Air from kitchens and bathrooms is best extracted separately to the outside and not recirculated. As said, this also prevents the raising of the house's overall humidity by vapour from these wet areas – you will remember the condensation point.

The emitters, i.e. grills
Again, grill areas must be large enough to deliver the adequate air volume without excessive velocity and, hence, noise. But, in fact, grills are fairly unobtrusive, the typical 200 × 150 mm size being roughly equivalent to a radiator of 3.5 m² surface area. Large or irregularly shaped rooms may need more than one grill for proper distribution. Grills in or near the floor are efficient from the mixing point of view but may get blocked by furniture: it is usually better to have high level grills throwing a downward air pattern for mixing.

Control systems

Automatic control is essential to any central-heating system, tailoring the output to meet the prevailing need and no more, so maintaining comfort without waste. Comfort and economy are cardinal aims and to skimp on controls would be fatuous. Control sophistication has increased enormously and the following is only an outline.

The thermostat. The thermostat is the most common device for maintaining a constant temperature, reacting to fluctuations away from the desired, preset level before a person would ever notice them. The simple type has a bimetallic strip, two metal strips of markedly different expansion coefficient, mated along their length. Temperature change causes differential elongation or contraction and, hence, bending, causing the strip to make or break an electrical circuit to the valves, pumps and so on.

Hot-water central-heating controls. Theoretically, there are two ways of regulating the heat output from the central-heating system, namely, varying the *temperature* of the water flow circulating to the radiators or varying the *amount* of the flow. Varying the temperature only would have drawbacks though. Assuming the boiler is also heating the tap supply to the hot-water cylinder, the temperature there would be undesirably fluctuating with the varying central-heating need. Also, the boiler would be obliged to operate at inefficiently low temperatures for much of the time.

The boiler thermostat. As a starting point, there will be a thermostat in the boiler: the water temperature falling a couple of degrees below the preset level, usually about 80 °C, completes the electrical circuit activating the gas valve or oil burner. Once the water temperature has risen, the thermostat circuit opens and cuts the fuel off again.

The room thermostat. The circuit flow is then controlled by a second, air-temperature thermostat. This is placed in a thermally 'typical' part of the house – not that there is such a place but, at least, it should be out of sunlight and away from local heat sources or draughts. This thermostat acts on the *heating circuit* pump, directly powering the flow to the radiators as needed. So, a typical sequence is – the room temperature falls; the air-temperature thermostat cuts in the central-heating circuit pump; the cool return flow of the circuit starts to pass through the boiler; the boiler water temperature falls so that the thermostat there cuts in the fuel pump; the boiler burners start to warm the central-heating flow; the room temperature rises; and the air-temperature thermostat cuts out the circuit pump again. Incidentally, when the circuit pump is off, some natural convective flow from the boiler may persist so, for a sharper cut-off response, 'anti-gravity' valves can be introduced which close unless there is positive pressure from the pump itself.

The three-way mixing valve. Alternatively, the air-temperature thermostat can act on a three-way mixing valve (3.11). The circuit operates in the normal way when heat is needed, i.e. as in the diagrams earlier but, once the house is warm enough, the bypass comes into play, i.e. the valve plunger lifts. The circuit pump carries on

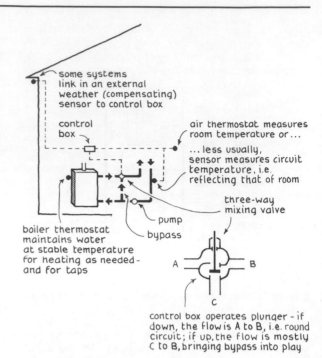

some systems link in an external weather (compensating) sensor to control box

control box

air thermostat measures room temperature or...

...less usually, sensor measures circuit temperature, i.e. reflecting that of room

three-way mixing valve

boiler thermostat maintains water at stable temperature for heating as needed - and for taps

pump

bypass

A   B

C

control box operates plunger - if down, the flow is A to B, i.e. round circuit; if up, the flow is mostly C to B, bringing bypass into play

3.11 Some typical heating control equipment, including the three-way mixing valve

working but most of the cool return from the radiators short-circuits straight back into the flow, without passing through the boiler. The boiler water, becoming isolated, will start to heat up and the burner will cut out. The three-way valve gives a more gradual modulation than the 'on-off' of the circuit pump.

Having more than one thermostat – motorised valves. The single thermostat has to assume that the needs in one place reflect the needs everywhere else. This is only loosely true, but works well enough in a small, open-plan house. However, two thermostats allow different settings in different parts of a house, for example, keeping a slightly higher temperature in the living rooms than in the bedrooms. It means having separate loops in the pipe circuit. There is still only one circuit pump and the flow modulation to the loops is achieved by electrically motorised regulating valves.

Thermostatic valves. Thermostatic regulating valves on the radiators themselves give selective, local control. They are pre-settable, opening and shutting in response to the temperature around them. Inevitably, the radiator's proximity can distort the reading slightly – remote linkage types with the actual sensor some way from the radiator reduce this problem.

Compensating systems – control by weather. This is probably beyond the requirements of the small house, but there are also 'compensating systems' in which an external sensor influences the heating system. It *antici-*

pates the inside needs by sensing the fluctuating weather temperatures and, sometimes, wind-speeds, outside. There are quite sophisticated compensating systems where external and internal sensors connect through a 'thinking' control box to the circuit flow controls.

Time switch. Whatever form the thermostatic controls take, they are generally subordinated to an overall time-switch. By this means, the whole system can be shut down automatically when it is not needed, such as when the family is out during the day, or when they are asleep at night. Some systems enable the time-switch to bring in alternative levels of thermostat setting – for example, a full evening rate and a lower night-time 'setback'.

Controlling the controls. Modern domestic boilers come with pre-packaged options. Typically, there is a boiler water-temperature control to set the temperature at a convenient level for hot water and heating. A second counter allows options such as 'hot water only' (for summer); hot water with house heating once a day, twice a day, or continuously. An incorporated 24-hour clock times the chosen heating periods – and a thermostat (or thermostats) elsewhere select the house-temperature levels.

Warm-air heating controls. The room thermostat and time-switch override are similarly essential controls in warm-air central heating. A demand for heat to the house starts the furnace burner and, once the second furnace thermostat senses that the selected air-flow temperature is available, it starts the fan. Conversely, once the heat demand is satisfied, the burner cuts first and then the fan. Usually, the fan can be independently switched to provide unheated air for warm-weather ventilation and the burner can be independently switched to pro-

vide heating for the hot-water supply only.

As said, room-by-room control is less easy (but, arguably, less necessary) than with hot-water systems. There is the point that excessive damping in one area will cause over-supply to others. But room grills can be shut off by hit-and-miss slotted covers or other dampers, and there can be dampers in the ducts themselves – for example, controlling the supply to bedrooms, as distinct from living areas. There are devices like thermostatically controlled motorised dampers but they are, generally, beyond the needs of the small house.

Notes on some recent developments

Solar panels

Much has been said and written about solar panels. 3.12a is a fairly typical solar panel arrangement. The panel is positioned to collect the maximum insolation annually, the optimum inclination depending on latitude but, once chosen, not usually arranged to alter seasonally. Typically, there is a glass (or plastic) outer sheet, a metal panel incorporating water channels and an insulation backing. The design exploits the 'greenhouse principle', the glass being highly transparent to the incoming short-wave radiation from the very hot sun but less transparent to the longer-wave infra-red radiation the warmed plate wants to reflect. Further, the plate can have a so-called 'selective surface', a special coating having a high absorption coefficient for the incoming solar radiation but a lower emission coefficient for the plate's infra-red. So the design is valve-like, receptive to the incoming energy but inhibiting its loss back outwards. The collected heat energy then passes via the incorporated water circuit to the house to assist in the hot-water supply and/or space heating. The circulation is normally pumped, meaning that the system is generically 'active' rather than 'passive'. 3.12b shows a simple system for hot-water supply assistance

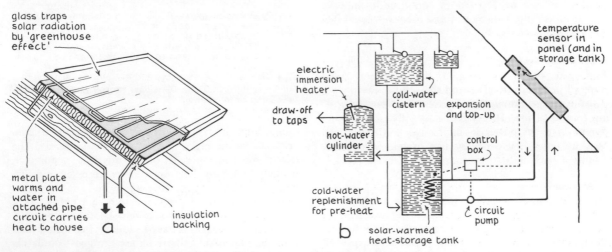

3.12 (a) The solar panel (b) Its simple circuit for assisting water heating only

only. Like any panel system, it has to overcome some fundamental problems. First, the circuit water will for most of the time be cooler than the 65 °C or so needed at the taps. So its intermediate storage cylinder is acting only as a back-up, pre-heating the cold intake on its route to the hot-water cylinder, where the temperature can then be topped up by an immersion heater, as required. Second, there has to be a means of shutting-off the circuit when the panel temperature is below that of the water in the storage tank lest the whole system start to work in reverse, transferring the heat back outside the building. So there are temperature sensors in tank and panel, respectively, connected via a control box to the circuit pump. Finally, there have to be safeguards against the circuit freezing in cold weather or boiling under the high sun, such as system draining and sealed-system pressurisation, respectively, or use of additives or other fluids than water to widen the operating temperature band.

There are all manner of more elaborate systems, notably those intended to back up the space heating as well and having further heat-exchange links with the boiler and central-heating circuit. And there are refinements such as the use of heat pumps – these will be described in a moment – to extract heat more effectively from panel water that is only moderately warm. However, there always remains solar heating's inherent drawback, especially for space heating, namely, that the supply-and-demand curves tend to be exactly out of phase – you collect least heat in winter when you need it most. The panels are better geared to heating swimming pools in summer than houses in winter and, in the latter role, away from hot climates, they are, at best, ancillaries to conventional heating installations. The payback period can look rather long when the equipment costs – total energy costs to produce and capital money cost to buy – are compared with the likely savings that result. But that is today and things could alter as panel technology continues to improve and production runs increase, and if conventional fuel costs continue to rise.

The Trombe wall

This, too, exploits solar gain, the insolation being trapped between the glass and the solid backing wall, with the wall's thermal capacity extending the useful heating period (3.13). The system is usually passive – no pumps or other power input. Convection carries the heat into the house or, by use of openable vents, dumps it back outside in summer.

Heat pumps

The heat pump, in simple form, is the basis of the ordinary domestic refrigerator, extracting heat from an ice-box already relatively cold and losing it to the surrounding room. The device can be similarly exploited to extract 'low-grade' heat from the relatively cool out-

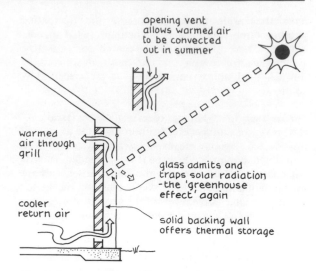

3.13 The Trombe wall – passive solar collection

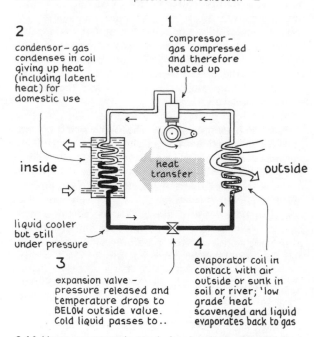

3.14 Vapour compression cycle forming the basis of the heat pump. Slightly modified, the circuit can be reversed to cool ventilating air in summer.

doors, from the air, soil or a river nearby, and to convert it to usable higher temperature heat indoors. The heat-pump principle relies on two physical phenomena, namely, that gases heat up when compressed and that liquids (condensed gases) eventually boil and evaporate when decompressed, provided they can extract heat from their surroundings to energise the boiling process. The sequence (3.14) is easy to follow – and note that, reversed, it could similarly be used to cool ventilating air in summer. Liquids suitable as refrigerants start to boil at room temperature, at pressures not far below atmospheric: ammonia was common but has been largely replaced by odourless synthetics.

Of course, heat can be more readily extracted when the outside surroundings are warmer so, again, the demand-and-supply curves are out of phase. Also, the pump uses up electrical energy. At present efficiencies, and today's fuel prices, domestic applications are still rather limited, perhaps to transferring heat to create a cold-storage room and benefiting the hot-water supply as a bonus, or to improving the heat transfer from the solar-panel circuit, as described. But the future for whole building heating and cooling looks promising. A services engineer consulted on this chapter was wryly talking of installing a heat pump to suck heat from that part of his house occupied by tenants into his own.

LARGER BUILDINGS

It was admitted in Chapter 2 *Enclosure* that any division between small and large buildings had to be arbitrary – that, in practice, any dividing line would be a wavy one. The same is true here in environmental services but, undoubtedly, as building size increases and as the activities inside become more varied and, sometimes, more environmentally demanding, providing indoor comfort becomes more complicated, too. The architect would normally be competent to specify the services for a small house but not for a large project. The services engineer's role in the 'design team' will be an important one right from the early stages, with the servicing proposals interacting with the building's organisation, structure, and costing. The architect, as head of the team – if the architect wants to stay head of the team – must be sufficiently competent in servicing as in the other disciplines to ensure that the team collaboration is effective.

Even with no particular services knowledge one only has to look at large buildings, at the basement plant space required and at the air-handling plant on roofs, and at the extended network of pipes and ducts and consequent routing provisions vertically through shafts and service cores and horizontally through hollow floors or voids over suspended ceilings, to appreciate that services have weightier implications on the form and cost of large buildings than they they do on small. But how far can one expect to find larger scale buildings still centrally heated by piped hot water and naturally ventilated through windows? Where does mechanical ventilation come in and where full air-conditioning?

Of the many factors at play, the first to consider is the effect of scale itself, which can bring a more than proportional increase in the services task. As explained, increasing building scale increases the volume by the cube but the enclosure only by the square, so the significance of volume increases faster than that of enclosure. Take ventilation. Natural ventilation is less likely to be effective in a large space, not only because it is hard to ventilate naturally from windows right into the deep-plan centre but, also, because the factors that produce the need for ventilation, people and processes, are volume-related, whereas the available window area is only enclosure-related. Take air-conditioning. We have explained the influence of these volume-related factors on the production of 'free' heat and the resultant thermal shift towards the need for cooling. Add to this the (often unavoidable) lightness and, hence, low thermal capacity of large enclosures and their (avoidable) high transparency and consequent high solar gain, and the potential overheating peaks can be acute, especially around the perimeter of the plan. Furthermore, the increased scale makes getting the services to each space that much harder, and makes control harder with different parts of the building needing simultaneous, different inputs, depending on their particular activity or orientation in respect of prevailing wind and sun.

From the strict point of view of scale increase, servicing solutions could be expected to pass through the spectrum of having heating and natural ventilation, then heating and an increasing amount of mechanical ventilation and, finally, jumping to full air-conditioning, i.e. introducing cooling. In practice, climate outside, activities inside and other factors mix in. Of course, if physical comfort were the *only* consideration, you could say that all buildings would be fully air-conditioned, but nothing is free and air-conditioning is expensive to install and run. There are many building types up to quite large scale where conventional heating and ventilating is all that is *practically* necessary and *economically* justified – provided, as ever, that those buildings are sensibly enclosed, shaded and shaped.

HEATING AND NATURAL VENTILATION

Ordinary heating and ventilation would be perfectly appropriate for a building like a medium-sized office or apartment block in a temperate climate. Also, there are those larger buildings where the volume-to-enclosure ratio penalty is less marked, owing to the plan's being spread out. For example, a school or hospital might effectively consist of a group of interconnecting medium-sized buildings rather than one large one – traditionally tight budgets may, anyway, reduce the case for air-conditioning except in select parts, the college lecture room perhaps, or the hospital operating theatre and treatment areas.

Direct heating

Gas and oil unit heaters
In fact, large buildings lend themselves more than ever to indirect, overall heating systems, but there are some applications for individual, direct heaters. A simple type is the hot-air blower, in factories and garages. Of course, such heaters can act as emitters off an indirect system's hot-water circuit, but those with their own burners have the advantage of being cheap to install, since they need only a gas or oil pipe to supply them and an electrical link for the fans. Units can be free-standing, wall-mounted or roof-hung, and usually need a flue. The burners and fans can be controlled by an air-temperature thermostat, obviously placed well away from the hot-air flow, or any hot processes or cold ventilation intake draughts. Generally, the hot-air delivery pattern is a bit crude from the comfort point of view and the fans are noisy, so the application to other building types is limited.

Incandescent ceiling panels (think of the gas-cooker grill, but moderated) are sometimes used in industrial buildings. They have quick response and the radiant component is pleasant, especially if offsetting radiant losses to cold skylights.

Electric heating
Today's costings make electric heating rare in larger buildings. Electric-element heaters would be costly in large spaces – in fact, even the electric fans in gas and oil unit heaters is a considerable cost factor – but they do have localised uses, like dispensing with visually intrusive pipes and radiators in a formal entrance lobby, or placed over external doors to create air-curtain draught barriers. Lobbies can have localised underfloor or wall panel heating. Off-peak electricity has been used with large water tanks as the heat store, but rarely. Underfloor storage is also rare – in addition to the points made in respect of the small house, the extra 50 mm needed in the floor slab adds up to a significant weight and height penalty over several storeys: also, the system would be unresponsive, too slow to cope with the weather/solar fluctuations around the perimeter band of floor area by the windows.

The 'in-between' category
There is also an 'in-between' category with no *overall* heating circuit as such, but where separate boilers and circuits heat subdivisions within a building – 'central' heating for each occupancy within an apartment block is the best example of this (3.15a). The fuel would normally be gas (oil piped around a muti-storey building could be a fire hazard). Each occupancy's distribution circuit, calorifier for tap-water heating and controls are then as described for the small house. Most modern systems back the boiler against an outside wall, using a balanced flue which makes a discreet appearance on the elevation but, alternatively, central flues can be run up the core or back of the building. This is less usual now but, in passing, the main types are the conventional low inlet and roof outlet arrangement, sometimes known as the SE duct, and the U duct with both inlet and outlet at roof level (3.15b, c).

Hot-air systems are less common in apartments. An unoccupied apartment gets less cold than a single house owing to the thermal inertia of the whole block, so the quick heat-up is less of a bonus. Also, the desirability of having the furnace in the centre of the space to cut duct runs is hard to reconcile with having balanced flues conveniently at the external wall.

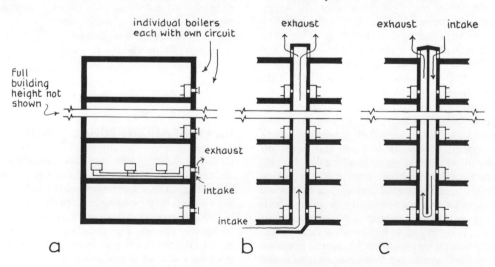

3.15 Individual boilers and circuits in e.g. an apartment block with (a) external balanced flues (b) SE-duct flue (c) U-duct flue

Indirect central heating

Most large buildings that are not air-conditioned are warmed by overall central heating, i.e. with a central boiler source and extended distribution circuit.

Larger-scale boilers

In contrast to the sooty boiler rooms in older buildings, the modern plant room, with its multiple boilers, air-handling equipment and so on, has a clinical air, clean, complex and highly organised.

We say 'boilers' in the plural because a simple scaling-up of the domestic boiler is not the most useful way of achieving higher outputs. Certainly, the boiler sizes increase to anything up to 1000 kW for sectional boilers and 6000 kW and over for shell – we will describe these types in a moment. But achieving the overall higher output by having *several* boilers has practical advantages. Apart from the scale limits helping the transportation of the boilers to the building and their installation, there are points on operating flexibility. Remembering that any boiler has its most efficient operating output level, it follows that the fluctuating heat demand in a building can be most efficiently met where there is the opportunity of varying the number of boilers working, as well as modulating the output of the boilers themselves. Also, having more than one boiler gives a back-up to cope with any one boiler's breakdown or need for routine maintenance. In very general terms, it would be common to find medium-sized buildings having two boilers each rated at around two-thirds of the likely peak demand. This accepts the chance of a slight deficiency in the unlucky event of a boiler's breaking down in very cold weather. Three boilers, each rated at 50 per cent, would avoid this, but two boilers are cheaper and, in any case, a building's thermal capacity will cushion it from the effects of any failure over the short term. Of course, large buildings can have three boilers or more. Interestingly, where there are many boilers, they are often sized unequally, so that judicious combination offers a graduated output scale, allowing the demand of the moment to be met by selected boilers operating very close to their optimum efficiency. This would all be automated.   be

Types.   As with domestic boilers, the main types are cast iron and mild steel. The *sectional boiler* is the commonest cast-iron type at large scale: 'sectional' because it is assembled rather as one would reassemble a sliced-up cucumber. The sections or slices are punctured with holes which line up when bolted together, making a combustion chamber to produce the heat and hollow water tubes to collect it. Sectional boilers are easy to install and can be up-rated by adding more sections later. Being cast iron, they do not corrode easily but are unsuited to high pressures. The *shell boiler* is probably the most common of the steel family. The outer shell or cylinder contains the water and the

furnace and hot gas tubes run through. Higher pressures are possible than with cast iron but, as explained before, corrosion is a problem if they are run too cold.

Fuels.   Again, fuel comparison is notoriously hard in view of changing trends and availabilities. Gas and oil are the most convenient and common. Coal may be about to make a come-back, though, in the 1980s, and there are improved systems for stoking (ram, sprinkler, travelling chain grate) and ash removal. Electricity is virtually to be discounted *at present*.

Location.   Ideally, boilers would be sited at the thermal 'centre of gravity' of the building, central in the distribution-pipe system. But other factors usually defeat this.

In a building with an extended plan layout, like a school, services plant needs to be zoned away from classrooms and other public areas, both for amenity and to avoid noise disturbance.

There has to be vehicle access from the perimeter road for maintenance, and for oil or solid-fuel supply. Solid fuel, especially, needs a lorry run-up to the bunkers, close to the boilers, whereas oil can be piped some way to the tanks and from tank to boilers.

There are other factors in high-rise buildings. Access, weight, fuelling and noise suggest plant location in the basement. As we shall see, it is often convenient to site air-handling plant on the roof, but gas- and oil-fuelled boilers can be sited up there as well should low-level siting be hard to achieve. Access is then harder and care has to be taken to mount the plant resiliently so that vibrations are not structurally transmitted to floors below (see Chapter 6 *Acoustics*) but, at least, fresh-air ventilation and, important, flue location are easy. On the other hand, planning needs may prevail. A couple of hotel towers come to mind where basement plant-siting was impossible (in one, there was an underground railway, preventing a basement) and where roof levels were given over to restaurants and penthouses producing high revenue. So, justifiably, the plant rooms were sited at mid-height, achieving the thermal centre of gravity criterion, but needing costly attention to acoustic separation and fire compartmentation and, inevitably, complicating the servicing later.

Size.   The factors influencing the required plant size are similar to those outlined for the small house, namely, heat loss through the fabric, ventilation losses, and hot-water supply for taps – and possibly for other processes or needs, for example, in factories, laundries and, rather obviously, swimming pools. The free-heat contribution in large buildings can be significant but its variability makes it an unreliable ingredient in the overall heating calculation – free heat from processes in a factory might be greater than the calculated heating need, but that would not mean the space heating could be dispensed with, for how, otherwise, would the building be heated in the morning before the processes started, and what if the building use and processes are later changed?

A more tangible point of difference from the small house concerns designing for 'acceleration', the heating-up of the building from cold. In larger buildings, over-capacity of the heating system is less effective as a means of providing a quick heat-up capability. There is more thermal inertia to be overcome owing to the building scale and, to a lesser extent, owing to the increased scale of the heating system itself. It would be unusual to find a heating system in a large building oversized by more than about 20 per cent or so: instead, reliance is placed on having a longer pre-heat time – a building like an office might well need a two-hour pre-heat before people arrived to work. Then, again, a related factor to consider is how the building's response speed is further influenced by its construction in general and enclosure in particular. A building with a fast response, owing to its lightweight and glazed enclosure, will need a faster response heating (and cooling) system, especially around its floor perimeter near the window areas, than will a building the enclosure of which has higher capacity and resistance insulation. It might just be the difference between opting for forced convectors rather than radiators but detailed decisions on plant size and system type are matters for the services engineer.

Larger-scale circuit

Clearly, the task of the distribution circuit is magnified in larger buildings, having to carry more heat and carry it further. Pipes have to be larger and, occasionally, there will be pressurisation to allow higher temperatures. But, although there are many options, the small-house principles persist.

Extending the plan. Take a building like a low-rise school – extending the small house sideways, as it were. In the extended central-heating circuit, the problem of getting a balanced flow supply to all emitters is that much greater. The usual layout will still include a two-pipe flow-and-return ring, but serving branch circuits to each zone rather than the emitters directly (3.16). (There may even be sub-branches off branches). The branch circuits can then be as for the small house, ordinary single or two-pipe, or microbore. The main-circuit piping will be larger than that in a branch, and a branch larger than any sub-branch – an obvious hierarchy of flow responsibility. But the problem of equal supply, equal pressure, to each branch still persists.

The basic measure is to make the main-circuit pipe sizes large, perhaps 150 mm, ensuring that the pressure drop away from the boiler – which, among other factors is inversely proportional to pipe diameter – is small enough to be insignificant. In effect, this helps to make

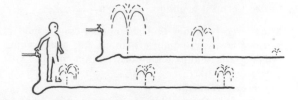

jets from hose same size only if holes progressively smaller towards tap. Having large-size main-circuit pipe can almost entirely avoid pressure drop away from boiler, but as with hose holes, 'balancing' valves are final adjuster of flow to successive branches

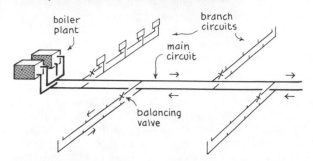

3.16 The larger central-heating circuit and the greater need to balance the flows

the system *self-balancing*, allowing all branches the same, or nearly the same, supply pressure. However, the extended circuit may need further, finer, balancing than this, which would be extremely awkward to achieve by gradation of pipe sizes alone. In any case, the engineer is obliged to work within the limited range of common pipe sizes manufactured.

The further measure is illustrated by the hose, where the sprays from the holes will be the same size only if the holes are made progressively smaller towards the tap. In the same way, the heating circuit can incorporate built-in resistances, *balancing valves*, allowing flow pressures to all branches to be appropriately adjusted.

But this does raise a supplementary point. It is accepted that heating systems, like all the building services, will have a commissioning period after they are installed, for adjustment and for the inevitable problems to be ironed out, so that everything works before the building is officially handed over. The trouble is that, even with a workable design by the engineer and a good installation by the contractor, the commissioning all too often happens late, inadequately or not at all, doing nothing for user satisfaction or relationships all round. A heating system that is designed, as far as possible, to be self-balancing, or any other services system that is designed to need minimum commissioning, is all the better for being so.

Incidentally, although the building layout will only occasionally be regular enough to favour it, the reversed return circuit (3.17) is interesting in that it is an inherently self-balancing system. The circuit is so

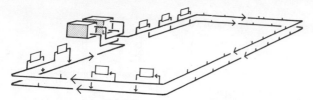

3.17 The 'reversed return' circuit is inherently self-balancing

arranged that, at any radiator, the *added* distance of the flow from, and the return to, the boiler is the same, meaning that the circuit resistance affecting the supply to any radiator is virtually the same.

Higher buildings. Returning to the apartment block, an overall central-heating system obviously involves an upward extension of the circuit (or downwards, if the boilers happen to be on the roof). Traditionally, multi-storey buildings used either ladder or drop circuits for this, as shown (3.18a), but these are really only single-pipe circuits, acting in parallel, and, as such, are now less used. The two-pipe circuit is more common, probably with the main flow-and-return ring located in the service core and with branches running off (3.18b). Again, these can be one- or two-pipe or, and this is perhaps simplest, they can be microbore, i.e. with manifolds off the main circuit and then spurs to the rooms and tap-water cylinder calorifiers. The balancing principles apply as before.

Now, suppose the high-rise building form is extended laterally as well, as in an apartment complex, for example, or an hotel or hospital block. The main supply can then take the form of a horizontal, two-pipe ring with vertical spurs at intervals, dividing the building plan into vertical, servicing zones (3.18c). This keeps the final

branches short and the pipes can be economically sized. In other words, there would be primary horizontal runs, secondary vertical runs and tertiary sub-branches, with decreasing pipe sizes in each case – all perfectly logical.

A note on pressurisation (and steam heating). Extended layouts are sometimes pressurised, accepting the impositions this places on system design – on strength and safety encasement of the hot emitters, but achieving smaller pipes (possibly) and usefully higher heat delivery. Either the whole system can be pressurised, or just the primary circuit which then serves the ordinary low-pressure sub-circuits through heat-exchange calorifiers. In practice, there will often be a combination, for example, a school might have high-pressure lines direct to the high-output unit heaters in large assembly spaces, but reducing through calorifiers to low-pressure sub-circuits for classrooms. A factory might reduce for the office accommodation but maintain high pressure for the main works space, to serve the unit heaters there and, possibly, to supply high temperatures for particular processes. In fact, the need for high-temperature supply, for processes in buildings like factories, laundries and hospitals, can be an added reason for choosing a pressurised heating system in the first place. Similarly, although steam-heating circuits are now very rare, they still crop up occasionally where there happens to be a need for process steam.

A note on district heating. The idea of having a central heat source and extended distribution circuit is carried to its fullest extreme in district heating. The 'district' – and this can mean a residential development, an industrial or hospital complex or, perhaps, a part of

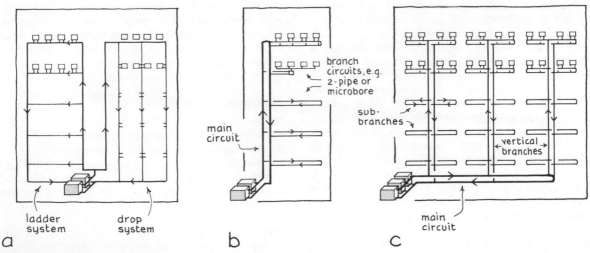

3.18 Heating circuits for multi-storey buildings (a) 'Ladder' and 'drop' systems (b) Vertical main circuit with horizontal branches off (c) Horizontal main circuit with branches and sub-branches off

a city – has a single heat source rather than individual boilers in each building. This can be a large boiler plant or, as in some of the most successful district-heating schemes, it can be the 'free' heat surplus from industrial processes or electricity power stations, or derived from natural sources, like hot springs. The main flow-and-return lines run underground or overhead, each line, typically, taking the form of a pipe within a pipe, concentric pipe rings with insulation sandwiched between. There will be spurs to each building's calorifiers.

Emitters

Radiators. Larger buildings tend to have larger rooms of course, but, unless a room is only intermittently used and requiring a fast heat-up, ordinary wall-mounted steel radiators are still the emitter.

Other 'radiant' systems (remember, radiant is a misnomer) include panels, where the circuit piping is threaded through the floor screed or wall construction. These can be useful in places where ordinary radiators might be found obtrusive, e.g. the entrance lobby again. Piped floor heating is also used as primary or background heating in a variety of buildings, a church, a warehouse. But, as with electric underfloor, the slow response is a limitation: you would not use underfloor in a factory where a build-up of process heat could be expected during the day. Equally, heated screeds are thicker and heavier, ruling them out for high-rise.

Radiant ceiling-panel systems have been used in commercial buildings. Essentially, they comprise a distribution-pipe grid incorporated in the suspended ceiling, the pipes being clipped to and, hence, conducting heat to, modular plaster or metal tiles (3.19a). They have faster response than underfloor systems and, agreeably, a higher radiant component. Temperatures are up to 130 °C or so. Room height is a factor, heights under 3 m tending to feel oppressive, and over 5 m starting to reduce efficiency, but the commercial floor normally lies within this band.

Convectors. Natural or forced convectors can be constructionally incorporated, occurring under windows as perimeter floor grills or built into spandrel walls (3.19b). Incidentally, the distribution branching in such a case, say, from the main flow-and-return in a multistorey service core, will often be routed through the suspended ceiling void in the level below, finally routing around the perimeter and having links punching up through the floor to the individual emitters in each window bay.

Perimeter convectors are particularly helpful in counteracting cold downdraughts, important where windows are high – the terminal building, library or sports facility.

A large, intermittently used space like a sports or meeting hall further suggests the use of *forced* convectors because of their quick air heat-up capability and

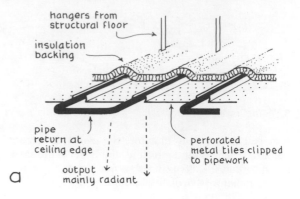

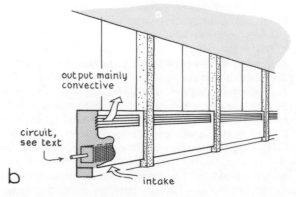

3.19 Further central-heating emitters include (a) Water-pipe heated radiant ceiling (b) Convectors incorporated in the spandrel area of the perimeter wall

because of the relatively good distribution across a wide space which the fan's 'throw' of air produces. In other words, we are really talking of unit heaters, as earlier, only now on a central-boiler water circuit rather than with their own elements or burners. Convectors, especially forced convectors, are the natural choice with pressurised high-temperature circuits. As already implied, the choice of high-output forced convectors for large spaces can be an added reason for pressurised circuits, because the convectors are so greedy for heat supply. The tail can wag the dog.

Summary

Boilers
- Multiple, possibly differently sized.
- Typically cast-iron sectional, or steel shell.
- Ideally at thermal 'centre of gravity' save for factors like access, flues, noise, and general planning.
- Capacity obviously varies with building size and purpose.

Circuit
- Types include drop and ladder (single-pipe), or two-pipe.
- Extended system requires flow hierarchy of main

and branch circuits – possibly, microbore branch circuits.

- Pressurisation common, e.g. of main circuit with calorifier to ordinary sub-circuits, or carried right to select areas and/or high output emitters.
- District heating is indirect system on grand scale.

Emitters

- Ordinary radiators – most applications, except where intermittent use makes fast heat-up important.
- Underfloor and wall panels – where radiators obtrusive; also underfloor system useful in large single-storey spaces as primary or background heating, provided fast response unimportant.
- Ceiling panels – in large spaces with 'ordinary' room height, e.g. office floors.
- Convectors – most applications, often built-in under windows, perhaps with pressurised circuits; conversely, pressurised, high-temperature circuits require convectors for safety.
- Forced convectors, including indirect unit heaters – suited to large, wide spaces, especially where intermittent occupancy; most effective with pressurised, high-temperature circuit.

WARM-AIR CENTRAL HEATING?

Duct systems circulating air around large buildings are generally more elaborate and expensive than the equivalent pipework in hot-water central heating and, where warm-air heating is chosen, it will usually be in cases where mechanical ventilation is required in any case – in other words, if you need ducted air to ventilate there is an argument for heating with it as well. Large spaces, like sports halls and other public facilities which are only intermittently used, can further benefit from the faster response, faster heat-up, that warm-air systems afford. Also, there are cases where parts of a building are mechanically ventilated/warm-air heated, and other parts, hot-water-circuit heated, both systems probably being served by the same boilers.

Certainly though, the convenience of using ventilating air to heat is to some extent offset, in as much as ducts for heating are normally larger than those required for ventilation alone. (We will come on to this point under full air-conditioning shortly and, also, to the typical plant provisions for air heating and recirculation generally, which air-conditioning necessarily includes in its other, further functions.)

And it is also worth noting that, even where the mechanical ventilation is *not* providing the basic heating, there may be a winter need for the incoming air to be, at least, pre-warmed to within a few degrees of the inside temperature, otherwise, it can cause discomfort near the inlets. A heat-recovery device (also to be described) can assist in this, once the building is already warm, but the usual thing is to have thermostatically-controlled electric elements or hot-water heat-exchange coils across the inlet air-stream, the latter possibly occurring as branches off the hot-water radiator circuit.

SO INTRODUCING MECHANICAL VENTILATION – AIR-DELIVERY SYSTEMS

Increased volume and plan depth are general arguments for having mechanical ventilation, the required system capacity being further determined by the occupancy – how far it is continuous or only short-term, and whether there are particular needs for limiting the build-up of heat, moisture, and odours or process fumes. Capacity is often measured in terms of air changes per hour, typical examples being; office floors, 2–6; classrooms, 3–4; restaurants, 10–15; kitchens, 20–40. Places for public gathering, like auditoria and dance halls, are usually measured in terms of the number of occupants to be served, say 30 m³/person/hour. A good system must achieve the prescribed rates, properly distribute the air and, ideally, create just enough air movement so that people have a feeling of freshness without perceiving draughts – about 30 m/minute, in fact.

Fans

The process starts with the fan (3.20). The propeller fan (a) is the simplest but is only used where there is no flow resistance to be overcome, where the extract is directly through an external wall or where the duct run to the final outlet is very short – inevitably, there is flow

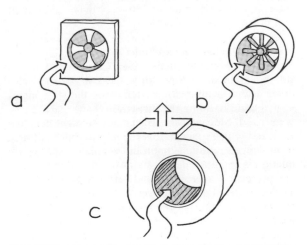

3.20 Fan types used in mechanical ventilation (a) Propeller (b) Axial (c) Centrifugal

resistance in ducts. The axial fan (b) is similar but much more powerful and more used nowadays. The domestic extracts for the small house would probably be axial and, in larger installations, axial fans can be neatly fitted within the air ducts themselves. The centrifugal fan (c) is the other common work-horse for powering air through ducted systems. It is more elaborate than the axial, but quieter at high power and the right-angled change of direction it allows is more useful than otherwise in designing system layouts.

Ducts

An important aspect of duct design is reducing flow resistance, especially in the extended layouts in larger buildings. For this reason, rectangular section ducts are kept as close to the square as the available space in voids above suspended ceilings and in other ducting cavities allows, thereby minimising internal surface area and friction. Constrictions and sharp changes of direction are avoided. Galvanised steel sheet is the usual ducting material, folded into the rectangular shape and encircled with stiffening collars to prevent the inside air pressure from drumming the walls outwards. (Circular-section ducts are becoming more popular: they make things a bit more awkward when it comes to stepping down the size at branches, or running through tight spaces, but they minimise the flow resistance and are inherently free from outwards drumming.) Plastic ducts have particular application where corrosive laboratory or factory fumes have to be expelled. Also, there are 'builders' work' ducts, where the actual construction of the building – brick shafts, suspended ceiling voids – provides airways without any inserted trunking as such. But they need care in construction if they are not to leak air in use.

Systems

It can be either the stale-air extract or the fresh-air inlet that is fan-assisted, or both.

Mechanical extract, natural inlet

We saw this in the small house extracts, but it can still be applied to the utility areas at larger scale, kitchen and lavatory areas, removing the stale air at source and providing the higher air-change rates these areas need. Mechanical extract causes a slight negative air pressure in a space so that replacement air infiltrates from the surrounding naturally-ventilated spaces via inlet grills and open doors, an inward flow that is inherent protection against the outward spread of odours. The extract fan can be direct through an outside wall but, in a hotel or apartment block, the grills are more likely to have 'shunt' ducts connecting to common vertical shafts with main outlet at roof level. Office blocks will normally have vertical shafts in their service cores (3.21). The shunts are shaped to facilitate exhaust flow

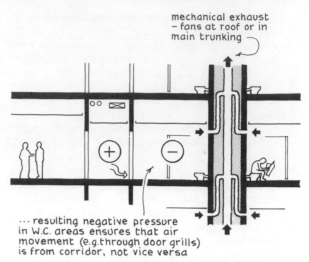

... resulting negative pressure in W.C. areas ensures that air movement (e.g. through door grills) is from corridor, not vice versa

3.21 Mechanical extract and natural inlet – one typical application

and discourage backflow from the main shaft – and they help protect against smoke spread via the duct should there ever be a fire on a lower level (see Chapter 7 *Fire safety*) and help prevent flanking transmission of noise via ducts (see Chapter 6 *Acoustics*). Direct, mechanical extract is also used to tackle process fumes at source, as where extract hoods over catering equipment remove steam and odours, or in factories where hoods directly over production machinery protect workers from harmful paint vapour, dust or chemicals. These extracts may be quite additional to the main ventilation system.

Mechanical inlet, natural extract

Mechanical inlet causes a slight positive pressure and, hence, outward leakage, sensible where it is to be the space that is protected from its surroundings. A simple example might be the office suite surrounded by a factory floor. In the more general case, where the enclosure is from the outside, a slight positive pressure inside the building is also helpful in preventing random and uncomfortable infiltration of cold draughts.

However, the really significant advantage of mechanical inlet over mechanical extract is the far greater power of the inlet to shape the character of the ventilation. It is better at shaping the airflow pattern through a space – random inlet will inevitably leave parts of the air volume unchanged – and is appropriate when the incoming air is also being used to heat. Extract only can never effectively serve a space of any size.

Mechanical inlet and extract combined – the 'balanced' system

In practice, though, either intake only or extract only leave the chance that the airflow will be short-circuited on its journey through a space by windows, doors or

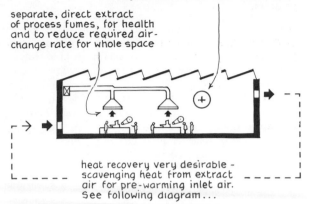

inlet power usually exceeds extract – shapes airflow pattern and creates slight positive pressure preventing draught infiltration from outdoors

separate, direct extract of process fumes, for health and to reduce required air-change rate for whole space

heat recovery very desirable - scavenging heat from extract air for pre-warming inlet air. See following diagram...

3.22 Mechanical inlet and extract combined

other openings in the external enclosure. Unsurprisingly, optimum, controlled air-distribution asks for both combined (3.22). Such systems are said to be 'balanced' in that the relative powers of the intake and extract fans are adjusted to give a slight positive or negative pressure in the space, as appropriate, intake normally being the more powerful, as shown.

The airflow pattern

Very broadly speaking, a long, low space can suggest horizontal airflow between inlet and extract grills in opposing walls. A high space might suggest a more vertical flow and, if so, there is an argument sometimes raised that this is best achieved by low-level inlet and high-level extract, since warm air inherently wants to rise. But, as mentioned earlier, floor-level inlets can get obstructed and also can be uncomfortable to people close by – so high-level inlet grills or nozzles are common, directing the air downwards for good mixing in the space. (In any case, the upwards delivery argument does not apply to full air-conditioning, where the air is as likely to be cool and wanting to fall). But there can be other considerations. Upwards flow to roof

extract might be better in a factory, even if it were long and low, in order to get process fumes away quickly. There can be specific extracts close over machinery or processes, removing fumes at source. A large sports arena or swimming pool might have nozzles throwing air inwards from perimeter walls – high air movements would matter less there, they might be rather pleasant, in fact. High glazed walls often have warm-air inlets set in the floor under them (as an alternative to the hot-water-circuit convectors mentioned earlier) to counter-act cold downdraughts and stop condensation forming on the glass.

But, having said all that, inlet and extract grill positions will as much be influenced by the need to find the most economic and least obtrusive routes for the associated ducting, depending on the particular building layout and the routing possibilities through its construction – the ability of the inlet to throw and shape the airflow is an enormous bonus in system design.

Heat recovery

Ventilation means heat loss and, as we might expect, there has been increasing attention in recent years to heat recovery, scavenging heat from the exhaust air-stream and using it to pre-warm the inlet air-stream (3.23). The *run-around coil* is one method and works by circulating liquid between heat-exchange coils in extract and inlet trunkings. Often, the heat pump principle will be incorporated, increasing the heat-transfer efficiency. Another method is the *heat-recovery wheel*. The wheel has special construction but, basically, it penetrates both air-streams and transfers heat by conduction when it rotates.

What is most likely, though, is that a need for extensive mechanical ventilation will align in any case with the need for full air-conditioning; may provoke the jump towards having it, in fact. Even away from hot regions, in temperate climates, the building which has extensive mechanical ventilation will tend to be air-conditioned, i.e. with cooling and proper humidity control. Factories and sports halls are among the more obvious exceptions to this, because the rapid air move-

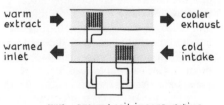

warm extract → cooler exhaust

warmed inlet ← cold intake

run-around coil incorporating heat pump: trunkings need not be close – circuit between heat-exchange coils could be longer than shown

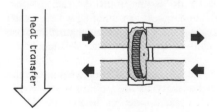

heat transfer

heat-recovery wheel is an alternative, transferring heat by conduction as it rotates through air-streams

3.23 Heat recovery from exhaust air-stream can be used to pre-warm the cold, fresh-air intake

ments associated with very high rates of mechanical ventilation happen to be welcome in these places. But, in large spaces where people gather, it has to be stressed that *unobtrusive* ventilation can seldom cope reliably with heat and humidity build-up. It must help, but would only be really effective when the atmosphere outdoors happened to be sufficiently dry and cool to give the incoming air an effective working margin, a hit-and-miss way of achieving comfort and, certainly, one least effective in hot weather, which is just when the cooling relief is most needed. The very action of the fans will heat up the incoming air slightly.

FULL AIR-CONDITIONING

'Air-conditioning' was first associated with humidity control in American textile mills earlier this century and, from the 1920s, started to be used for comfort more generally. The rapidly increasing scale of public and commercial buildings demanded effective ways of cooling and dehumidifying the deep-plan interior (remember the effect of scale on unwanted heat gain) and the air-conditioning industry boomed.

In the fully air-conditioned building, the windows are fixed shut. The air indoors is enclosed and separate from the air outdoors, and its temperature and humidity are controlled to close limits by circulating it through special plant.

Factors leading to the choice of full air-conditioning

Climate. This is an obvious factor. Hot, perhaps humid, regions have the greatest need for cooling and dehumidification. But in more temperature regions there is:

Scale. As explained, this must be an underlying factor, the scale effect of increased volume and, hence, of lighting and other 'free' heat, and the related thermal impacts of the lightweight enclosure. And, within this scale idea, 'aspect ratio' is important, the ratio of plan length to width. The large, bulky building with a squarish plan will find it harder to ventilate naturally deep into its centre than will the narrow-plan slab block.

Activities. The inside activities can contribute to the overheating and humidity to the point where straight ventilation becomes inadequate. In the very deep-plan office, it will be heat from lighting, equipment and people; in the crowded cinema, it will be, primarily, people; in the office, television studio or factory, it can be processes – albeit mechanical ventilation is the more usual way of cooling factories in summer. There are also those activities calling for a finely controlled environment –

computer suites, and certain laboratory or engineering processes, may require temperatures and humidities to be kept within close tolerances. Cold stores need refrigeration.

The need for sealed windows. Openable windows get impracticable as height increases. For one thing, tall buildings are prone to 'stack effects' in cold weather. Warm air wants to rise and, if there are windows open, air tends to rise in stairways and lift shafts, leaving the building at higher levels and drawing in colder air at lower levels to replace it. It is generally a marginal effect, but it can be wasteful, make balanced temperatures that much harder to achieve and contribute to draughty conditions in the lobbies below.

Windows at high level are subject to high wind speeds and the effects of local turbulence, and of pressure on the windward side of the building and suction on the leeward side. This makes it impossible to ventilate naturally in an even, controlled way – imagine doors slamming and sheaves of paper blowing off desks. And it is more costly to make openable window frames adequately strong and weather-tight. So, ironically, those higher levels surrounded by the cleanest air are least able to take advantage of it.

Windows may have to be sealed as protection against excessive noise or pollution outside. Hospitals are now tending to air-condition to avoid ordinary urban pollution levels – if not the wards, at least the treatment and X-ray areas, and operating theatres. Conversely, the outside may have to be protected from the building where laboratory experiments or other factory processes make free connection with the atmosphere undesirable.

Finally, opening and shutting windows is a crude comfort control in a large building with many occupants. With ordinary heating, and natural ventilation for cooling, there is always the person who feels hot and opens the window instead of turning the heating down (actually, to be fair, there is often no means of turning the heating down) – farcically wasteful. Increased scale increases the case for putting the indoor climate entirely under automatic control, and this increases the case for sealed windows.

Quality and prestige – and fashion? The prevalence of air-conditioning is inevitably linked to affluence and, taken with today's higher expectations of comfort, opulent offices, banks, luxury hotels, public spaces and such are more likely to have it included in their budget. And, even if air-conditioning manufacturers do labour the point somewhat, 'optimum' comfort can increase people's concentration and contentment – the more productive office worker or contented shopper may mean more money back for money spent!

And yet, with air-conditioning so expensive to install and run, and with the energy crunch so evident, it is curious how far the choice for air-conditioning can owe – has so often owed – more to fashion, even fad, than to real, environmental need. Buildings with no clear candidature for air-conditioning, perhaps with modest bulk in a moderate climate, and which could have been well served by proper enclosure, ordinary heating and natural ventilation, have had it only because it was arbitrarily assumed to be desirable or, as earlier implied, have had to have it owing to lacks in the enclosure's thermal performance – just a cautionary note.

Plant

3.24 is a typical air-conditioning plant. The sequence can be followed through:
1) The air is extracted from the space.
2) A proportion of the return air is exhausted to the outside.
3) A proportion of fresh air is admitted. The inlet grill must be well separate from the exhaust and, as far as possible, in a gust-free, unpolluted air-stream – witness an Aberdeen office building where the inlet had been faced towards a fish-filleting works next door not, it seemed, with good results. Energy conservation is an important influence here. Clearly, a heat-recovery link between the exhaust and inlet trunkings is a valuable

measure. Also, the air-intake proportion is very important. When the temperature outside is towards the unkind cold or hot extremes, energy conservation asks that the air circulating inside the building intermarry as little as possible with the air outside, as little as is consistent with the need for ventilation. Assuming there are no particularly contaminating processes, the fresh-air-intake proportion can reduce to 20 per cent or less of the whole, most of the circulating volume being economically recirculated within the building. Looking at it the other way, the volume of fresh air needed for breathing and ventilating generally is much less than the volume needed for unobtrusive heating and cooling – the total volume there being determined by the need to deliver enough air for the heat input or extraction required, allowing that practical and comfort constraints place upper and lower limits on the delivery temperature. Actually, slightly more air is needed for cooling than heating, because heating air can be hotter than the ambient room air by a greater margin than can cooling air be cooler, while still remaining unobtrusive and comfortable. This is one reason why ducts are usually sized to meet peak needs for cooling not heating – the other being that the large building's peak summer

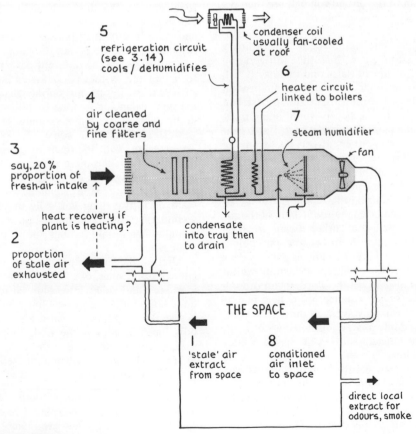

3.24 A typical air-conditioning plant sequence

cooling needs are often greater than the winter peaks for heating.

However, the proportion of fresh, 'ventilating' air intake can be increased when the outside temperature is more helpfully temperate, it then either helping make up the heating load the plant is being asked to supply or, in the overheated condition, reducing the cooling load by carrying away internal heat gains. Optimum system design acknowledges a quite complex interrelationship between heat recovery, heating or cooling load, inside/outside temperature and intake proportion.

4) Filters remove bacteria, dust and other pollutants from the return and fresh air mixture, maintaining the quality of the air in the building and protecting the air-conditioning system itself, which would otherwise clog up in time. The type of filter used depends on what is to be filtered and, often, there will be a combination of types of series. *Impact filters* consist of solid plates, for example, metal, corrugated to increase the area and coated with viscous oily liquid to trap the dirt. *Fabric filters* trap with their weave and fibres. *Electrostatic filters* trap by imparting opposite and, hence, attracting, charges to the particles in the air-stream and metal filter plates they pass. Of course, the filters themselves have to be cleaned. Some types of fabric filter progressively wind from the clean to the dirtied roll and are then replaced, some impact types continually pass through a cleansing bath.

5) The features, so far, could occur in ordinary mechanical ventilation plant but the refrigeration equipment brings us into full air-conditioning. It will only be working when cooling and/or dehumidification is needed. As explained, most refrigeration is based on the heat-pump principle. In the air-conditioning plant or rather, the cooling plant that is part of it, the expansion (or 'evaporator') coil either extracts heat from the passing air-stream directly, or from water which, once chilled, is circulated through a heat-exchange coil in the air-stream. The condenser coil, or a water circuit linked with it, loses the heat outside the building, usually by a forced-draught air cooler on the roof (think of the car radiator) but, occasionally, by other means, such as linking the water circuit with an ornamental cooling pond or fountain. Or, where the climate has hot days but cold nights calling for cooling and heating within the 24-hour period, it may well be worth incorporating a heat-storage facility. For instance, using heat pumps and calorifiers, heat can be scavenged during the cooling mode and passed into a large, insulated water tank, the water's thermal capacity storing the heat for later extraction, as required.

Cooling the air raises its relative humidity. At its particular dew-point temperature (depending on its vapour content), it becomes saturated and starts to condense – the condensation is run to a drain. The cooling is, therefore, a fortuitous means of dehumidification and the final humidity can be accurately controlled, since the more the cooling drop, the more the condensation and, hence, the drier the air when it emerges from the reheater at the desired delivery temperature.

6) The heater, usually supplied by a hot-water circuit from the main boilers, but it could be an electric element, is for slight reheating as necessary in the cooling cycle and, of course, for the ordinary heating of the building in colder weather.

7) Conversely to 5), the steam jet (or 'air washer' humidifier in an alternative layout) is for humidifying air used to heat if it would otherwise emerge too dry from the heater.

For convenience, we have shown the plant as a single package but, in practice, the various items may be separated, for example, having the main plant in the basement by the boilers but locating noisy pumps and condensers on the roof. This will be clearer when we come to look at the various system types.

Distribution throughout the building

There are three fundamental aims in designing air-conditioning distribution. The first is to seek economy by achieving the minimum duct layout that will do the job. The second is to retain, nevertheless, the capability of serving the different spaces according to their differing requirements, to have flexibility in other words. Obviously, these requirements vary with building type and a variety of distribution systems has evolved in response. The third is that the fan pressures, and duct and grill sizes, to the different parts of the building be such that the system is, as far as possible, balanced, each part getting its appropriate share of the supply – the final balancing is done by adjustable air dampers in the ducts at the commissioning stage.

Most systems have some sort of flow-and-return arrangement, with the ducts progressively reducing in size away from the main plant: in fact, the air-conditioned building will generally have a duct hierarchy analogous to that of the extended heating-pipe circuitry – main ducts travelling vertically in core or shafts and secondary, possibly tertiary, branches radiating horizontally, commonly through the voids over suspended ceilings.

Altering the air temperature at the plant is simple enough – the snag, of course, is that most large buildings are subdivided into many spaces, often having quite different heating or cooling needs at the one time. We spoke of domestic warm air's control snags, but how much greater is the design challenge when the scale is larger and there is to be heating *and* cooling. First though, two cases where the selective distribution problem is not confronted.

The self-contained 'packaged' unit

These are compact boxes placed in or near the single space to be served. They are mainly for small spaces, the small restaurant or discothèque or, in warmer climates, the familiar cases of small office, hotel bedroom or apartment, where a measure of cooling is wanted without the problem of ducts. The units are often sited under windows, recirculating the room air, together with a smaller proportion of direct fresh-air intake and exhaust through the wall. There will normally be no heating capability, nor humidity control specifically – though inevitable condensation on the cooling coils (requiring a drain) will provide some drying in humid weather. They can be quite noisy, though.

The single-zone, 'central' plant

This is the simplest of the ducted systems. It is used for single, undivided spaces, like auditoria, where the air conditions and, in consequence, the required air-delivery temperatures, are virtually uniform throughout. All that is needed are flow-and-return ducts connecting grills to plant and, possibly, an ancillary smoke extract direct to the outside – all as in the simple circuit already shown.

The dual-duct system

Most buildings are multi-spaced, posing varied demand. In the dual-duct system (3.25), the main plant supplies to separate hot and cold flow ducts, distributing around the building – there is a common return. Each space has its own thermostatically controlled mixing box. It is a theoretically appealing but very expensive solution. Having ducts large enough for heating and cooling is costly, but having one for heating and one for cooling is costlier still. The system may be jus-

tified where there is a considerable ventilation requirement and where fluctuating heating/cooling demands ask for fast, effective response, for example, a laboratory. But the layout must be simple for even remote economy.

The variable volume system

Another way of catering for fluctuations is to have a single-flow circuit with air temperature set to the *average* demand, and then have local thermostats controlling the final volume delivered to each space. They do so by activating dampers at inlet grills or in the branch ducts themselves. Again, this is for where there is a high ventilation need anyway, where there is a high heating/cooling internal load and – since the single-flow duct cannot provide simultaneous heating and cooling – for where there is an all-heating or all-cooling demand. Central service cores are a possible case but, in truth, the effective applications are coming to be seen as rather limited.

The zoned system

This is the usual system for larger, multi-space buildings. The building is subdivided into zones, groups of spaces or rooms likely to have similar demands at the same time. For example, an office block, the plan of which was elongated on the east-west axis, might sensibly have two vertical zones, one facing north and the other south. A square plan might have four zones. Areas like core conference rooms and staff restaurant could also be zones (3.26).

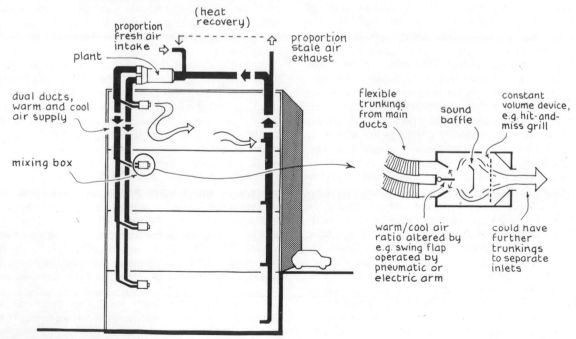

3.25 'Dual-duct' system for air-conditioning distribution

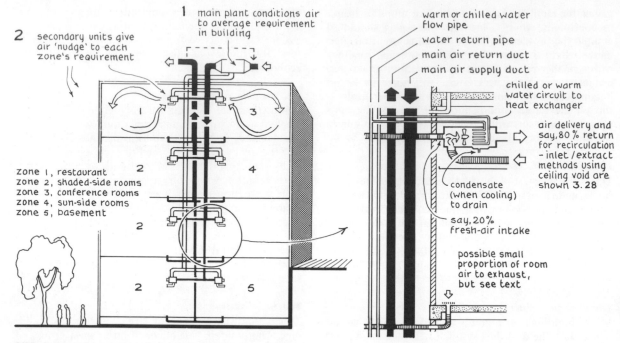

3.26 A form of 'zoned' system for air-conditioning distribution, using main (so-called 'central') plant and secondary fan coil units

In simple form, the zoned system can be taken to mean central plant, with integral design and ducting capable of satisfying different ventilation and heating/cooling loads in different zones. Often, there will be local reheat batteries in the zones – think of central-heating convectors – giving the delivery air a final temperature nudge. But, for interest, the system illustrated is going a stage further, being a common form of 'zoned' system where the central plant is conditioning to the *average* need in the building and local, zone units are adjusting the heating or cooling to each zone's or, indeed, room's, *particular* requirement. There are many versions of this approach but, generally speaking, when zone spaces are large the secondary units are referred to as *secondary air-handling units*; where the spaces are small, as here, we have the ubiquitous and popular fan coil units. Essentially, the central plant is delivering both conditioned air and water circuitry in parallel, the air ventilating and taking the bulk of the heating or cooling load and the water enabling the fan coils to supply the final heating/cooling temperature adjustment. In the simple *change-over* version shown, there is a two-pipe flow-and-return water circuit, heated in winter and 'changed over' to chilled in summer. But multi-pipe versions are admittedly more common, having both heated and chilled flows, and a common return – or, indeed, separate returns. This allows greater flexibility, including the ability to heat in one zone while cooling in another.

The added water circuits are a complication, but then the single air-flow duct is a simplification. And, as can be seen from the diagram, there is the very considerable advantage that the higher proportion of air that is recirculated – for efficient, unobtrusive heat transfer, as distinct from fresh-air ventilation – only has to go back to the secondary unit, not all the way back to the main plant. The main vertical ducts in the service core, now only having to handle the smaller amount of fresh-air inlet and exhaust, can therefore reduce in size.

As to the exhaust from the zone back to the main plant, this will be less than the fresh-air inlet by the amount of the air leakage to the outside that occurs – that is, unless the enclosure is to something like a laboratory or hospital operating theatre and specifically sealed against this. The desirability of balancing the inlet and extract to give a *slight* pressure over (and, hence, isolation from) the air outside has been described. To be sure, some smaller-scale systems rely entirely on this leakage for exhaust, scrapping the exhaust return to the main plant. But at anything away from small capacity, this would represent an unacceptable heat loss, missing the chance of heat recovery from a controlled exhaust at the main plant. And, on the subject of energy management, the water circuitry, in taking a share of the heat transfer between zone and plant, also offers opportunities for heat recovery and heat transfer between zones. Surplus solar gains on the sun side of a building, and heat gains from the lighting and other activities, can be a considerable cold weather bonus overall.

Incidentally, local *induction units* are often used instead of fan coils. These have no fan. Instead, the fresh-air intake from the main plant (or, perhaps, directly from the outside, where units are on the external wall) is supplied at high pressure through nozzles and this, and the casing design, are such as to induce room air to recirculate and join the flow over the heat-exchange coil, again making up an adequate volume for efficient heat transfer. Heating or cooling is modulated by varying the water flow or by having a flap adjusting the proportion of air flowing around the coil or bypassing it. The units are compact and having no fan reduces electrical wiring and later maintenance.

The above systems offer only an outline guide in the widest range of options. There are many varieties and different energy-saving policies within the systems described, and larger buildings may be served by different systems in combination. Building scale may reach the point where it is worth having separate air-conditioning systems to limit the duct travels involved. In tall buildings, this separation will often happen on height, but it is fair to say that general plant separation – say, having separate plant floors at basement level and, again, at mid-height – has more to do with the water services (see next chapter) than with air-handling. Air ducts do not develop 'pressure heads', as such, so the break-even point where it is worth having quite separate air-conditioning systems has more to do with the total floor area served, i.e. the total air volume to be handled. Where there is parallel heating/cooling water circuitry, it can be designed to cope with higher pressures than would be tolerable in the ordinary water-supply plumbing, and the more pressure-sensitive elements, like boilers, can avoid the pressure problems by being sited at roof level. A single air-handling system up to 30 storeys would be perfectly feasible.

Distribution in the spaces themselves

The airflow in the spaces are governed mainly by the inlet and only slightly by the outlet arrangements, and the factors – good distribution, noise control, unobtrusiveness – are similar to those for straight ventilation, except that when cooling, a downward flow makes more sense from the natural convective point of view and also because it is more comfortable to have a cool head than cold feet. System balancing is essential. For example, in the zoned systems, the proportions of air that are exhausted or recirculated will depend on the size of the grills, the secondary-unit fan pressures and so on and, even with the mixing proportions right, a bad layout might be leaving room air in one place stagnant and uninvolved in the circulation, with a perverse 'bundle' of air in another place, endlessly recirculating.

As you will infer, the type of space and system influences the grill arrangements. An auditorium with single-zone plant might have inlet slots incorporated in the acoustic ceiling construction and extract outlets around lower wall or under seats (3.27). Delivery in large spaces, generally, may be by nozzles, say for throw from a high ceiling perimeter or, more commonly, by circular or linear *diffusers* which provide good mixing with the ambient air and, hence, avoid cold draughts.

Diffusion is especially important with delivery from low ceilings, for example, in office buildings: in fact, the office suspended ceiling is usually involved in the distribution arrangements. 3.28a is the *plenum* system: this has the duct delivering air into the ceiling void, where the resulting slight surplus pressure causes the air to permeate evenly into the room below, either through perforations in the ceiling tiles themselves or through slots in the supporting metal channels. Since the whole ceiling is effectively one big inlet grill, the system is particularly suited to situations where delivery air volumes are large, for example, to cope with the high cooling load in computer suites. Extract grill

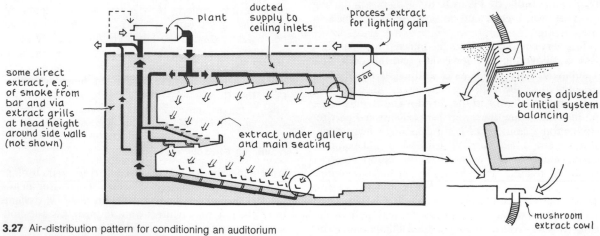

3.27 Air-distribution pattern for conditioning an auditorium

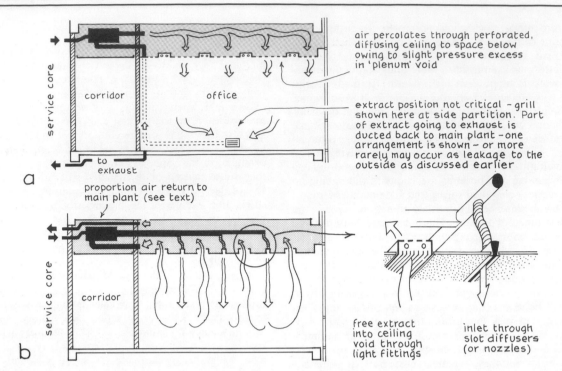

3.28 Common ways of distributing conditioned air in office zone. (a) The 'plenum' ceiling (b) Slot diffuser inlet and (heat removing) extract through lighting fittings

positions, you will remember, are less critical, except that they should neither be near the door, lest there be short-circuiting from the corridor, nor placed where they are likely to be obstructed. The wall extract shown returns the room air to the fan coil for immediate treatment and recirculation. And, as said, unless the space (or, more specifically, the heating/cooling load) is very small, then energy-wise the exhaust should be by extract back to the main plant rather than just by ordinary leakage through the enclosure. In fact, this could occur at the fan coil by having a proportion of the return room air diverted back to the main return ducting or, as shown, there could be separate extract back to the ducting.

It is very common to have the fan coils to each space located above the corridor ceiling and this does raise a supplementary point about extract. It might seem easy to reduce the ducting by having the office air leave through extract grills to the corridor and, thence, pass up into the ceiling there, for part exhaust and part recirculation – and, indeed, this has often been done. But airflow direct into the corridors could certainly carry smoke in the event of a fire outbreak, disastrously violating their integrity as escape routes. Ideally, corridor ventilation should involve positive pressurisation (see Chapter 7 *Fire safety*).

Another approach is to *extract* through lighting fittings (3.28b). Even at quite moderate illumination levels

(around 500 lux but see Chapter 5 *Lighting*), supplementary day-time or straight night-time lighting can be contributing over half the heating gain, adding massively to the cooling load in summer and, possibly, outweighing the heat loss through the enclosure in winter. Extract through lighting fittings traps this heat at source, carrying it to exhaust in warm weather or allowing its recovery to warm the fresh-air intake in colder weather. And cooler lighting fittings operate more efficiently. The diagram shows a secondary unit above the corridor, with the adjacent suspended ceiling void affording an input route to the supply nozzles and extract route back through the lights. Clearly, one or other of the routes must be ducted and here we show the former. Because inlet is the effective determinant of the airflow pattern, the delivery nozzles adequately serve the space without short-circuit to the extract points, even though the latter are quite closely adjacent. Note, also, the separate exhaust back to the main plant – there certainly would be an exhaust in a case like this, helping the heat extract and, hence, recovery, i.e. rather than relying on wasteful leakage to the outside. It makes sense to site the exhaust in the ceiling void, since that is where the air is hottest.

A note on controls

An engineering colleague was called to a small office building, just commissioned, where the environmental services seemed to be hardly working at all. The clients were blaming the architect who, in turn, was questioning the services engineers who, in their turn, were blaming the installers who, to complete the circle, were claiming the plant had been designed too small in the first place. Litigation seemed possible but, in fact, it eventually transpired that the trouble lay with quite simple faults in the control system. The room thermostats had been sited too close together and were fighting each other. There were defective humidity and duct airflow sensors. There were valves mounted upside-down, so the system was doing the opposite of what it was told, that is, where it was being told anything at all. For months, the heating and cooling plant had been working simultaneously and at full blast, the boilers producing their full output and the cooling plant fighting to chuck it all out at the roof!

This was an extreme and ludricous case, but it is true that the services are only as good – as effective and as economic – as their controls will allow them to be. Preoccupation with comfort and, at the same time, energy conservation has enormously advanced control design over recent years. Briefly, the control task is similar to the small house's, in as much as there has to be an appropriate average output from the plant and a specific output from each outlet but, of course, there is more plant – multiple boilers, refrigeration, humidification – and there are many more outlets. The controls must sense the temperatures and humidities, and act on the dampers and fans in the air system, and the pumps and valves in the water circuit. The common electrical sensor is the on-off temperature thermostat described earlier, but there are others. There is the electronic sensor, in which sensitive elements alter their resistances in response to changes in temperature or humidity, alterations which are magnified by transistors into usable current for control operation. There are also pneumatic types, such as those where similarly sensitive elements progressively open or close an orifice, releasing or building-up the air pressure in what is called a pneumatic actuator, to give a working control arm. Pressure sensors can be used to monitor the airflow in ducts. Humidistats monitor humidity. There will be time-switch overrides and, possibly, outside compensating sensors. The controls have an added role to play in directing heat recovery and its appropriate transfer between zones. Increasingly, the control operation in larger modern buildings is computerised.

Summary

Factors for air-conditioning choice are:
- Hot/humid climate.
- Large scale and deep plan.
- Heat or pollution – processes, people.
- Windows sealed owing to height, external noise or pollution, or for comfort.
- Quality and prestige (as distinct from fashion).

Plant
- Extracts from ducting system or from space directly.
- Exhausts a proportion of stale air.
- Intakes a proportion of fresh air.
- Filters.
- As necessary, cools (and dehumidifies) or
- Heats (and maybe humidifies).
- Delivers the conditioned air.

System types
- Packaged units – compact, for e.g. individual rooms in existing buildings.
- Single-zone plant – air ducts from plant to single large space, for e.g. an auditorium.
- Dual-duct – parallel hot and cold ducts allowing blended supply locally; high performance; for e.g. a laboratory.
- Variable volume – single-duct only, measure of control by altering air flow quantity, for e.g. central core.
- Zoned system – central plant delivering air to suit average requirement in different zones, with re-heater batteries or secondary units making local adjustment; for any large multi-space building.
- Induction and fan coil units – secondary units making final zone or room adjustment; served by parallel heating/chilling water circuitry from central plant.

Inevitably, the very fact of separating the technical functions into chapters presents them in a more isolated way than they would ever be treated – or, at least, *should* ever be treated – in the design process, and this is nowhere more true than with the 'Climate services'. Not only are there all the basic choices just described, shaping the system to suit the particular building's function, plan, size and location but, also, there is the need for mutual accommodation between the climate services and the other technical functions of the building. How are the *services* and *structural* solutions to be reconciled? Are the pipe and duct vertical runs to be in a service core and are the horizontal runs to be between suspended ceiling and structural floor slab? How does the kind of enclosure affect the required capacity and flexibility of the system? How are the distribution runs to be integrated with the other utility services' runs, such as water supply and drainage? Will the artificial lighting system be such as to require extract at the fittings, possibly with heat recovery? How are the needs of fire safety, to avoid fire and smoke spread through the building, to be reconciled either with having service ducts passing through and potentially violating compartment divisions, or with a ventilation system that passes air from one room to another? Some of these questions we have covered, some we have yet to cover. The need to integrate the technical functions in design may be self-evident but it does bear stressing.

4

UTILITY SERVICES

The *utility*, or 'useful', services bring working convenience to the building. They include:

- Water supply
- Drainage
- Refuse handling
- Electrical supply
- Telecommunications
- Gas supply
- Mechanical conveyance

The comparison of piped services with arteries, telecommunications with nerves and so on, is a bit overworked sometimes but it does remind us of the services' essential role in linking the building with the material support of its surroundings and bringing it operationally to life.

The 'utility services', like the 'climate services', will sensibly tend to collect into common runs through the building; in a large building, for example, they will usually travel vertically through duct spaces, possibly in a service core, and horizontally through floor constructions or through the voids over suspended ceilings. And, as stressed in the last chapter, the building's internal organisation must help minimise and simplify these runs – particularly, those of the water

services to bathrooms and kitchens, and the drainage services therefrom. Servicing can amount to 30 per cent of a building's total budget, in fact up to perhaps 50 per cent in a large, highly serviced building like a hospital, and they absolutely cannot be threaded through the design as an afterthought – not, at any rate, with any logic or economy.

WATER SUPPLY

THE SOURCE

The natural cycle that waters the land is pretty familiar and the process of side-tracking some of this water for sanitation in buildings is the oldest and most elaborate utility service (4.1). Above ground, the rainwater sheds into streams and rivers. Below ground, it tends to percolate through permeable strata and collect over impermeable strata, sometimes re-emerging as springs but, generally, forming the underground water table,

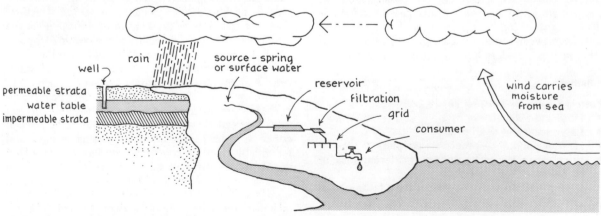

4.1 Tapping the natural water supply cycle

whose depth below the surface will vary, depending on topography, subsoil and, of course, the prevailing rainfall.

Isolated places

In isolated places, private wells or boreholes can be sunk to reach the water table. Water seeps in and (it is hoped) provides a constant source for pumping up to a building's water-storage tank. A float switch in the tank triggers the pump when the water content drops below a predetermined level.

Cities

The classic urban supply system has rivers in the hills serving large reservoirs which, in turn, supply an extending piped network right to the water mains under each street. The delivery pressure from a pipe is proportional to the 'head' or height of filled pipe above it, save for losses owing to friction. The convenient mains head in the street is between 30 and 70 m, enough to push up to the header storage tanks in most buildings, and for adequate fire-hydrant pressure, and yet keeping within the strength limits of practical plumbing. Since level changes between reservoir source and distant consumer can amount to hundreds of metres, there have to be subsidiary reservoirs, breaking the primary flow and supplying local areas. These can be service reservoirs at or below ground level or, occasionally, water towers supplied by pump, all depending on local topography – San Francisco must have posed interesting problems.

Purification

Purification includes filtration to remove sediment, disinfection to remove those bacteria that are harmful and, possibly, softening to remove hard-water salts. New private sources in outlying districts have to be analysed. Well-water in these cases is preferable to surface sources because it is less open to pollution and will have been pretty well filtered in passing through the ground. Further filtration, if needed, is normally through a bed of sand. The sand itself removes sediment and, in time, develops a gluey film known as the 'vital layer' which screens off some bacteria as well. Chemicals may be added to promote this layer's formation and to disinfect the water. Today, the process is usually contained in a compact steel chamber (about 2 m³ for a house), pressurised to speed the flow. Periodic washing-back cleans the filter. Often, there will be a second-stage fine filter on the drinking water tap.

Public supplies from reservoirs are filtered through large-scale sand beds and disinfection is normally by controlled amounts of chlorine.

Soft and hard water

Water is 'soft' or 'hard', depending on the amount of dissolved salts it contains. Some hardness makes it more palatable, but excessive hardness prevents soap lathering, can increase corrosion of the plumbing system and, if the hardness is 'temporary' rather than 'permanent', meaning that the salts are precipitated on boiling, there can be further damage owing to scaling. Scale, the whitish deposit that can build up round the element of an electric kettle, can equally 'fur up' the plumbing installation. The risk is, arguably, reduced on the boiler and central-heating side, assuming the water there is recycling unchanged and therefore harmless once its initial attack has been exhausted – but, in fact, there are occasional top-ups, as the last chapter explained. And, in any case, the problem remains on the hot tap-water side. The temperature there may be lower but there is regular fresh water replenishment. Where needed, water softening is usually done at the point of use, i.e. at the building. There are several methods but, essentially, the troublesome salts are precipitated or converted by passing water through a cylinder containing appropriate chemicals, the chemicals being recharged periodically. (The central-heating circuits in larger buildings may incorporate 'dosing pots', allowing corrosion and scaling inhibitors to be periodically added).

DISTRIBUTION IN THE HOUSE

Cold supply, 'direct' and 'indirect' systems

4.2 shows the two cold distribution systems for a house, *direct* and *indirect*. In the direct system, the supply to all the *cold* water taps is directly off the mains. There is a mains-supplied storage cistern but only for the hot part of the system. The indirect system has the cold as well as the hot supply via the storage cistern, save for the direct-mains drinking-water tap, usually in the kitchen. Each system has its merits.

The main advantage of the indirect system is that each building has its own storage rechargeable over the 24-hour period, meaning that the public mains is excused from the immediate responsibility of meeting peak demands. Mains pipes can be smaller and water resources are optimised wherever supplies are restricted. Buildings are less vulnerable to any temporary cut-back in the mains supply. On the other hand, there is the cost and nuisance of the larger storage cistern and extra pipework. Drinking from an indirect cold tap (most of us occasionally do) can be harmful if the cistern ever becomes polluted – a rat can fall in (although the cistern should be covered) or the water can stagnate if people are away for a long period. The risks are remote though.

The direct system reduces piping, the more so since the storage cistern, now only supplying the hot-water system, is smaller and can be located, say, above the

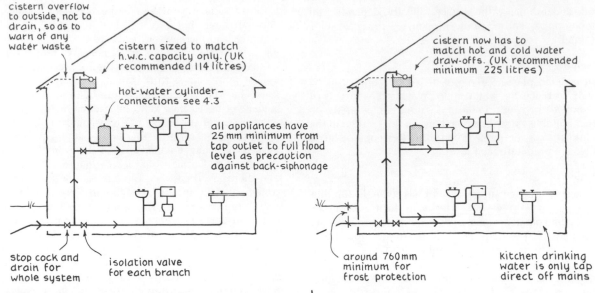

cistern overflow
to outside, not to
drain, so as to
warn of any
water waste

cistern sized to match
h.w.c. capacity only. (UK
recommended 114 litres)

hot-water cylinder –
connections see 4.3

all appliances have
25 mm minimum from
tap outlet to full flood
level as precaution
against back-siphonage

cistern now has to
match hot and cold water
draw-offs. (UK recommended
minimum 225 litres)

stop cock and
drain for
whole system

isolation valve
for each branch

around 760 mm
minimum for
frost protection

kitchen drinking
water is only tap
direct off mains

a

b

4.2 Domestic cold-water distribution systems. (a) Direct (b) Indirect

- All cold taps for drinking.
- Greater peaks on mains.
- Risk of back-siphonage.

bathroom airing cupboard, rather than in the attic. If there is a bathroom shower, though, the cistern will still need to be in the attic for adequate head. All the cold taps are drinking water. On the other hand, peak demands on the mains are greater. The plumbing under the higher mains pressure is likely to be more noisy, and taps and valves are more stressed. But the chief worry is the risk of contamination of the mains supply by back-siphonage. This is where the mains supply happens to fail and water already in the domestic plumbing starts to drain back out again. There can be trouble if a tap is infected or has its outlet under the flood level of an appliance and able to back-siphon the impure contents towards the main. The risk is less severe in today's surer water supplies but, for safety, all tap outlets have to be above the flood level of the appliances they serve, and appliances like bidets, where this is not possible, have to have their own in-direct storage tank.

Summary

Indirect system
- Cold and hot supply off storage cistern.
- Reduced peak demand on mains.
- Building cushioned from mains failure.
- More plumbing, bigger cistern.
- Cold taps not for drinking.

Direct system
- Cold supply direct off mains.
- Neater cistern (hot only) and plumbing.

Much can made of the arguments either way. Obviously, a remote place supplied by a pump from a well or other source requires header cisterns in the buildings and is inherently indirect. Areas with stretched resources, or extended supply, as in cities, have tended to be indirect also. But the tendency is now towards direct. The UK is beginning to accept direct. The US and the continent tend towards direct – some countries actively prohibit indirect storage tanks on sanitary grounds, despite the back-siphonage worry. Either way, regulations have tended to confirm the early preference and, as elsewhere in the building industry, this polarisation is not always useful.

Hot supply

Hot-water supply can be from a central boiler to all taps or from point-of-use heaters to particular taps. In either case, there may be a hot-water storage facility or water may be heated 'instantaneously' on demand.

Central boiler and storage cylinder
This is now the most common thing for a modern house, especially the combined system, where the boiler has the combined role of heating the water for the central-heating circuit as well as for taps. So we have an interface here with the last chapter – see 3.6. 4.3a shows the *separate*, hot tap-water side of the heating system. Ideally, the boiler is close to the hot-water cylinder to minimise heat loss from the primary circuit (which, like the cylinder itself, should be lagged with insulation) and therefore, by implication, the

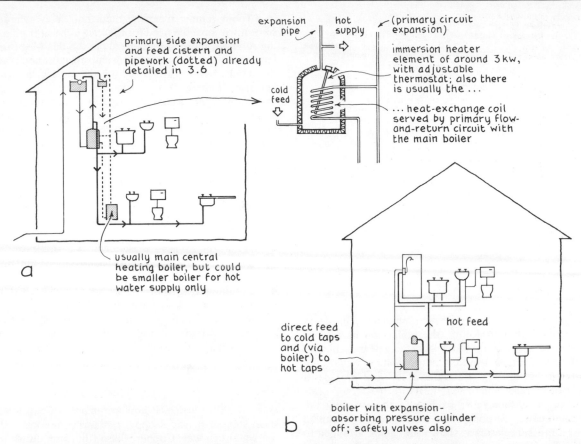

4.3 (a) Hot-water supply system incorporating (indirect) storage cylinder and central boiler heating (b) Direct, 'continental' hot-water system

boiler should be fairly central in the system to minimise pipe runs to taps. The excessive pipe-run, or *dead-leg*, is a double cause of waste: it wastes cold water run-off before the hot arrives and, more important, wastes heat from the water inevitably left in the pipe after the tap is turned off.

The hot-water cylinder provides a vehicle for the heat exchange to take place and, in providing storage, it allows a reasonable time for this. Water has a high specific heat – it takes quite a lot of heat to heat it up. 'Instantaneously' heating a bath-tap flow, say around 10 litres/minutes raised from 5 °C to 55 °C, would take about 45 kw no less, around three times the output of the average domestic boiler, let alone that boiler's other house-heating responsibilities. Storage heated on a continuous basis allows a reasonable recovery time, excusing the boiler output and heat exchanger from having to keep direct pace with a draw-off as it occurs. This is very sensible. Expansion is back into the cold-water tank.

Other heat sources for the storage cylinder. There are other heat sources. The electric immersion element (rated around 3 kw) is simplicity itself, if rather expensive to run. It is often used as a back-up to the main boiler, e.g. in summer when the boiler is off. Where there is no main boiler, smaller, purpose-made boilers can be linked to the cylinder's heat exchanger. They are compact but need a flue. Fires and stoves with back-boilers circuiting to the exchanger are coming back into vogue. There are boilers with an incorporated water-storage facility, i.e. allowing recovery and good response, but without a separate storage cylinder, as such.

The direct 'continental' hot-water system
In many parts of Europe and the USA, direct hot as well as direct cold is the rule (4.3b), i.e. taking things a stage further than the direct system of cold tap supply. Both cold cistern and hot cylinder have now disappeared and, with them, of course, the arguable advantages of cold and hot storage already described. But the plumbing simplicity is immediately apparent. Helpfully, also, the hot and cold are now at the same mains pressure (or below mains if the mains pressure is high and there is a pressure-reducing valve in the initial feed) and this inherent balancing makes the working of single-lever mixing taps and shower mixing

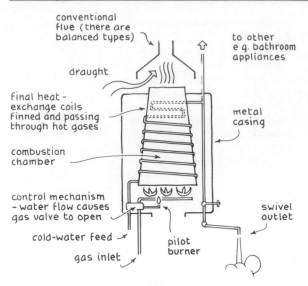

4.4 Gas 'multi-point' water heater

valves that much easier. Where temporary hardness might cause boiler scaling, the boiler can include an intermediate 'primary' water jacket as a heat exchanger with the cold flow – in fact, many systems include the water jacket anyway, since the water mass provides a kind of storage which, together with the larger heat-exchanger surface, goes some way to retrieving the 'instantaneous' response. The system expansion can be into a pressure vessel as shown, or can be arranged to occur back along the mains or run to waste. Both boiler and hot distribution have back-up safety valves.

The multi-point heater – 'instantaneous' flow

The 'multi-point' gas heater, or geyser (4.4) provides hot water on demand. It is called 'multi-point' because it is capable of supplying two or more taps – for instance, it might be mounted in the kitchen but have a run to the bathroom appliances as well. It needs no header tank, since it can operate off the mains and, of course, there is no hot-water cylinder – all very simple. Turning on a tap releases the gas flow to the burners, where it is ignited by the continuous pilot flame (there is an automatic cut-out if the pilot fails). The water passes through finned heat-exchange coils in the combustion chamber. But, as earlier implied, the output is restricted by the difficulty in heating a fast water flow, so it is really only for a small house where no central boiler is planned.

Single-point heaters

Point-of-use heaters, *geysers*, operate similarly. They are no alternative to a proper central system, but are a cheap-to-install heat source for a single basin, bath or shower. And, even if there is a central boiler, it can make sense to use a geyser, say, where a basin is added in a room remote from the main hot-water cylinder, since it will save the need for the long hot-water plumbing run and, more important, avoid the dead-leg problem thereafter. Gas types are usually instantaneous and the small size avoids the need for a flue. Electric types, because of their limitation to around 3 kw to avoid overloading the domestic wiring, and because of the slower heat exchange from electric elements, have more restricted flow, and most incorporate enough storage for a single use of the appliance they serve.

Summary

- Central boiler with storage cylinder – 'combined' boiler for central heating and hot water; hot-water cylinder allows recovery period and copious draw-off.
- Other boiler types – these includes gas/oil purpose-made for hot-water cylinder; back boilers; and boilers with incorporated hot-water storage.
- 'Continental' system – direct hot as well as cold; no cold-water tank or hot-water cylinder.
- Single or multi-point heaters – usually 'instantaneous' if gas, 'storage' if electric; restricted flow but cheap to install; remote single-point avoids long dead-leg.

Materials

Domestic cold-water cisterns are mainly of galvanised sheet steel or plastic. Copper remains common for the hot-water cylinder. Copper is also still common for the pipework, being easily worked and durable with a high tensile strength, allowing the pipes to be thin-walled and light. Mild steel pipework is stronger still, but this virtue has more application to high-pressure heating systems than to domestic plumbing. The cost of lead and, indeed, its toxicity, debar it from most modern plumbing, except that its easy workability is occasionally useful in making awkward connections, e.g. cistern to WC and so on. Plastic pipework is tough, light, flexible and workable. It was formerly restricted to the cold-water side, owing to plastics' generic tendency to soften around 70 °C, but special 'hot-side' plastics – pioneered in the USA principally – are now finding wide acceptance.

DISTRIBUTION IN LARGER BUILDINGS

Scale increase brings further factors into play. For a start, peak demands from large buildings are going to be that much greater, immediately favouring indirect supply via storage tanks rather than direct supply, which might starve the mains. Building height brings pressure problems, both in feeding water up the building and in distributing it down from high-level storage cisterns. Hot distribution has, somehow, to avoid overlong dead-legs.

Cold-water distribution

Boosting by pump

Building height may well outstrip the 30–70 m supply head of pressure available from the street mains. A 30 m head is only equivalent to eight or so storeys, let alone the need for a two-storey margin to allow for frictional losses in the upward feed pipe and to give a sensible working pressure at the highest appliances. Some countries, notably North America, tend to pump-boost both appliance- and drinking-water supplies. Others, including the UK, boost to high-level cisterns from which there is a gravity down-feed to appliances, and boost directly only for drinking water. 4.5a shows such a system. Note that the drinking water is taken directly from the mains as high as pressure allows. Above that, in the boosted section, it is supplied by gravity from the drinking-water header pipe – this is just an enlargement of the supply pipe en route to the main storage cistern, and its capacity (at around 4.5 litres/dwelling in a block of flats) is small enough to ensure enough turnover to keep the water fresh but large enough to avoid having the booster pumps switching on and off all the time. A system of float switches in the main cistern and in the header pipe cuts in the pumps when the level in either falls low.

Boosting by pneumatic cylinder

This is similar to the above, except that the boost to the higher levels is maintained by a pressure cylinder connected to the initial feed (4.5b). This comprises a small reservoir with an air cushion over the water. The water pumps and air compressor operate intermittently, the first to replenish the water when it falls low and the second to maintain the pressure and replace any air dissolving away. Since the system is supplying the high-level drinking water as well, the cylinder capacity is kept small to ensure rapid turnover, and the air entering the compressor is filtered against dust and insects. The continuous nature of the head avoids the need for, but does not preclude, high-level storage tanks.

Break-cisterns, 'indirect' supply

The use of high-level storage tanks makes a system indirect but, actually, the term 'indirect', applied to larger buildings, can imply something more. Even with high-level tanks, peak-demand pumping from the mains can cause an unacceptable drop in the head available to buildings further up the line, and this is even more true with pneumatic boost and no storage tanks. Many authorities (in Europe at least – US cities tend to have copious street mains) ask for a *break-cistern*, a reservoir between mains and building, on which the pumps can draw. Only the fresh drinking water is then off the mains direct, boosted as required. This is illustrated for the high-rise system (4.6).

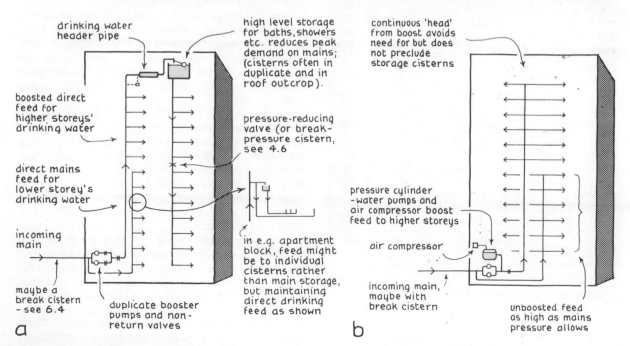

4.5 Cold water distribution in multi-storey buildings. (a) Pump-boosted (b) Pneumatic-cylinder boosted

High-rise systems

Further height increase brings further pressure snags. Over about 30 m (say, up to 10 storeys), the pressure at lowest taps supplied by roof-level tanks becomes unmanageable and, over 70 m (up to 20 storeys), the pressure at the foot of the boosted initial feed and on the pumps becomes unmanageable. 4.6 shows a typical system for ordinary and drinking-water supply, using 'break-pressure' storage cisterns at intervals on height, restricting the unbroken initial feed to rises of 60 m and the gravity distribution to taps to drops of 30 m. The low-level break-cistern reservoir cushions the mains and, incidentally, it reduces the water loads to be carried higher in the building structure. Actually, common practice in the UK is to provide a day's supply of storage, with $\frac{2}{3}$ in the break-cistern and $\frac{1}{3}$ higher in the building.

The drinking-water system shown is appropriate to the sinks and basins in residential blocks and hotels. In an office building, the drinking flow may only be to drinking fountains and, consequently, small, and could warm up unpleasantly in transit up long vertical pipes, especially were the pipes sharing routes with other warm services. Instead, the drinking points can be supplied off a pumped flow-and-return circuit, lagged and passing through a chiller.

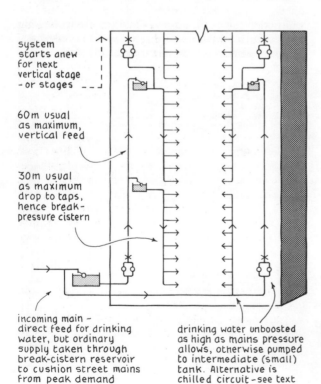

system starts anew for next vertical stage - or stages

60m usual as maximum, vertical feed

30m usual as maximum drop to taps, hence break-pressure cistern

incoming main - direct feed for drinking water, but ordinary supply taken through break-cistern reservoir to cushion street mains from peak demand

drinking water unboosted as high as mains pressure allows, otherwise pumped to intermediate (small) tank. Alternative is chilled circuit - see text

4.6 Vertically staged cold-water distribution for very high-rise building

Summary

- Break-cistern reservoirs at ground level are often required to cushion mains from peak demand – bulk supply is then truly 'indirect', only drinking water being kept direct; low-level storage makes added structural sense.
- Available mains head has to reach building height plus working margin; otherwise, boosting needed by pump to high-level storage tanks, or by pneumatic cylinder, possibly without storage tanks.
- Very high buildings require break-pressure cisterns at intervals, restricting continuous upward feeds to around 60 m and continuous gravity drops to taps to around 30 m.

Hot water distribution

Local systems

Hot water distribution in large buildings faces the added problem of limiting dead-legs. In a building like a factory, where washing facilities tend to be grouped in only a few select zones, it makes sense to have local systems particular to those zones. The options are then similar to those for domestic hot water, i.e. having either single or multi-point geysers, or a hot-water cylinder with its own boiler or electric immerser. Occasionally, local cylinders are heated by a heat exchanger looped off the ordinary heating circuit, in which case, there would be an electric immerser for the summer months when the heating was off. In factories, they may be off a process steam circuit.

Flats can be localised, in as much as they often have their individual boilers for both heating and hot water, rather than there being a central service supply from a main plant. The systems are then, again, as ordinary domestic, in the UK usually having the hot-water cylinder venting into the flat's own cold-water header tank.

Central boiler systems

The majority of large buildings – schools, hospitals, multi-storey hotels and offices – have extended hot-water distribution from a central boiler plant. The boilers will usually be providing the space heating also, so again, we have the interface with the last chapter. Centralising the plant offers scale economies and is certainly cheaper to fuel than any system using many local boilers and, possibly, relying on electrical back-up. But there are three principal problems requiring solution. First, as already explained, scaling risks require that the water recycling through the boiler plant be kept separate from the regularly replenished supply to the hot taps. Second, any extended system must, obviously, accentuate the wasteful, dead-leg problem. Third, there are the pressure-head limitations applicable to high-rise. The system (4.7) deals with all three.

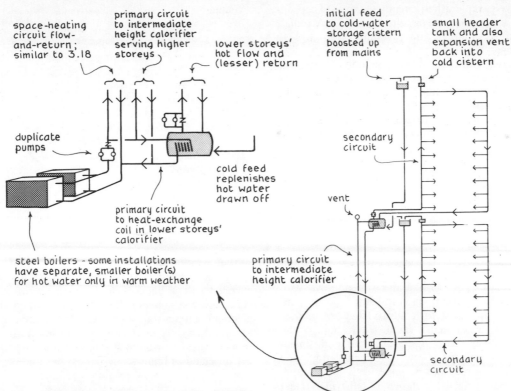

space-heating
circuit flow-
and-return;
similar to 3.18

primary circuit
to intermediate
height calorifier
serving higher
storeys

lower storeys'
hot flow and
(lesser) return

initial feed
to cold-water
storage cistern
boosted up
from mains

small header
tank and also
expansion vent
back into
cold cistern

duplicate
pumps

secondary
circuit

cold feed
replenishes
hot water
drawn off

vent

primary circuit
to heat-exchange
coil in lower storeys'
calorifier

steel boilers - some installations
have separate, smaller boiler(s)
for hot water only in warm weather

primary circuit
to intermediate
height calorifier

secondary
circuit

4.7 Hot water distribution at high rise: at left is the relationship between the 'central' boiler plant, primary circuits and hot-water calorifiers; and at right, the intermediate height calorifier separates the system into stages, limiting pressure heads and, together with the use of secondary circuits, reduces the 'dead-leg' problem

As at domestic scale, the boiler water is kept separate by having primary circuitry transferring the heat to the replenishment hot-water supply by means of heat-exchange coils in indirect storage cylinders, or 'calorifiers', as they are more properly called. The primary circuitry can be low-pressure hot water or, for greater efficiency at large scale, pressurised hot water or steam. The dead-leg problem (whether in a low-rise school complex or a multi-storey building, as here) is reduced, in that the calorifiers can be situated remote from the central plant, acting as distribution outposts to the taps in their particular zones. In a small zone, the calorifiers could serve the taps directly, but large zones, like the multi-storey stages here, call for further, *secondary* flow-and-return circuits with short branches to taps off them – circuits are not dead-legs by the very virtue of their being circuits and, of course, they will be well lagged with insulation.

The pressure-head problems are answered in that the intermediate-height calorifier is inherently acting as a kind of break-pressure cistern, halving the pressure head with which the boilers have to cope, and the associated, vertical staging limits the pressure head at the lowest taps. Actually, the system shown is quite modest as pressures go – high-pressure steel boilers and pipes can push to calorifiers some 100 m and more above them, i.e. some 30 storeys. And, as described in the last chapter, ultra-high-rise will divide into quite separate, vertical stages from the servicing point of view, each with its own boilers, air-handling plant and so on. Incidentally, if the boilers happen to be on the roof, pressure-reducing valves can be used to limit the head in the drops to calorifiers. They also allow more than 30 m drops *within* a zone, i.e. from calorifiers to the lowest taps. Choosing the optimum system is a complex matter for the services engineer – whether the system is central or local, low-level plant or high, single plant, or horizontally or vertically zoned. Building type, shape, height and other factors are involved.

Summary

- Large system design must solve dead-leg and pressure-head problems.
- Local systems (single or multi-point, with or without storage) are suited to isolated groups of appliances, e.g. in factory – they limit dead-legs. Apartments can be 'local' in having their own domestic-type systems.
- Central boiler plant is used for most large buildings – interface with space-heating systems.
- Primary boiler circuits to calorifiers and, thence,

secondary, replenished circuits to taps a) keep primary and secondary circuits separate, b) reduce dead-legs, c) break the pressure heads in high buildings.

- 60 m rises and 30 m drops common, except that high-pressure boilers allow rises to e.g. 100 m; services for ultra-high buildings in completely independent vertical stages.
- If boilers at roof level then pressure-reducing valves possible in drops to calorifiers and from calorifiers to taps.

DRAINAGE

Anyone's relationship with drains will improve with closer aquaintance, the thing is to learn to love them! Despite any prejudice to the contrary, they are not boring and only occasionally smell. And, though the distinction between 'two-pipe', 'one-pipe' and 'single-stack' systems, and the operation of vents, traps and manholes, may sometimes be imperfectly understood, the logic underlying drain design is satisfyingly clear and simple.

Essentially, the discharge progress is through lateral runs or *branches* from each sanitary appliance, a main downward collecting *stack*, the underground lateral run or *drain* and the public collecting run – the *sewer*. Discharge from basins, baths, washing machines is *waste*, and that from WCs is *soil*. For the purposes of discussion, soil is euphemistically said to contain 'solids'.

Now, if avoiding blockage were the only worry, drainage design would be child's play, only requiring pipes of a suitable material, size and progressive fall to the sewer. To be sure, there would still need to be removable covers above ground, and rodding eyes and manholes below ground to allow access in case the system ever did become blocked, but these do not affect the layout, as such. The inherent snag is that, while there must be a clear route for discharge *outwards*, some means has to be found to prevent sewer and drain smells passing back *inwards* to the building. This prospect was once thought a most awful menace to health (almost obsessively so in Victorian England – Prince Albert died of typhoid) and, while we do not take quite the same view today, smells are unsavoury nonetheless. Mechanical valves are out since they would become blocked. The solution, gratifying in its simplicity, is the trap.

The trap
The trap (4.8) works by giving the pipe a dip in its run,

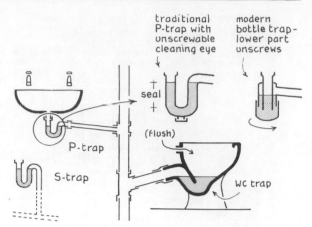

4.8 Sealing traps (branch waste and stack connection outlines assume plastic system)

so that a water seal is always left behind after the discharge has passed. All sanitary appliances are trapped, except where a bank of appliances, e.g. urinals, discharge into a common branch which would then be given a common trap. Some authorities require drains to be trapped at their entry to the sewer. But there remains a snag: without special precautions, traps can get unsealed as follows:

Suction and pressure threats to traps

Suction owing to self-siphonage. Waste traps from appliances are normally assumed to flow 'full bore' and there is the risk that, at the end of the flow, the body of discharge will have enough impetus to pull the seal after it by siphonic action. Baths are less at risk than appliances like basins because their shape results in a slower tailing off at the end of the flow, tending to leave the trap sealed. And WCs are less at risk, since their branches do not flow full – their 100 mm outgo (large-sized to prevent blockage) is very much larger than the pipe supplying the flush.

Induced suction and compression. Also, flow passing down the system from one appliance can create enough partial vacuum in the air behind it to draw traps of other appliances. The common case is where flow from a higher floor passes down the stack past adjoining branch inlets on a lower floor, but it can also occur where a discharging branch connects with another branch. And, in the stack, the flow can create enough air compression in front of it to blow adjacent traps. Looking at the hydraulics more closely, the vertical stack is also sized at 100 mm to prevent blockage and will not suffer full-bore flow even when appliances discharge simultaneously, but research has shown that, at around 30 per cent capacity, there comes the risk that the flow will bridge the stack and form a *plug* (4.9). In the lateral drain, gravity holding the flow on the pipe

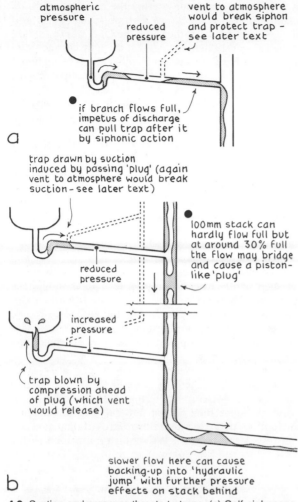

atmospheric
pressure

reduced
pressure

vent to atmosphere
would break siphon
and protect trap –
see later text

if branch flows full,
impetus of discharge
can pull trap after it
by siphonic action

a

trap drawn by suction
induced by passing 'plug' (again
vent to atmosphere would break
suction – see later text)

reduced
pressure

100mm stack can
hardly flow full but
at around 30% full
the flow may bridge
and cause a piston-
like 'plug'

increased
pressure

trap blown by
compression ahead
of plug (which vent
would release)

slower flow here can cause
backing-up into 'hydraulic
jump' with further pressure
effects on stack behind

b

4.9 Suction and pressure threats to traps. (a) Self-siphonage by discharge in the branch itself (b) Induced suction and compression in branches when main stack is bridged by 'plug' or 'jump'

floor must hinder bridging but, on the other hand, the flow is slower and may back-up in the section immediately after the stack into what is known as a *hydraulic jump*. Plugs and jumps are like pistons, producing considerable suction and compression in the stack and, sometimes, in branches too.

Protecting traps

As we shall see, it is possible to size and shape drainage pipework carefully so that threatening suctions and pressures do not develop but, first, some of the more traditional methods of protecting traps:

By their design. To some extent, unsealing can be prevented by the design of the traps themselves. Making them deeper helps. There are also 'anti-siphon' traps. Some of these incorporate either an air bypass

pipe or a relief valve capable of venting either way when negative or positive pressure threatens. Others have a wider chamber at the outlet containing enough water in reserve to fall back and reseal the trap after it has discharged or been otherwise threatened. In fact this tends to be the approach in many countries, the European continent for example, the argument being that some suction is helpful in scouring and ventilating the pipe. So philosophies vary. Anti-siphon traps are useful where no other remedies are practicable but, depending on type, tend to be more expensive and noisy, or have their efficiency progressively reduced by blockage.

By venting. A classic and still widely used precaution is to connect a vent pipe from the point immediately behind each trap to the outside air, i.e. as shown dotted in the diagram. This prevents direct siphonage, as would a hole in a siphoning hose, and relieves any induced suctions and compressions – with the pressure either side of the trap kept stable, i.e. at atmospheric, the seal *must* be safe. The branch vents can connect to a main vent pipe running to the roof (smells up there are not really 'smells', if no one is there to smell them) or, in small installations, they can connect back into stacks above the highest point of inlet – stacks will, in any case, be carried up open to roof level, to improve their own venting.

Summary

- Simple discharge progression is: appliance – branch – stack – drain – sewer. But traps are needed to prevent back-smells.
- Trap seals are threatened by self-siphonage in branches and, also, suction/compression induced by flow in stack, especially where stack bridged by plugs or jumps.
- Traps are protected to some extent by their design (including seal-depth increase and anti-siphon air bypasses/relief valves).
- Fail-safe, traditional protection is by vent pipes.

. . . and so the domestic systems

Two-pipe and vents

The *two-pipe* or *dual-pipe* system just merits a mention, although outdated. It is so called because it has two separate stacks, one for WC soil branches and one for waste – what with branches, vents and all, this duplication could add up to quite a clutter of pipes threading down the backs of old buildings. The idea must have been partly generated by the drain-smell phobia and the consequent desire to get the soil quickly and separately to the sewer. Today, it would only be used where WCs and waste appliances were well apart on plan, i.e. where a combined stack would involve long branch lengths. A distinctive feature is that the waste

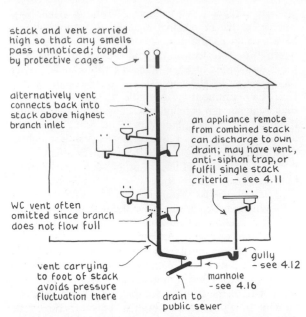

4.10 'Combined' stack, i.e. combining soil and waste, with vertical vent pipe connecting to branches

stack, being soil-smell free, can be open at points, characteristically where first-floor basin and bath branches and possibly rainwater downpipes, collect into an open, bucket-like 'hopper' at the stack head on the wall outside. Hoppers reduce plumbing connections but are always prone to blockage by leaves or freezing.

Combined or one-pipe (and vent)

The *combined* or *one-pipe* system (4.10) largely superseded the two-pipe, it being altogether neater and cheaper to combine the soil and waste. The open tops of the combined stack and the branch venting pipe are at roof level, where any smells escaping will pass unnoticed, but the system is otherwise closed, with all inlet branches securely trapped.

Single-stack (no vent at all)

The most recent innovation, the single-stack system, is 'single' in that there is *no vent stack at all* and, granted, this will at first seem a complete contradiction of all that has been said. Traditional practice, with no accurate knowledge of the way discharge behaved in pipes, quite understandably evolved a set of standards by trial and error, rules on pipe sizes, falls and, of course, venting. This guaranteed good performance in all circumstances and, if the provisions were more than were actually required in some circumstances, well, that was a fact of life. Work, principally at the British Research Station and the US National Bureau of Standards, has allowed a very much better understanding of drainage hydraulics. An unvented system was being monitored in the USA in 1934 but wider adoption was well post-

war and, curiously, considering the early research, its main adoption has been in Europe (Scandinavia and the UK principally) rather than the USA where venting remains widespread. The view there may change – the sight at the NBS of water and leather stools being timed as they hurtled down transparent plastic pipes might, on the face of it, seem amusing, but improved guidelines in drainage, or anything else applied throughout a national building industry, are a massive saving potential.

4.11 outlines the factors ensuring that unvented traps are safe. They are essentially these. The stack is kept straight below the highest branch inlet and the bend at the foot is kept gradual, to minimise disruption and the chance of plugs or jumps. Branches connect separately into the stack, again, to reduce disruption and to prevent one branch from inducing suction or compression in another. They are kept short enough, and with shallow enough fall, to prevent their flows from developing enough impetus either to pull seals by self-siphonage, or significantly to assist in plug formation should the stack already happen to be under flow from an appliance above. Actually, the length and fall values can be traded, shorter runs allowing steeper falls and vice versa, and generous pipe diameters are a mitigation – the diagram shows typical limits. Note that basin branches are more critical than baths and sinks, since their narrower diameter is more likely to flow full and the flow tail-off is more abrupt. WC branches are at little risk since they do not flow full but, to minimise disruption, they are swept downwards at their junction with the stack. Also, a WC branch must not join the

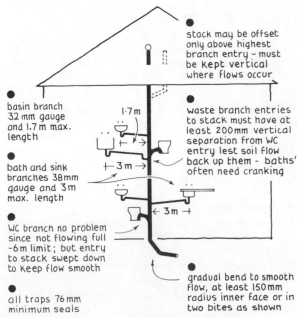

4.11 'Single stack' (i.e. unvented) system is possible provided certain criteria are met – UK values shown here

stack at the same level as waste branches, otherwise, soil flow might back up them, blocking them and blowing their traps.

Linking-in rainwater to the drain

Rainwater collection into roof gutters and downpipes was illustrated under Chapter 2 *Enclosure*. Picking up the story, the system for disposing of the flow from these, and from other surface areas like patios, to the drains can be *combined* or *separate*, the terms, in this case, referring to whether the flow combines with the ordinary sanitary drain and so to the sewer, or is to a separate rainwater-only drain leading to its own surface water sewer, river or other point of disposal. If it combines with ordinary drainage, it must do so in such a way that its flow (which can be torrential in a storm) does not interfere with the sanitary flows and their traps. Connection is, therefore, below the lowest appliance inlet and the usual thing is to link it and the sanitary drain into a common manhole from which the combined drain then carries to the sewer – 4.12 shows a 'back-inlet' gully, collecting both rainwater downpipe and surface water flows. Logically, any open gully must be trapped against back smells if it leads to a combined drain. But no vent is needed, since the surface-water flow will tail off slowly after a shower and leave the seal safe. Incidentally, waste from a remote appliance, such as a kitchen sink, can connect in, as shown dotted, either directly to the pipe or at the grating, in the latter case carrying through the grating to bypass any surface blockage by ice or leaves.

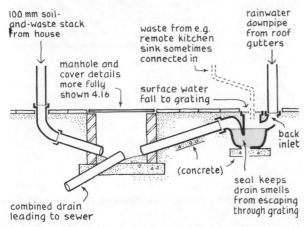

100 mm soil-and-waste stack from house

waste from e.g. remote kitchen sink sometimes connected in

rainwater downpipe from roof gutters

manhole and cover details more fully shown 4.16

surface water fall to grating

back inlet

combined drain leading to sewer

(concrete)

seal keeps drain smells from escaping through grating

4.12 'Combined' sanitary and rainwater drainage; back inlet gully at right collects rainwater downpipe and surface-water flows

Summary

- Two-pipe (and vent) now virtually obsolete.
- One-pipe (and vent) has largely superseded two-pipe and is common except for recent increasing use of:

- Single-stack (no vent). This demands tight appliance grouping, so that required branch and stack configurations are met.
- Rainwater drainage. Either 'combined' with sanitary drainage (hence, surface rain gullies etc., trapped) or 'separate' e.g. to own surface-water sewer.

THE SYSTEMS APPLIED TO LARGER BUILDINGS

Sanitary drainage principles at increased scale are conveniently similar. It might be thought that plugs and extreme pressure fluctuations would arise, owing to large flows from numerous appliances and, in tall buildings, owing to high flow velocities down long stacks. In fact, it does not work like that. For one thing, while the daily discharge from a large building is obviously proportionately greater than that from a small one, the volume of any given discharge as it occurs will be more or less the same. Even with 100 WCs, or many more, the chance of two flushes coinciding at the same time and place in the stack are remote. There will be more frequent flows but they will tend to be the same size as or, by a rare coincidence of flows, only slightly greater than, the domestic flow. Increased scale eventually finds the 100 mm stack giving way to 150 mm, but rarely more. For another thing, the idea that excessive velocities might develop in a long stack is now known to be false. Stacks were once offset at intervals down a tall building, a loony idea actually, increasing the chance of blockage and turbulence but, presumably, intended to prevent 'solids' rocketing down and smashing out of the bottom bend of the pipe. In practice, the friction between the discharge and pipe's air and walls results in a gentle terminal velocity after only two or three storeys – solids descend elegantly and stacks are now kept straight!

The drain stacks carry down inside a large building in ducts – ducts probably shared with other services. Now, a prime planning aim is to have as few stacks as possible, while yet avoiding overlong branches on each floor and the possible need for venting, in consequence. It is a question of finding the right balance but, clearly, both needs will be served if sanitary spaces can be closely grouped on each floor and vertically zoned through the floors. The terms 'two-pipe', 'one-pipe' and 'single-stack' are less useful now – there may be many stacks and whether they are waste, soil, or waste and soil combined, and whether vented or unvented will, again, depend on the appliances and their grouping. Rationalising the sanitation planning to simplify the service runs is mainly a matter for the architect, obtaining optimum systems within the resulting plan is for the services engineer.

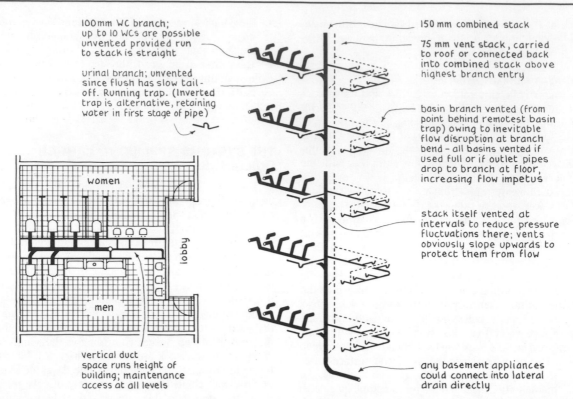

100mm WC branch; up to 10 WCs are possible unvented provided run to stack is straight

urinal branch; unvented since flush has slow tail-off. Running trap. (Inverted trap is alternative, retaining water in first stage of pipe)

women

lobby

men

vertical duct space runs height of building; maintenance access at all levels

150 mm combined stack

75 mm vent stack, carried to roof or connected back into combined stack above highest branch entry

basin branch vented (from point behind remotest basin trap) owing to inevitable flow disruption at branch bend – all basins vented if used full or if outlet pipes drop to branch at floor, increasing flow impetus

stack itself vented at intervals to reduce pressure fluctuations there; vents obviously slope upwards to protect them from flow

any basement appliances could connect into lateral drain directly

4.13 Multi-storey drainage – partially vented system in e.g. an office building

Office blocks. The office (4.13) usually collects most of its sanitation in the service core. The main provision is the male and female lavatory accommodation and this will generally need venting, although not necessarily for the branches themselves. WCs are safe, even in banks up to 10 or so, provided the common branch runs straight to the stack. If not, then the flow disruption at the bend, and consequent risk of induced siphonage, generally asks for a vent back from the furthest appliance to the vent stack. Urinal branches are usually unvented – there is a safe tail-off after the flush – and basins can be unvented where hand washing is briefly under a running tap, thereby avoiding filled basins and full-bore emptying. The stack itself, however, is shown vented regularly into the parallel vent stack to prevent occasionally coincident or closely consecutive flows from developing pressure fluctuations sufficient to threaten branches.

At the very least, such duct spaces will be additionally carrying the hot- and cold-water pipework and, quite possibly, the environmental and other services too. All these will require adequate access for maintenance. This is generally through removable panels, allowing either adequate reaching coverage into the duct, or better, full walking access to support gangways within the duct.

Hotel blocks. The hotel block is generous in its sanitary provisions but fortunately can have a very regular and vertically repetitive plan (4.14). Pairs of bathrooms flank a common, vertical duct space and branches can be arranged to fall within single-stack limits quite easily – single-stack plumbing has been found workable in quite tall buildings, up to 25 storeys or so; beyond that, the longer travel and greater chance of multiple flows might lead to excessive air turbulence.

Apartment blocks. Here, again, there is the chance of a vertically repetitive plan, ideally. with adjacent dwellings handed back to back so that their kitchens and bathrooms flank common duct spaces. Achieving single-stack limits may be hard and venting may be needed, especially for the vulnerable basin trap. In the proprietary European development (4.15), appliances at each level are connected into a special stack chamber whose design prevents soil backing up the branch entries, smooths the flow and allows venting into an integral cavity ring rather than to a separate vent stack. It is a hybrid between the single-stack and fully vented system.

Buildings with dispersed appliances. Some building functions make sanitary zoning virtually impossible. A laboratory block might ask for sinks dispersed on each floor. Hospital sanitation is extensive and varied – general-utility sluice rooms, patient and staff facilities, treatment areas and so on. The choice between having

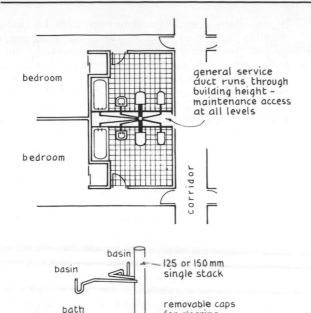

bedroom

bedroom

corridor

general service
duct runs through
building height –
maintenance access
at all levels

basin

basin

— 125 or 150 mm
single stack

bath

removable caps
for clearing

WC

WC

bidet

bidet

branches cranked to
achieve 200mm vertical
separation from WC entry

4.14 Single-stack (i.e. unvented) drainage at multi-storey, in common duct space between hotel bathrooms

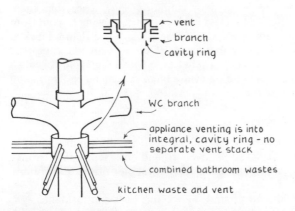

vent

branch

cavity ring

WC branch

appliance venting is into
integral, cavity ring – no
separate vent stack

combined bathroom wastes

kitchen waste and vent

4.15 Proprietary fitting – a hybrid between single-stack and fully vented systems, useful in e.g. apartments where single stack limits exceeded

few stacks but longer branches, or many stacks but shorter branches, then arises sharply. Longer branches are, arguably, more prone to blockage and, falling outside the single-stack limits, will need venting and vent stacks. The alternative of having more down-

stacks will certainly come into its own where the dispersed appliances can, at least, be arranged to occur regularly on each successive floor, allowing the stacks to be straight and short unvented branches to be achieved – but that is not always possible. The solution will, like as not, be a partially vented compromise, i.e. between branches and stacks, and venting only those branches which fall outside the single-stack limits.

Summary

- Hydraulic principles are mostly as for domestic scale, since individual flows have similar magnitude.
- Rationalised sanitation planning is most important economic factor.
- Design compromise is between limiting the number of stacks and avoiding long, possibly vented, branches.
- Single stacks successfully used to 25 storeys or so, otherwise, systems partially or wholly vented; also hybrid types.

OTHER DETAILS – FOLLOWING THE SYSTEM TO THE SEWER

In the further points on pipe construction, and the routing and access below ground as the system leads to the sewer, the provisions for small and large buildings are similar enough to be discussed together.

Pipe size and fall

Obviously, pipe sizes vary with the type and, to a lesser extent, number of appliances served. But they are of the following order. Waste pipes are usually 32 mm from basins and 38 mm from sinks and baths, either connecting direct to the stack or to a 51 mm branch waste. Branch vents are upwards of 32 mm and vent stacks are 75 mm with a 100 mm drain stack, and 100 mm with a 150 mm drain stack. As explained, avoiding blockage dictates the 100 mm soil stack in domestic work, with buildings having to be quite large before requiring 150 mm stacks. In fact, *over*sizing can cause problems in horizontal runs above and below ground. Deposit can build up on the upper internal surface of a pipe that is too infrequently flushed and the tendency to shallow sluggish flow can strand solids. More specifically, the minimum flow speed on lateral runs to avoid stranding, the 'self-cleaning velocity', is normally taken as 0.7 m/sec. Obviously, fall as well as flow determines whether or not this will be achieved and for a typical WC flush travelling in a 100 mm drain pipe this should be 1 : 200 – *theoretically*. In practice, slightly sharper falls are needed to allow for the chance of small inaccuracies in laying or of ground movement later. A suggested rule-of-thumb here is a 1 : 40 fall for a 100 mm pipe, and 1 : 60 for a 150 mm pipe.

Materials

Ideally, pipework should be cheap, light and quick to assemble, tough, non-corrodible and smooth-surfaced. Plastics fulfil all these things and, remembering that the effluent is neither under pressure nor likely to be significantly hot, the move to plastic systems is hardly surprising. This is particularly true of branches. For stacks, the once common cast iron is a further alternative which is tough but requires corrosion protection, and, for rainwater, there is asbestos cement (unfortunately persisting despite its health risk in manufacture). Underground alternatives include cast iron, earthenware and pitch fibre. That there *are* alternatives is only partly due to the building industry's resistance to change. There may be local code requirements for cast iron, or other materials tougher than plastic, in vulnerable locations close to ground level, and under concrete ground slabs, where breaks would be costly to repair. But then again, there are contrary opinions. The initial idea that drains passing under footings and so on should be laid in rigid concrete protection is giving way to an argument for using more flexible piping and joints, e.g. plastic and synthetic rubber, respectively, which will accommodate movement rather than trying to resist it. Where reactionary local building codes unrealistically require cast iron in all cases, one can always use plastic and paint it black?

Access above and below ground

As 4.16 shows, there have to be access points in case the system ever becomes blocked, either allowing direct entry to the blockage or allowing it to be dislodged by extendable rodding. The siting of the points reflects two things, namely, that blockage is most likely at bends and junctions and that rods, although flexible, can only get round gentle bends. Drain-clearing rods are made up of whippy wood or plastic sections about a metre long, with male/female screw connections allowing them to be successively joined and fed into the system. The feeding is done with a continuous turning motion to keep the screw junctions tight – a disconnected length of rodding, stranded up an already blocked drain, is not much help. The first access then is at the appliance trap, older traps having a removable bung and modern traps being demountable or unscrewable. The head of a branch needs rodding access, and so does any offset in the stack, including the bottom bend.

Access below ground is more expensive but is absolutely critical. Gullies, of course, afford simple access at a drain head. *Rodding eyes* are also a relatively simple provision at intermediate points along a straight run or where the drain changes level. Modern US practice

is particularly keen on rodding eyes, even for drain direction changes or junctions. Most European practice, in these latter cases, prefers a *manhole*. This should either be at the point in question or at least close enough to afford decent rodding access to that point. Similarly, straight runs need intermediate eyes or manholes if they would otherwise outreach rodding range – the maximum interval allowed will, again, vary with local legislation, but 40 m is about the limit. Manhole depth varies with topography but will, in any case, tend to increase as the drain falls to the sewer. Really deep manholes (actually, if over 900 mm deep, they are called 'inspection chambers' in the UK) may have access shafts leading down to wider working chambers. And, apart from the brick manhole shown in (4.16), there are recent proprietary systems too numerous to illustrate. There are plastic drum manholes, where the drum is sawn off on site to give the required depth, and the base of which has preformed channels for use as inlets and outlet as required. There are precast concrete types with similarly preformed bases, and with the drum body built up to ground level using successive precast rings.

Connection to the public sewer.

The final connection to the sewer is generally preceded by a manhole, (4.16). Traditionally, this included an *interceptor trap*, the idea being to keep the main sewer gases, and indeed rats, from reaching the building drain. In fact, it is now realised that the ventilation tendency is into the sewer rather than from it, i.e. going with the flow, and there are less rats. Authorities will only ask for an interceptor where the street sewer happens to be in poor condition and, hence, blockage-prone – a blocked interceptor trap, owing to sand from builders' work, say, being a lesser evil than a blocked sewer under the street. If there is an interceptor, it should include a relief pipe with removable cover, so that any flooding over a blockage can be easily drained before the system is cleared.

Summary

- Direct or rodding access needed for unblocking.
- Achieved above ground by removable traps, branch heads, and covers at pipe offsets and bottom bend.
- Below ground, by rodding eyes and manholes.
- Usual to have final manhole before sewer, and vented to outside if there is an interceptor trap.

Sewer

Public sewerage ('sewerage' is the pipes, 'sewage' the content) is really outside the province of the architect, except in so far as its type affects the drain systems leading to it. As said, sewers may handle rainwater and sewage in combination or separately. Traditionally,

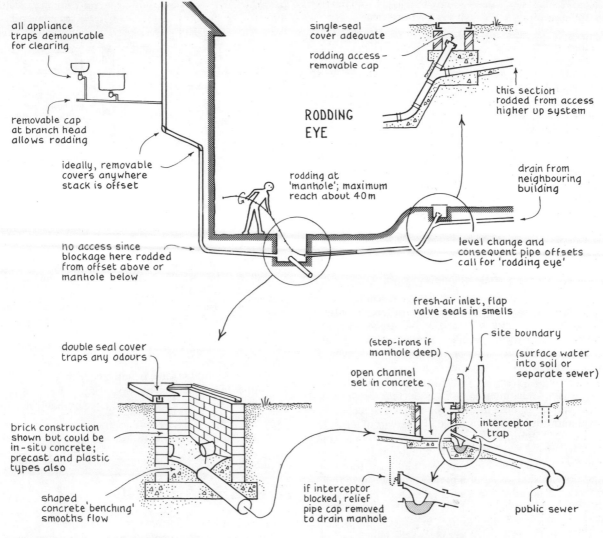

all appliance
traps demountable
for clearing

removable cap
at branch head
allows rodding

ideally, removable
covers anywhere
stack is offset

no access since
blockage here rodded
from offset above or
manhole below

single-seal
cover adequate

rodding access-
removable cap

this section
rodded from access
higher up system

RODDING
EYE

rodding at
'manhole'; maximum
reach about 40m

drain from
neighbouring
building

level change and
consequent pipe offsets
call for 'rodding eye'

double seal cover
traps any odours

brick construction
shown but could be
in-situ concrete;
precast and plastic
types also

shaped
concrete 'benching'
smooths flow

fresh-air inlet, flap
valve seals in smells

(step-irons if
manhole deep)

site boundary

(surface water
into soil or
separate sewer)

open channel
set in concrete

interceptor
trap

if interceptor
blocked, relief
pipe cap removed
to drain manhole

public sewer

MANHOLE

INTERCEPTOR TRAP

4.16 Access to drains for unblocking – typical provisions
above and below ground

they were usually combined, requiring only a single
pipe under the street and a single drain from each
building (or group of buildings, depending on the site
layout). But, of course, rainwater flow includes run-off
from streets and pavements, let alone from roofs and
impermeable areas around buildings and, in a storm,
it will be massively greater than the ordinary foul flow.
So the design and capacity of the traditional sewer had
to guard against fouled flooding, and the sewage-
treatment plants had to handle considerably greater
flows in wet weather, often to the point of having to
discharge untreated sewage direct to the sea or other
outfall. This is usually unacceptable today and it is
now considered more practicable to separate the

flows. The extra cost of duplicated pipework is offset
both by the savings at the plant and the fact that the
surface-water sewer can be directly discharged to rivers and other natural watercourses.

There are intermediate solutions, particularly in
existing areas. Any new buildings may be required to
have separate drains, albeit temporarily combined into
a single sewer, in anticipation of the addition of a surface-water sewer later. Also, to relieve the load on a
combined sewer, surface water from a site can be
passed to a *soakaway*. This is a sunken pit – often filled
with rocks or crushed brick – and large enough to hold
storm run-offs without flooding, i.e. until the water has
had a chance to seep away into the surrounding land.

Disposal where there is no sewer
To return to our starting point for water services – out-

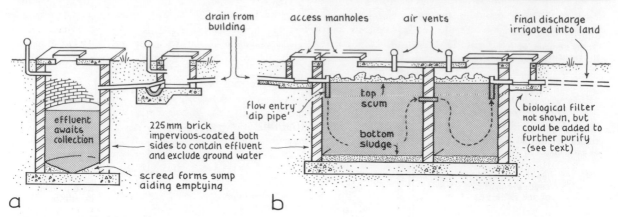

drain from building

access manholes

air vents

final discharge irrigated into land

flow entry 'dip pipe'

top scum

bottom sludge

biological filter not shown, but could be added to further purify - (see text)

effluent awaits collection

225mm brick impervious-coated both sides to contain effluent and exclude ground water

screed forms sump aiding emptying

a

b

4.17 Effluent disposal where there is no sewer. (a) A cesspool can be used to hold the sewage pending collection by disposal vehicle or (b) A septic tank can purify the sewage for safe disposal to the surrounding land

lying districts may have no public sewer at all. Surface water is little problem, since it can be discharged to a river or soakaway, but sewage is a different matter. Obviously, it cannot be directly discharged onto the land, nor into rivers, where it would be immediately insanitary and, in decomposing, would deoxygenate and, hence, stagnate the water.

The simplest thing is to store it in a *cesspool* (4.17a), a ventilated chamber in brick, concrete or plastics, large enough to hold the sewage, pending collection by local-authority pumping vehicle at intervals of a month or so. The alternative is to treat it. In the *septic tank* (4.17b), the sewage decomposes under the action of 'anaerobic' bacteria, breaking down into gas, top scum and bottom sludge. After a period of between 16 to 48 hours, the liquid is innocuous enough to be irrigated away into the ground, usually by underground pipes with open joints – field drains, in effect. The process happens in the absence of oxygen, so the tank is sealed, and the inlet and outlet pipes are so arranged as to avoid breaking the scum, as well as to prevent sewage short-circulating across the surface untreated. A local-authority vehicle will come every year or so to desludge the tank.

Rightly, there are strict controls nowadays on the purity of discharge into rivers and other watercourses, and a septic tank *alone* will seldom meet the standard. A *biological filter*, sometimes called a *percolating filter*, is in fact not really a filter at all. It is an additional chamber in which any remaining impurities are exposed to air and bacterial action, thereby allowing their removal by oxidation. The chamber contains crushed rock or other granular material, ventilated from above and underneath, thereby providing a large surface area for aerating the liquid as it seeps through. Together with the septic tank, it provides a complete sewage treatment plant at small scale.

Summary

Public sewers
- In combined sewerage, foul and rainwater flows are in common pipes – cheaper at building, but less efficient for whole system, especially owing to storm-flow problems.
- Separate systems becoming more common – rain-flows passed separately to watercourse or soak-away.

Where there is no sewer
- Cesspools hold sewage pending collection.
- Septic tanks act as mini-plant, treating sewage before discharge. Discharge to river probably needs biological filter as well.

REFUSE HANDLING

In recent years, more consumption, more packaging and less incineration, owing to the turn from open fires to central heating, all have helped to increase the refuse that buildings produce. A five-person house produces, perhaps, a cubic metre a week. And, if recycling of materials and energy recovery are to play their necessary part in refuse 'disposal', there will inevitably be collection complications in separating paper, glass, metal and the like, whether at the building, collection vehicle or city plant – this has only modest impact, as yet, but the expression 'refuse disposal' is already being replaced by 'refuse management'.

HOUSES AND SMALL PREMISES

The time-honoured dustbin

The dustbin is still the familiar thing for small buildings.

Its siting is plain common sense – adjacent to the kitchen, but discreetly outside, and sealed to keep smells in and insects away. Distances up to 50 m are quoted as a maximum for the nearest vehicle access but, remembering that the person collecting has four journeys, i.e. there and back for the full bin and there and back to return it, this would mean walking over a mile to service just ten houses. Existing urban terraces are often landed with having bins carried through the house to the front for collection off the street, an obviously tedious and unsightly arrangement. It is better if there can be service alleys up the side of buildings or, in the case of terraces, trolley or service vehicle access in the site behind.

There are various arrangements allowing an outside bin or disposable sack to be filled from inside the house, essentially consisting of an openable flap and through-wall chute or hopper. Disposable bags are becoming more common. They are cleaner and need only one journey to the house from the collection vehicle.

Waste grinders

Waste grinders or 'garberators' located under the kitchen-sink outlet can be used to cope with the putrefactive content of refuse. As you will know, these comprise an electric-powered shredder which reduces the refuse to the point where it can flush down the ordinary waste pipe to the sewer. Of course, there still has to be ordinary disposal for tins, paper and so forth, but it is then less bulky and cleaner. The system is useful where ordinary bin disposal is tedious, for example, from an apartment block or where refuse collection is infrequent – in remote places perhaps. A caution on the latter case, though: care has to be taken to ensure that cess-pools or septic tanks have enough spare capacity to handle the added sewage load a grinder imposes.

LARGER BUILDINGS

Apartment blocks

At worst, older apartment blocks may require bins to be carried laboriously down to large public collecting bins, 'paladins'. Better, but still not ideal, bins can be carried or trolleyed along external walkways or central corridors to service lifts.

Refuse chutes

The common thing now is the refuse chute, the essential features of which are shown (4.18a). Access can be off a common landing on each floor but, if apartments are handed back-to-back on plan, they can share common refuse chutes with access direct from kitchens or utility rooms – we have already seen how mirror plans allow common runs for water services, drains, flues and the like and, clearly, refuse disposal favours the arrangement as well.

The 'Garchey' system

As said, putrefactive waste can be shredded into the ordinary drains. In the *Garchey system*, developed in France, there is no shredder. Instead, the kitchen-waste bowl discharges to a waste pipe and common stack, large enough for solid matter, even small tins and bottles, to pass without blockage (4.18b). Initially, the solids and waste water are allowed to collect in a receptacle under the sink. Lifting a central plunger then discharges the contents cleanly away – the waste water is the flushing agent. A single vertical stack can serve up to about 20 storeys, further storeys needing a further stack. At the bottom of the stack, or stacks, the water is normally strained off from the solid refuse and passed to the public sewer. This is either done by local plant or by the specially equipped collecting vehicle – the latter case calling for a large holding receptacle. Actually, the idea behind the system was that the strained-off refuse could be burnt for useful heat recovery but there has been only limited success in this and the whole installation is costly, more so than refuse chutes and, in any case, still needs an ordinary disposal system for larger items. But it is very clean and convenient.

Other large building types, office, hospital

In the office block, the waste paper and other dry refuse from each floor will normally be bagged and manhandled or trolleyed to the service lift and, from there, descends to a central collection point in the low-level service area. Putrefactive articles from occasional kitchens are but a small proportion of the total refuse and it is usually more economic to route them via the service lift than to install a separate refuse chute.

Incidentally, sanitary tampons can fall under either the drainage or refuse headings. Domestically, discharge is through the WC but the greater number from large buildings risks blockage. Incinerators in female lavatories are one solution but these require a flue. Alternatively, there can be an automatic shredding receptacle and, thence, ordinary soil discharge, or they can follow the ordinary refuse routes in hygienically sealed bins.

Trolley and service lifts are the refuse arteries in most large buildings, hotel bedroom blocks, public buildings and the like. But some buildings need special solutions. For example, a hospital has a multiple disposal problem – ordinary refuse including catering waste, soiled linen, medical waste – and a common thing is to route them in colour-coded bags. Refuse and linen can be trolleyed to the service lift for transfer to central refuse collection and laundry, respectively. Lift access on each floor is often adjacent to the 'sluice

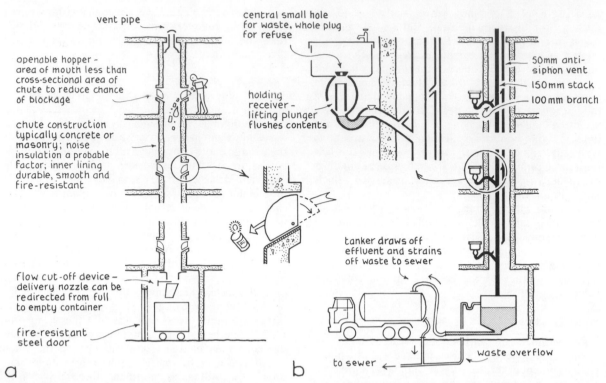

vent pipe

openable hopper -
area of mouth less than
cross-sectional area of
chute to reduce chance
of blockage

chute construction
typically concrete or
masonry; noise
insulation a probable
factor; inner lining
durable, smooth and
fire-resistant

flow cut-off device -
delivery nozzle can be
redirected from full
to empty container

fire-resistant
steel door

central small hole
for waste, whole plug
for refuse

holding
receiver -
lifting plunger
flushes contents

50mm anti-
siphon vent

150mm stack

100mm branch

tanker draws off
effluent and strains
off waste to sewer

to sewer ←

waste overflow

a

b

4.18 Refuse disposal from multi-storey apartment block. (a) The ordinary refuse chute (b) The less usual but effective 'Garchey' system, flushing refuse to a ground-level container

room', the general utility and cleaning space. To avoid cross-infection, medical waste from operating theatres and treatment areas is bagged and either ordinarily routed or, in some modern systems, is chuted down to intermediate collection points – i.e. rather than being laterally routed on the immediate floor and, thence, passed via lifts to the hospital incinerator. Body parts and other sensitive items are bagged and separately routed (carried by porter, usually) to the incinerator.

Ground-level handling arrangements in large buildings are affected by the refuse management policy in the region, for example, dictating whether waste paper is kept separate for later recycling at the city plant, and whether the refuse is compacted before collection, or incinerated. Bulk collection, such as from shopping precincts, calls for special attention to fire safety, preferably with collection points protected by automatic sprinklers. Industrial wastes may call for special arrangements with the collecting services owing to their bulk, recycling potential or hazardous nature.

Summary

- Dustbins and disposal bags – general application but particular reference to small buildings; main design factors are user-access at house and collection access from road.

- Waste-grinders – domestic buildings, especially multi-storey. They reduce disposal bulk remaining to bins and chutes by shredding putrefactive part of waste down ordinary drain.

- Refuse chutes – multi-storey buildings and particular reference to high-rise apartments.

- Continental 'Garchey system' – as above; sink refuse flushed by waste water through large-diameter pipes to lowest floor container.

- Trolley and service lift – general application to larger buildings, especially, offices and other public buildings.

ELECTRICAL SUPPLY

Electricity is an almost unbelievably convenient deliverer of power around buildings, power for heat, light, machinery and communications, all at the flick of a switch. But, for all that the electrical services are vital, they route their cables so easily that they have little effect on building form. This is, perhaps, one reason why electrical installations in buildings tend to be poorly understood by those members of the design team not directly concerned with then – notoriously so.

What are substations and transformers? What exactly is single and three-phase supply? What, in fact, is electricity?

THE NATURE OF ELECTRICITY

What is electricity? In fact, this apparently easy question defies an easy answer. We cannot see it, only observe its effects, but its explanation lies at the sub-atomic heart of matter. All atoms contain particles which convention describes as being positively or negatively charged and it is the magnetic or other energy effects of these particles, either at rest in matter or in fast motion through it, that we call 'electricity.'

As we know, matter attracts matter, and electrically charging matter increases its attraction power. Electricity and magnetism are interrelated, and a magnet is just a metal bar electrically treated to create permanent, opposite, charges at either pole.

Electrical current – conductors and insulators

A charged object has an electric energy potential or 'potential difference' over an uncharged one and, if the objects are connected by a conductor, an electric *current* of charged particles will flow and discharge the imbalance. A *conductor* is a material in which the charged particles can be dislocated from their atoms and flow freely, e.g. metals, water and, in consequence, animal tissue. Conversely, *insulators* hold the particles within the bounds of their parent atoms, e.g. most non-metals such as plastics, glass, dry wood and most gases.

If a conducting coil is rotated through the 'lines of force' in a magnetic field (4.19), current is generated and will flow in the circuit, as shown. This phenomenon is the basis of the power-station alternator and electric energy transmission. The current peaks when the coil is cutting through the lines of force most rapidly, and is zero when moving parallel to them. Also, by geometry, the current must reverse in the coil every half revolution or half cycle. This is *alternating current*, AC. The simple switch, also shown, would convert the output to *direct current*, DC. Incidentally, the output from the chemical reaction in batteries is a form of DC. To be sure, AC is not so much of a current as a kind of 'pull and push' – very broadly speaking. But, whether AC or DC, the concept of current 'flowing round' a circuit serves our purpose.

The water-flow analogy – volts, amps, ohms, watts

Current in a wire can be usefully likened to water flow in a pipe. The potential charge or electric 'pressure', *volts*, corresponds to the head of water; and the current, *amperes*, corresponds to the water flow. The

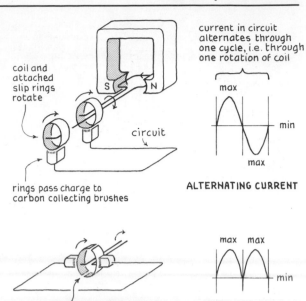

4.19 Generating alternating and direct electric current by rotating a conducting coil in a magnetic field

volts' potential is there whether or not amps are flowing, just as the water head is there whether or not the tap is turned on. The resistance of the circuit, *ohms*, corresponds to the pipe resistance, i.e. it is less when the wire or pipe is thicker. Electrically

$$\text{volts} = \text{amps} \times \text{resistance} \ (V = AR)$$

i.e. for given voltage, higher resistance means less current and vice versa. In a circuit of given resistance, more volts means more amps and vice versa.

The main concern, of course, is the available power. This is measured in watts and in 'single-phase' supply (to be described)

$$\text{watts} = \text{volts} \times \text{amps} \ (W = VA)$$

Increasing either volts or amps increases the power transmitted.

At the consumer end, electrical power can be converted back into mechanical power or heat. Mechanically, the electric motor is just the alternator principle reversed – if current is supplied to a coil in a magnetic field, it rotates and, in fact, dynamos and electrical motors are very similar in construction. Heating-wise, it can be imagined that electrical particle movement in a circuit is associated with increased atomic vibration and, hence, temperature.

Now, clearly, there has to be some way of ensuring that the energy release is at the appliance and not in the circuit supplying it. This is easily arranged. There is always a voltage drop across a circuit proportional to its resistance, just as there is a pressure-head drop in a water pipe proportional to the pipe's resistance.

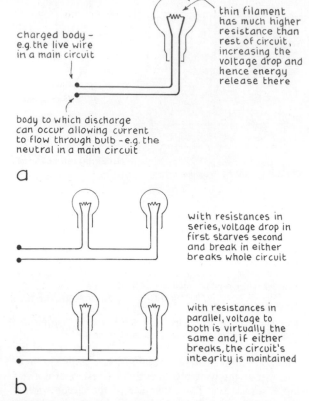

thin filament has much higher resistance than rest of circuit, increasing the voltage drop and hence energy release there

charged body – e.g. the live wire in a main circuit

body to which discharge can occur allowing current to flow through bulb – e.g. the neutral in a main circuit

a

with resistances in series, voltage drop in first starves second and break in either breaks whole circuit

with resistances in parallel, voltage to both is virtually the same and, if either breaks, the circuit's integrity is maintained

b

4.20 (a) Arranging for the useful energy release to occur at the appliance rather than in the circuit itself (b) The advantage of connecting resistances, i.e. appliances, in parallel rather than series

Giving light-bulb filaments or heater elements a much higher resistance than the rest of the circuit serves to concentrate the voltage drop at these places (4.20a). The current is constant but volts have been lost. Watts = volts × amps, so watts have been lost – watts that can only have gone to heat up the appliance. Local 'constrictions' give local energy releases where they are wanted.

Appliances in series and in parallel

It happens that the bulb shown is connected *in series* with the circuit. This is more obvious if we show two bulbs (4.20b). The problem with having appliances in series is that one will tend to starve the voltage available to the next, i.e. the appliance 'constriction' is enormously adding to the voltage drop in the circuit. Appliances in series cannot be separately switched and one failing breaks the whole circuit – the old Christmas-tree light exasperation. The easy solution is to connect appliances *in parallel*, as shown, so that the circuit is continuous and virtually independent of the appliances off it. There will always be a slight voltage drop in the

circuit itself but, practically speaking, the same voltage is now available at all points. Almost analogously, one is reminded of the advantage of the two-pipe over the single-pipe central-heating circuit.

DISTRIBUTION – POWER STATION, TO 'GRID', TO BUILDING

Source and grid

Most power-station alternators are driven by steam, generated by the heat of fossil-fuel combustion or nuclear reaction. In hydroelectric schemes, they are turned by water pressure. The power is pooled into the *grid*, a countrywide network of cables. Since the cables are very long-distance, some means has to be found of minimising the power losses within them. The essential thing is that the voltage drop and, hence, power lost to heat, be insignificant compared with the total voltage and, therefore, power transmitted. Very thick, low-resistance cables would achieve this but would be prohibitively expensive. The alternative, since the voltage drop is virtually independent of the total voltage applied, is to have very high voltages, because the drop is then an insignificant percentage of the whole.

Transformers

The snag, though, is that high voltages at the consumer end would be dangerous. In an accidental electric shock, it is the current flowing through you to the ground that harms you, and this is proportional not only to the resistance (your body and what you are touching) but, also, to the voltage received.

The conflict is resolved by use of transformers (4.21). The transformer comprises primary input and secondary output windings around a common metal loop, and the respective number of windings determines whether the voltage is stepped down or up. In other words, volts and amps can be swapped either way, although the power flowing (V × A) is virtually unchanged. In the UK, for example, the power station output is stepped

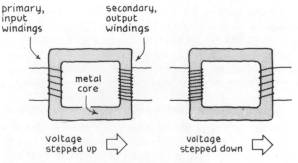

primary, input windings

secondary, output windings

metal core

voltage stepped up

voltage stepped down

4.21 'Step-up' and 'step-down' transformers

up to 132 000 volts, or 132 kilovolts (KV), into the grid and, indeed, up to 400 and 750 KV in the supergrid. Neighbourhood substations step this down again for delivery along streets and to buildings. In buildings, the final voltage must be high enough for efficient transmission without resort to uneconomically heavy cables – and yet, safe. The US compromises at 110 volts, parts of Europe and the UK at 240 volts.

Supply to building

The UK supply up to buildings, then, is AC (alternating current at 50 cycles per second), '3-phase', 415/240 volts. It is AC rather than DC because it is better generated and transmitted that way, because it better suits the operation of transformers and electric motors, and because it then suits 3-phase supply. 3-phase supply (4.22a) has three live *phase* wires, or *live* wires, and a common neutral return earthed back at the substation. Each phase carries AC alternating one third of a cycle out of step with its neighbours. The available voltage is 415 between any two of the phase wires and 240 between any phase wire and the neutral.

Summary

- Electricity and magnetism interrelated; a rotating conductor coil in a magnetic field generates current.
- Potential difference measured in volts, current in amps, conductor resistance in ohms, power in watts. V = AR and W = VA.
- Appliance resistance 'constricts' circuit, locally increasing voltage drop and, hence, energy release.
- Appliances normally in parallel, not in series.
- Need for long-distance transmission efficiency, and yet consumer safety, reconciled by use of transformers.
- Voltage stepped up into grid and down to buildings.
- UK supply at street is AC, 50 CPS or Hertz, 3-phase, 415/240 volts.

DOMESTIC BUILDINGS

It is unusual to carry 3-phase into small buildings. Typically, it runs along the street (or up an apartment block) with each of the phases serving every third house in rotation (4.22a). This balances the load of the phases and usefully minimises the current in the common neutral. The supply to the consumer unit in each house is then single-phase, 240 volts, using a live 'line' and neutral 'return'. The purposes of the fuses and earth connection at the consumer unit (4.22b) will be explained in a moment; first, the supply runs in the house itself.

The modern practice is to run simple *radial* circuits to individual, heavily-loaded appliances, like cooker and immersion heater, and *ring-circuits* for power. The power routes either way round a ring, effectively doubling the load the circuit can take, avoiding the need for unworkably heavy wiring. Lighting circuits, more lightly loaded, are usually in simple radial form. This separation facilitates routing for isolation and repair, *and allows each circuit's wiring to be sized according to its maximum likely load*. The design loadings, shown in amps, are common in the UK.

Ring-circuit

The power ring (4.22c) starts and finishes at the consumer board. It is a multi-core cable carrying live, neutral and earth, looping in and out of the sockets in each room. Appliances plugged into the sockets are then automatically in parallel. Spurs off the ring can be used to serve occasional outlying sockets. Now, for the given size of cable there are limits on the work a single ring can be asked to do. The loading, i.e. maximum number of appliances, is one constraint, and the allowable voltage drop (for a given resistance, this is proportional to length) is another. Actually, loading and voltage drop can be taken as interrelated. There are also limits on spurs, partly because they have only one cable to carry the load and, also, because too many spurs would defeat the economy of the simple ring layout. The accepted UK guide, and it is only a guide, is that a ring should serve no more than 100 m^2 of floor area. There can be any number of sockets, since that will not increase the number of appliances in use at any one time. But there should be no more spurs than sockets, and no more than two sockets per spur. A large house might have a power ring to the high-load appliances in the kitchen and then separate power rings to each floor, or to different sides of the house.

Fusing

If a circuit is overloaded, it will heat up and eventually fail – it can cause a fire. A fuse is a deliberately inserted weak link designed to fail and break the circuit well before the safe amperage is exceeded. The term 'fuse' used to be literal, there being a metal wire link with resistance and melting point such that it melted, fused, at the specified amperage. Its operation was inexact, though, and a nuisance to rewire, and the modern fuse is as often a clip-in cartridge, which encases the fuse wires in sand and is very exact, or a cut-out switch designed to throw by magnetic or thermal input. Fuses are graded from cable entry to wall socket, according to the stage of the system they are protecting. Note that the fuse is placed to break the circuit on the live side not the neutral. Were it on the neutral side only, a faulty appliance could fuse the circuit, making it apparently safe but, in reality; live. Some unsuspecting person dismantling an appliance could complete the

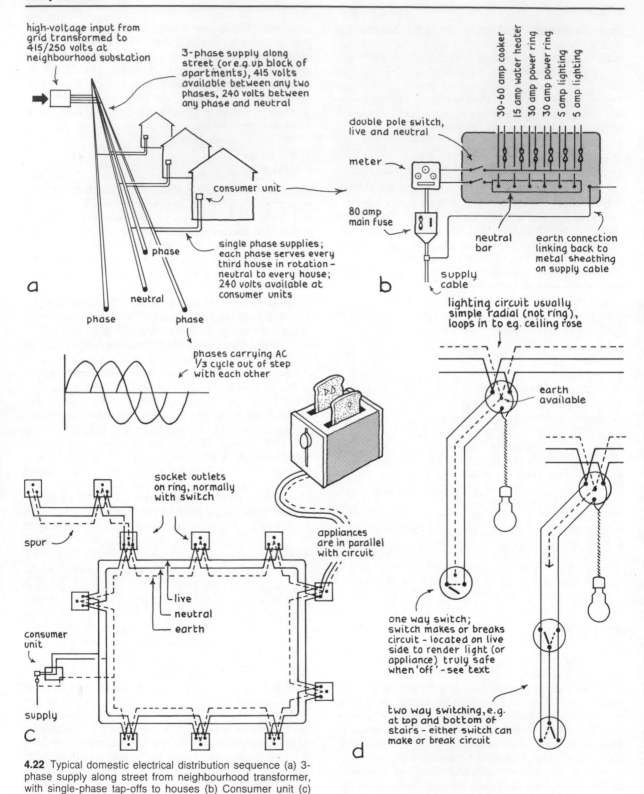

high-voltage input from grid transformed to 415/250 volts at neighbourhood substation

3-phase supply along street (or e.g. up block of apartments), 415 volts available between any two phases, 240 volts between any phase and neutral

consumer unit

phase

neutral

phase

phase

a

single phase supplies; each phase serves every third house in rotation – neutral to every house; 240 volts available at consumer units

phases carrying AC ⅓ cycle out of step with each other

30-60 amp cooker
15 amp water heater
30 amp power ring
30 amp power ring
5 amp lighting
5 amp lighting

double pole switch, live and neutral

meter

80 amp main fuse

neutral bar

earth connection linking back to metal sheathing on supply cable

supply cable

b

socket outlets on ring, normally with switch

appliances are in parallel with circuit

spur

live
neutral
earth

consumer unit

supply

c

lighting circuit usually simple radial (not ring), loops in to e.g. ceiling rose

earth available

one way switch; switch makes or breaks circuit – located on live side to render light (or appliance) truly safe when 'off' – see text

two way switching, e.g. at top and bottom of stairs – either switch can make or break circuit

d

4.22 Typical domestic electrical distribution sequence (a) 3-phase supply along street from neighbourhood transformer, with single-phase tap-offs to houses (b) Consumer unit (c) Power ring-circuit (d) Remote switching loops off lighting circuit

circuit by themselves becoming the alternative neutral return.

Earthing

The fuse protects the circuit but the earth primarily protects the user. Suppose a loose wire makes the outer casing of an appliance live. Provided the casing has its own *low-resistance* connection back to earth, the circuit will immediately complete, short-circuiting and blowing the fuse on the live side, and making the system safe. *The earth ensures that the system fuses.* Another development is the earth-leakage circuit-breaker, where a small flow of current to earth, insufficient to blow the fuse, will, none the less, trip a switch on the live side.

Earths used to be connected to the building's metal plumbing but this is unreliable nowadays, with the chance of insulating plastic pipe insertions breaking the connection. Instead, the earth is taken back to the consumer unit and, thence, to the metal sheath of the supply cable.

Switching

Like fuses, switches are always fitted on the live side to prevent the appliance they control from being apparently 'off' and yet potentially live. Switches are just make-and-break metal contacts. The power socket normally incorporates the switch directly, whereas the lighting fixture has it remote on the wall, with the live line diverted to include it. 4.22d shows single and two-way light switching, the latter allowing control from two places, say, from the top and bottom of a staircase. For safety, bathroom switches are either mounted outside the room or are operated by an insulating length of pull-cord. Remember, the severity of a shock is proportional to current and, hence, inversely proportional to resistance – a 240-volt shock received while washing or while standing barefoot on a wet floor would probably electrocute, that is to say, kill. Sockets are never mounted within reach of plumbed appliances in bathrooms, except the safety type for electric razors, which contain an isolating transformer.

Cables

Multi-core domestic cables include plastic-insulated types, i.e. metal conductors plastic-insulated from each other and plastic outer-sheathed; and mineral-insulated types, i.e. metal conductors outer-sheathed with metal and with mineral insulating compound packed in between. Where cables run under floor screed or behind wall plaster, they are normally housed in metal or plastic conduit pipe. This allows them to be easily drawn through during installation and at any later

rewiring, and gives them mechanical protection throughout the building's life. Conventionally, these hidden runs are made horizontal or vertical for easy location later and to reduce further the chance of accidental damage. Underground cables, like the 3-phase supply in the street, are 'armoured' by steel-wire sleeving in their outer composition and are run in conduit.

Summary

- Domestic supply is AC, single-phase, 240 volts, i.e. each house tapped-off the 3-phase street supply in rotation.
- Consumer board serves individual circuits for e.g. cooker, ring-circuits for general power, and lighting.
- 100 m² floor area is rule-of-thumb limit for ring-circuit, to limit total load and voltage drop. Maximum of two sockets per spur.
- Fuse inserted to protect each stage in distribution hierarchy – deliberate weak link. Always on live side of circuit, sometimes on neutral also.
- Earth ensures circuit fuses e.g. if appliance casing becomes live.
- Switches at least on live side, or on neutral also, i.e. 'double-pole'.

LARGER BUILDINGS

Obviously, increased scale means that the electrical loads are heavier and further distributed. An office building might impose 150 watts/m² averaged over its many floors, and up to double that if it is air-conditioned – such is the difference air-conditioning makes. As we might expect, the economical solution is, again, analogous to piped water systems, namely, having a reducing hierarchy of cable sizes economically reflecting the loads to be carried in each stage. Logically, there will also be a corresponding, reducing hierarchy of fuse sizes, so that a local appliance fault only takes out its own local circuit rather than a large part of the building, enabling the fault to be easily located and remedied.

For example, an office block

4.23 shows a system for a medium-rise office block. The total load may well be of the same order as for a street of houses, so it may pay to maintain the 11 000-volt supply (again quoting UK ratings) right to the building's own transformer, where it is stepped down to the required 415/240. (In fact, groups of offices, industrial buildings and so on are often served by an

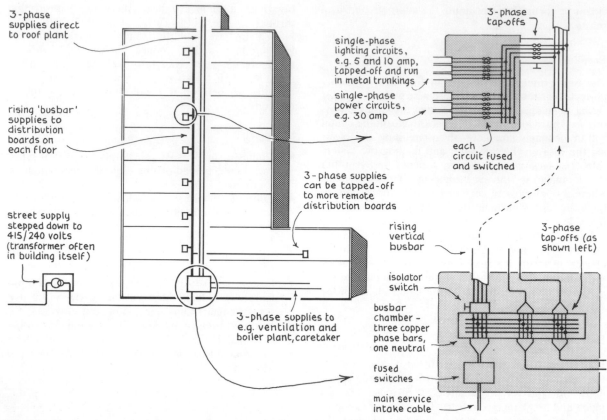

3-phase supplies direct to roof plant

rising 'busbar' supplies to distribution boards on each floor

street supply stepped down to 415/240 volts (transformer often in building itself)

single-phase lighting circuits, e.g. 5 and 10 amp, tapped-off and run in metal trunkings

single-phase power circuits, e.g. 30 amp

3-phase tap-offs

each circuit fused and switched

3-phase supplies can be tapped-off to more remote distribution boards

rising vertical busbar

3-phase tap-offs (as shown left)

isolator switch

busbar chamber – three copper phase bars, one neutral

fused switches

main service intake cable

3-phase supplies to e.g. ventilation and boiler plant, caretaker

4.23 Electrical distribution sequence in multi-storey office building

underground 11 000-volt ring-main direct from the sub-station). The entry supply passes through meters and master cut-out switch to the *busbar chamber*. Busbars are uninsulated metal conductors from which the various 3-phase circuits can be 'tapped-off' e.g. for basement plant, roof-level plant, lifts, and general power and lighting. Each has its own initial fuse-switch for safety and to allow separate isolation for repair.

Our particular interest is in the general power and lighting and, for this, there is a 3-phase *rising busbar*. Each floor has its own distribution board tapping-off this busbar and distributing to the power and lighting sub-circuits. The illustration shows this simplified. In practice, many sub-circuits may be needed if the wiring is to be economically sized and yet excessive loads and potential drops avoided. The 100 m² rule serves for houses, but economic distribution in larger buildings calls for proper calculation by the electrical engineer. Large plans may have separate, remote boards distributing to outlying areas and/or separate busbars rising in different parts of the plan. Very tall buildings may have busbars exclusive to different batches of floors on height.

Other building types

Arrangements are similar for other building types. For example, an apartment block might have taps off the initial busbar for lift, plant and caretaker, and single-phase taps off the 3-phase rising busbar, metered and switched in to the consumer boards in each occupancy. A factory might have scattered boards for the ordinary sub-circuits and, typically, a high-level, horizontal busbar allowing 3-phase taps direct to machinery, as required.

Cables and trunkings

Large-scale distribution, as, for example, in the office above, has the insulated cables running through conduit pipe or through various kinds of metal and plastic trunking. Trunkings often come prefabricated with detachable sides for continuous access and with fused tap-off boxes attachable along their length, as required. The vertical supply is in its own builders' work duct. The duct is fire-stopped at intervals – in fact, electrical supply is generally kept clear of the other building services for both fire and electrical safety. Horizontal lighting circuits are easily run in conduits in the void over suspended ceilings. If there is no suspended ceiling, they might be taken in conduit through the screed on the floor above and dropped through the slab at intervals. Fluorescent lighting circuits are often continued at 3-phase, in which case the separate phase wires have to be kept a minimum safe distance apart to avoid the risk of any maintenance worker getting a 415-volt shock. For horizontal power supply, small office spaces can be adequately served by sockets off skirting-board trunking running round the room perimeter. Larger spaces needing additional sockets towards the plan centre need conduit or, more likely, trunking, through the floor screed – possibly, there may be a regular underfloor trunking grid.

Summary

- 11KV (UK) possibly maintained to large building's own transformer; thence, 415/240 volts to meters and main distribution board.
- 3-phase busbars in distribution board; 3-phase individual supplies tapped-off to plant; 3-phase rising busbars to local distribution boards for power and lighting.
- Single-phase power and light circuits tap-off boards; typically conduited under floors or over ceilings, with tap-offs to sockets and lights respectively; fluorescent light circuits sometimes continued at 3-phase, in which case cables well separated to avoid 415-volt shock danger.

TELECOMMUNICATIONS

Telecommunications include main telephones, internal communication telephones, paging and alarm systems, radio and TV rediffusion and other specialist equipment like teleprinters. For the most part, they concern large buildings and, while the complex nerve routes these systems demand again have but minor effect on building form, forethought is needed if the installations are to be made conveniently and discreetly.

An office block

Main telephones

Again, we can take the medium-size office as our central example. The main telephone system works at around 50 volts. As with all telecommunications wiring, safety in use and during maintenance requires insulation and separation from the mains electrical wiring. The main entry cable from the street leads to a *distribution frame*: this is usually located in the basement and needs a ventilated cupboard with a plan area of around 1 m². The frame interconnects with the controlling switchboard which, being manned, preferably has day lighting, nearby toilet facilities and so on.

The frame then radiates the various speech and bell wires to each telephone. As we might expect, it does so via a system of vertical cables in rising ducts, distribution cases on each floor and horizontal secondary ducts. The maximum sensible run from distribution case to telephone is 30 m, so large-plan offices have several risers. These are pretty easy to accommodate – a 100-pair vertical cable is around 35 mm diameter and it, and the cases off, call for a rising duct space around 150 mm × 600 mm. But their location must allow maintenance access off corridors rather than intrusively off rooms.

The horizontal runs are more tricky. The sheet metal or plastic ductings can take the form of hollow skirtings, or wall or floor channels, all with junction boxes and terminal outlets. The ductings are either conventionally conduited through the screed or, construction permitting, routed through a hollow floor, or through the cavity floor decking that is increasingly common in highly serviced buildings. In terms of overall floor layout, the risers can connect with central feeder ducts along corridors, with branches either to window bays or running around room perimeters. Spurs then lead to telephones. But ultimate flexibility in large open-plan offices may call for a comprehensive underfloor grid on a 2 m or so module. Ducts must be accessible through a continuous removable cover to allow connections to be altered – office layouts are far from static and telephone positions will constantly change.

Internal telephones

Similar provisions are needed for the building's internal telephone system. Types of private automatic branch exchange (*PABX*) depend on office size and whether it has single or multiple occupancy. The design must take early account of the switchboard, batteries and other apparatus involved, of the spaces they require, and of the concentrated floor loads they impose. Early thought for duct routes can result in simpler, cheaper installations.

Other systems

Early thought is also needed on the plant and ducts for paging systems, and fire and security alarms, teleprinters and other specialist communications equipment. The central telecommunications control room in a large building is now common.

Other building types

Telecommunications are fairly similar in other large building types. A modern hotel block will need a telephone and possible radio and TV wiring to each room, plus alarm and paging systems. But, at least, the regular bedroom planning, flanking the corridors, allows easy feeder runs up the plan centre (typically, above the corridor suspended ceilings) and later flexibility is less of a consideration than with offices, since bedside installations tend to remain fixed.

Blocks of apartments will have telephone rising ducts and, depending on the floor layout, corridor feeders. Horizontal routing there is often minimised by having a riser for every flat on a floor. As in a house, the internal wiring from lead-in terminal to the skirting level connection will preferably be concealed and in conduit.

The likes of hospitals, schools and factories, often comprising groups of buildings, will have a central telephone switchroom, with distribution to adjacent buildings by overhead cable or, more usually, albeit more expensively, by protected cable underground. Distribution within the buildings is as already described, except for the advantage that telephones are fewer and more permanently sited than in offices. True, hospital wards may have telephone outlets and radio connection to each bedside but, then again, bed positions tend to remain fixed and are helpfully sited around the room perimeter.

Summary

- Telephone entry cable runs to distribution frame, with controlling switchboard off; there are design points on imposed loads, ventilation and staff amenity.
- System has vertical cables in rising ducts, then horizontal runs (up to 30 m) from floor distribution cases to each telephone.
- Cross-floor system design acknowledges varying needs for flexibility.
- PABX has similar provisions.

GAS SUPPLY

Piped gas supply is, perhaps, the simplest utility service. Of course, the piped national grid is a comprehensive enough undertaking – pressures are up to 1000 lb/in² to minimise the volume for maximum calories transmitted, and, again, there is the hierarchically sized network of pipes from source to consumer. But accommodating the small low-pressure pipes in buildings is straightforward enough.

Source
Gas used to be derived from coal and oil mainly, often known as *town gas* or, more generically, *manufactured gas*. Today, in the UK, most of Europe and the USA, this has given way to *natural gas*, more directly derived from natural sources and mostly oil-related. The delivery pressure and calorific value are higher than with town gas, tending to allow smaller, neater pipes. Natural gas has its manufactured equivalents. A common one is *SNG*: this is often taken to mean 'synthetic' natural gas – a curiously self-contradictory idea when one thinks about it – 'substitute' natural gas is, surely, a better term. *There is also liquid petroleum gas, LPG.*

Installation in building

Pipe entry is usually at lowest level through a metal or stoneware sleeve set into the brick or concrete construction. A typical sequence is then stop-cock, pressure governor, meter and piped distribution. In a small house, distribution pipes might be 25 mm, reducing to 15 mm or less, depending on the appliances served – boiler, cooker, unit water heater, fire. A block of apartments will have a main riser with domestic-sized branches off, possibly with a main meter at entry and/or individual meters to each occupancy. Vertical routes up stairwells and so on are fairly easy to achieve. Horizontal branches can be laid in the floor screed or, pipe sizes and construction permitting, chased into the walls. Where pipes have to punch through walls or floors, it is, again, better they do so through loose-wrapping outer sleeves, both for neatness and to avoid their being restrained and, possibly, damaged by thermal movement. A point in passing is that, with town gas, there was the risk of condensation in the pipes and possible blockage. Branches were given a slight fall to ensure drainage back out of the system, either to the street mains or to low-level receivers or bungs. It was bad practice to have meters and pipes in cold basements. However, natural gas is much drier and the risk of condensation is negligible.

Safety

Safety is, obviously, a crucial factor in the design of gas installations. Most modern gases are less toxic than the older varieties of town gas, but can still asphyxiate, or produce carbon monoxide if incomplete combustion occurs in a faulty appliance. Supplies should not be routed through bedrooms, since a leak at night could be fatal. Air-gas mixtures are dangerously explosive. Meters and pipes are never housed in unventilated cavities where a small leak could build up to a high explosive potential, quite undetected. Pipes are never routed in contact with the electrical services. Flues, discussed in chapter 3 *Climate services*, have the added safety value of helping vent away any gas leaking from a faulty or unlit appliance.

Summary

- Town gas now being replaced by various, safer, gases.
- Typical sequence is entry, stop-cock, pressure governor, meter and pipes.
- Traditionally, pipe falls and bungs removed condensation, but less needed with most modern gases.
- For safety, runs neither in unvented cavities, nor adjacent to electric services.

MECHANICAL TRANSPORTATION

LIFTS

At the 1853 New York exposition, E. G. Otis cut the supporting ropes of his passenger elevator car and reputedly said 'All safe, gentlemen'. He was able to say this simply because he had not plunged to his death – safety clamps had immediately engaged on the car's vertical guide rails, an innovation that virtually heralded the arrival of the passenger lift or 'elevator'. High-rise buildings and lifts arrived hand-in-hand, each facilitating the development of the other. And now, as then, lifts are as costly as they are essential, both in capital outlay and, very important, in terms of floor space consumed. In fact, most of the development thrust has been towards increasing efficiency, so reducing the number of cars for a given passenger volume. Sophisticated control systems automatically give optimum response by a bank of cars to a given pattern of demand. And speeds increase. Otis' original lift barely reached 1 mph, modern lifts reach 20 mph and more.

Design and construction

4.24 shows the simple installation essentials. Usually, the shaft is concrete and forms part of the service core. Desirably, the motor room is directly over the shaft (actually forming a visible outcrop in the roofs of many buildings). There are safety overruns above the top-landing level and in the basement pit. The counter-weight balances the car weight and ensures that the hoist-rope's friction grips the driving sheaves. The compensating cables are there to offset the weight of the hoist cables, transferring to the counterweight side as the car rises, keeping the load on the sheaves balanced. Slower lifts may be driven by an AC motor but vari-

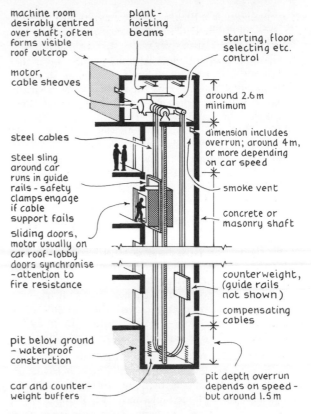

4.24 Simple passenger lift installation

able-voltage DC motors give a better, smoother performance for high-speed lifts.

The car

The typical modern car comprises a box with sliding doors, all in sheet metal, suspended in a structural 'sling' of metal channels. It will have fire resistance to satisfy the particular authority, as will the outer shaft and its doors. True, lifts should theoretically not be used for escape in a fire, but they may be and, in any case, the shaft enclosure is as much for 'compartmentation', preventing its becoming a disastrous route for the spread of fire and smoke up through the building (see Chapter 7 *Fire safety*). The car should be well lit and ventilated, and have inner finishes hard-wearing and cleanable.

The architect will be advised on the number of cars needed and, hence, floor space to allocate. Factors here are:

- Building size and type (type affects peaks to be handled, more in offices than, say, apartment blocks)
- Each car's capacity
- Car speed

- Maximum acceptable waiting time (offices, say, 30 secs, apartments, 90 secs)
- Mode of control/operation

As regards planning, lifts should, ideally, be at the centre of gravity of the building's circulation. In a deep building, this may remove them some way from the ground-level main entrance but they should, at least, be visible from that entrance. There should be adjacent lobbies large enough to prevent waiting people from obstructing corridor circulation. Preferably, opposite banks of lifts serving the same floors should not be separated by through corridors or stairs, otherwise conflicting circulation patterns may result as people rush for the doors.

Control systems

The original and most basic control has an attendant operating floor-selector buttons from within the car. This is inefficient and nowadays rare, suited only to the occasional 'posh' circumstance, like a smart shop. Modern systems are user-operated, with call buttons on each lobby, and floor-selector buttons in the cars. There is a variety of common systems, including:

Simplex collective. Here, a single lift answers all calls in sequence up the building and then, in sequence, returning down. This is adequate for the medium-size block of flats or public building, or small office.

Duplex/triplex collective. These operate similarly to the simplex but have two/three cars sharing the calls between them.

Group automatic. Here, two or more cars are controlled as a group. Control sophistication can range from merely intelligently spacing the cars, to the computerised operation where the cars are worked as a team, giving a constantly optimum response to the varying demand. There may be a timer override facility, homing the empty cars down for entrance-level availability during the morning peak, and upwards for high level availability during the evening peak.

Double-deck lifts

There are other innovations, increasing capacity. Two cars can be mounted one above the other in the same shaft so that two floors are served simultaneously at each stop.

Intermediate and express lifts

As explained, ultra-high buildings may divide their services into two or more separate, vertical stages, possibly having complete plant floors at intervals on height. The need to save floor space applies similar logic to lifts. For example, the World Trade Centre in New York has an installation with three vertical stages, each having four banks of six lifts. At the 44th and 78th floor divides, there are 'sky lobbies', each linked to the entrance levels by eleven express shuttle lifts. It is rather like having local and express trains.

Special lifts

There are specialised lifts, like the hospital lift sized to take a stretcher, or the fire-service access lift (again, see Chapter 7 *Fire safety*). And, concerned with quality as well as quantity in travel, there is the transparent observation car, affording spectacular views as it climbs up inside or outside the building. The cabins in the San Francisco Hyatt Hotel afford a novel perception, as one shoots up through the large central, galleried atrium.

Of a totally different breed are industrial freight lifts. Often, they are heavily loaded but low-rise, introducing a different gearing problem, best suited to operation by hydraulic ram. Slow speeds do not matter when loading times are long, anyway. There are unmanned industrial lifts with an automatic loading and unloading facility.

ESCALATORS

The escalator (4.25) is suited to moving large numbers of people through a limited number of floors – in public concourses, airline terminals and department stores. They are convenient and, certainly, circulating people through shops encourages impulse-buying more than do lifts. They are even fun – witness the external tube escalators at the Pompidou Centre, Paris. Reversibility to cope with commuter tidal flow is an advantage.

Of course, escalators arranged like an ordinary staircase allow only one-way traffic flow – as can be imagined, a more elaborate double cork-screw arrangement

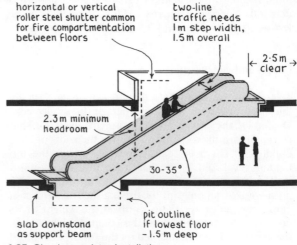

4.25 Simple escalator installation

is needed for two-way traffic. Capacity is principally governed by width, 1 m allowing standing and passing room. Speed is less significant, since a queue of people can only transfer to an escalator at a given rate – $\frac{1}{2}$ m/sec is common. Escalator flights are heavy pieces of equipment, for which the immediate floor structure must allow, and they are, preferably, delivered in complete units, for which the building's access must allow. The escalator well's open nature is a clear violator of fire compartmentation, calling for special safety precautions.

PATERNOSTERS

A paternoster is an endless, continuously moving chain of open-fronted cars, passing up and down a shaft. Its speed has to be slow enough to allow people to step in and out conveniently, around $\frac{1}{3}$ m/sec. This is much slower than lifts but, of course, there is no waiting time. They are best suited to buildings of modest rise where there is a continuous interflow of people between floors. They are not encouraged under some countries' legislation, for example they do not comply with the current British Standards, but this may prove to be only a temporary hindrance to their development.

Summary

- Lifts, ideally at circulation centre of gravity and probably part of central service core; waiting lobbies should not conflict with ordinary circulation.
- Shaft details as drawn; main constructional and design implications are overrun needs, i.e. relating to machine-room outcrop in roof and depth of basement pit.
- Capacity and efficiency increased by speed, number of cars, various sophistications of group operation, and use of intermediate and express lifts.
- Escalators – suitable for large numbers of people through few floors, e.g. public buildings, shops.
- Paternosters – slow but also suitable for buildings up to medium-rise with heavy interflow between floors.

5

LIGHTING

Good lighting design in buildings, daylight or artificial, is a matter of both *quantity* and *quality*. The architect – in collaboration with the lighting engineer in a large project – is concerned not only with providing enough light for the given tasks in each space but, also, with providing it in such a way that proper advantage can be taken of it, allowing visual efficiency and comfort. And around all this, there is the need for the lighting properly to 'reveal' the building; for lighting, how we see, and aesthetics, how our spirit reacts to what we see, must surely go hand-in-hand?

LIGHT

What is light?

The nature of light has been the subject of considerable speculation down the centuries. Plato thought we saw by spraying surrounding matter with particles from our eyes. Newton, rather nearer the truth, thought our eyes responded to particles emitted by the matter. Even today, our knowledge is incomplete. The particle or 'corpuscular' idea retains a place in explaining light's properties but, essentially, light is accepted as comprising what we choose to call *electromagnetic radiation* – energy transmitted by electromagnetic waves. Electromagnetic radiation's properties vary with wavelength and only a small band is capable of stimulating the sensitive retina at the back of the eye as visible light (5.1). 'Colour' is the quality the eye perceives in light of marginally different wavelengths within the band. We see daylight as predominantly white, although it is really a mix of the spectral colours from red, through yellow and green, to violet. It may be that Newton only added the other spectral colours because he happened to like the number seven!

Properties of light

Light travels at 300 000 km/sec. Gases are transparent to it and solids are usually opaque to it, notable exceptions being translucent glasses and plastics. It travels in straight lines except when 'refracted', bent, at the boundary between media of different densities. Different colours refract by slightly different amounts, which was why Newton was able to use the air/glass refraction through a prism to split sunlight into its component colours.

Illuminance – concentration of lighting
'Illumination' is the general process of lighting. 'Illuminance' is the resulting illumination level, luminous flux density, achieved on the task.

Most, though not all, light sources are incandescent, the sun, the oil lamp, the electric filament. The illuminance from a source reduces away from the source in inverse proportion to the square of the distance, i.e.

varying as $1/D^2$. This is the inverse square law.

The Imperial unit of illuminance is the *lumen*/square foot, the illumination on a surface from a wax candle one foot away – *1 foot candle* in the USA. The metric equivalent which we shall use is lumens/square metre, or *lux*. 1 lumen/ft^2 is approximately 10 lux. The light from a bright overcast sky is about 10 000 lux. A good working level inside is about 300 lux.

Luminance – brightness

The *luminance*, or brightness, of a surface depends both on the incident illumination and on the surface's *reflectance*. A white surface can have a reflectance upwards of 0.9, i.e. 90 per cent, a black surface, less than 0.02. Assuming the surface is matt, avoiding the chance of direct reflections of the source itself, then

$$\text{Luminance} = \text{illuminance} \times \text{reflectance}$$

The Imperial luminance unit in the USA is the *foot-lambert* – lumens/ft^2 × reflectance. The metric unit, is the *apostilb* (asb) – lux × reflectance

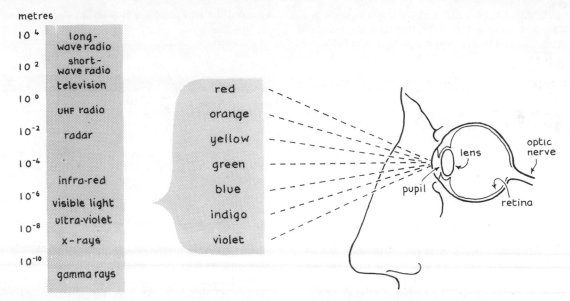

5.1 Only a small band in the range of electromagnetic wavelengths is visible light, capable of stimulating the retina at the back of the eye. The wavelength range within the band further divides into the various spectral colours, the mix being the 'white light' we see

Colour

The colour of a material is its property of preferentially absorbing selected wavelengths in the incident light and, in consequence, preferentially reflecting others. *It is, therefore, a property of the incident light as well as of the material.* A surface can only appear green if there is green to reflect. A green-painted table would simply appear dark grey, in light from which the green wavelengths had been filtered out.

An international colour standard is the *Munsell System*, originating from A. H. Munsell's colour atlas of 1915. It classifies colours three ways – hue, value and chroma – each having a graded scale. *Hue* is the wavelength quality, red, yellow, green, blue and purple; *value* grades from light to dark and is, therefore, related to reflectance; and *chroma* is the colour intensity.

Properties of the eye

Of course, vision is also a function of, and therefore influenced by, the eye and brain. The eye is often likened to a camera, the lens focusing the image on to the millions of photo-sensitive cells in the retina, whence the optic nerve carries the 'picture' to the brain. This is a useful shorthand except that the received stimuli have to be sorted out by the optic nerve and the brain, meaning that seeing is linked to our intelligence and experience, making it a psychological as well as a physiological process. We perceive better than dogs not because we have better eyes but because we are cleverer.

Visual acuity

Acuity is visual sharpness, one measure of which is the ratio between the size of a just discernible detail and its distance away, i.e. the angle the detail subtends at the eye. As we might expect, acuity increases with illuminance, but only up to a certain level, and other factors, such as contrast, play a part. Research has produced agreed standards on the illuminance required for various tasks (5.2).

Adaptation

The brightness of the surroundings in direct sunlight can be some 10 000 times more than in the softly-lit domestic interior, and yet the eye can adapt to either.

task difficulty	example	lux
prolonged, acute visual difficulty	local 'task' lighting in operating theatre, watch-making	⬆ 1500
prolonged task with fine detail, display lighting	precision factory work e.g. electronic, car paintwork retouching, supermarket	1000
fairly detailed	drawing office, accurate industrial	750
tasks of ordinary visual difficulty	art room, general office space, reading, laboratory, factory. Classroom	500 300
tasks of little visual difficulty	dining and living spaces, rough industrial work, changing rooms, circulation routes	200

5.2 Suggested illuminance levels in lux for various tasks. The text will later distinguish between local and general levels, for example, watch-making here is assuming locally high illuminance, office assumes general

The dilation and contraction of the pupil is the immediate response to a change in illumination level, backed up by slower and more complex processes in the retina behind. Developing full night vision can take half an hour or more. But, while the eye can adapt to the luminance prevailing, it cannot adapt to different levels *simultaneously*. This is fundamental to lighting design. The apparent brightness of a task or, indeed, the illuminance needed to see it clearly, very much depends on the relative brightness of the whole visual field, i.e. on the eye's prevailing state of adaptation at the time.

Contrast

Contrast is, therefore, the eyes' perception of the different brightnesses within the visual field, i.e. the brightnesses in relation to each other and, indeed, in relation to the average brightness to which the eyes have had to adapt. Black print contrasts with the white page and the page helpfully stands out from the grey desk top. The shading on a sphere tells us it is a sphere. Contrast is part of seeing. To be fanciful for a moment, if the eyes *could* simultaneously adapt, we would see no differing values in light and shade – only hues – and our sight would be heavily impaired.

Glare

Glare, on the other hand, is excessive contrast, bad adaptation conditions to the point of visual discomfort or disability. In buildings, it principally occurs where windows or light fittings appear too bright compared with the average brightness of the interior. We shall see that its avoidance is highly significant in lighting design.

Constancy

Seeing is not always the same as perceiving. *Constancy* is one example of our *psychological* reaction to lighting. A white ceiling appears greyer as it recedes away from the window wall but we perceive it, correctly, as uniformly white because we *know* the cause of its apparent non-uniformity. We accept a yellow tinge in the ceiling if we can see the room has a yellow carpet. Such subtleties are usually intuitively allowed for in design, but are coming to be better understood and quantified today.

Summary

- Illuminance on a surface is measured in lumens/ft^2 (Imperial) and lumens/m^2, i.e. lux (metric).
- Luminance, the brightness of a surface, is the product of illuminance and surface reflectance; units are foot-lamberts (Imperial) and apostilb (metric).
- The Munsell System grades colour in hue, value and chroma.
- Acuity is visual sharpness.
- Adaptation is the eyes' adjustment to prevailing light intensity.

- Contrast and glare are functions of adaptation.

DAYLIGHTING

A principal reason for having daylighting in modern buildings is that we like it! In strictly visual terms, it is less predictable and smoothly controllable than artificial lighting, and it is not necessarily 'free', as is sometimes suggested, owing to the thermal penalties that windows in the enclosure can incur. It is sometimes argued that daylight's spectral make-up is visually beneficial, but there is no real substantive evidence for this. The truth is that the varying quality and intensity of daylight entering a building, and our view out, bring a diverting and welcome sense of the world outside, a very valuable amenity to our psychological well-being.

GEOGRAPHICAL INFLUENCE

Daylight, of course, varies with place, time and weather. Direct sunlight is brightest, then the bright overcast sky and only then, perhaps unexpectedly until you think about it, the clear blue sky. Place, i.e. climate, gives rise to two very broad geographical classifications for the designer, *overcast maritime regions* and *clear non-maritime regions*.

Take the non-maritime tropics, the hot dry climate. The intense sun prevailing in a clear blue sky is an awkward lighting source, the sun being brilliantly luminous and the sky much less so, and yet windows encountering both during the course of the day. The sun is the overriding factor since, unchecked, it has teeth, causing crunching heat gain and glare (5.3a). Windows are kept small and, in modern buildings, may be reflective-tinted. There need to be shading devices outside, as described in Chapter 2 *Enclosure* and, inside, highly reflective room surfaces are avoided lest they become secondary sources of glare. However, the very fact of the sun's brightness allows a compensating strategy, at least at domestic scale. Reflective patios and courtyard walls can be used as indirect reflectors, bringing a more diffuse light to the building.

Conversely, in maritime regions, the tendency towards the overcast sky leads to larger windows (5.3b), which happily concurs with the greater need for ventilation in hot humid climates but conflicts with the need for heat conservation in cooler, higher latitudes.

This is all very general. The hot dry climate can be

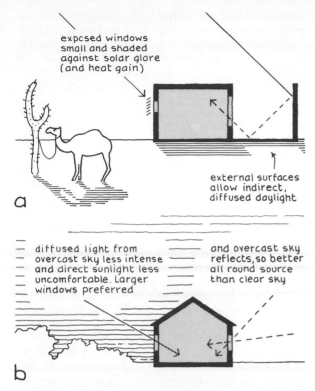

a

b

5.3 Daylighting design is normally taken as being for either the clear non-maritime region, of which the hot dry climate (a) is an extreme case, or for the overcast maritime region (b)

hazy and the maritime Mediterranean climate can be bright. And the effect of latitude on light intensity is an influence – the renaissance façade conceived in bright Florence was obliged to adopt a more open aspect when translated to its northern European derivatives. But the point remains that, today, the office-block fenestration conceived in the overcast maritime lighting strategy of London or New York cannot be simply transplanted to the Middle East – as has too often happened. Our main discussion will centre around design for overcast maritime conditions.

DESIGNING FOR DAYLIGHT QUANTITY

Influence on building shape

Given that natural lighting is to be the day-time source or, at any rate, the main source, it inevitably has its influence on building shape – sometimes a profound influence. In the small house, there may be the chance to orientate the principal fenestration away from adjacent obstructions and – climate depending – towards or away from the maximum sky brightness. This is in addition to the considerations of thermal performance

and view described in Chapter 2 *Enclosure*.

Large buildings must, obviously, consider the effect of plan depth on light penetration (5.4). Assuming a 3 m storey height and glazing to the full height and width of the external wall, around 7 m is the maximum plan depth that will still allow an adequate illumination at the back of a room (a) – one definition of 'adequate' being a 2 per cent daylight factor, as described in a moment. Greater depth will call for supplementary daylighting to the back, perhaps a roof-light (b) or clerestory, a band of glazing high on the back wall. In the multi-storey block (c), there can be a narrow plan depth or, more likely these days, supplementary artificial lighting. A deep low-rise building that needs daylighting in most of its spaces, like a school or college, can vigorously step its section (d) – an extreme case of design for daylight influencing the whole building form.

There may be the question whether to go for a single-storey building allowing roof-lighting but entailing a large site, or whether mainly to dispense with daylighting in favour of having a more compact multi-storey building, better thermally and, possibly, cheaper to construct. In practice, the decision will usually involve the experience of the lighting and environmental engineers, and cost consultants, as well as that of the architects.

Windows or roof-lights?

Windows and roof-lights merit a moment's comparison.

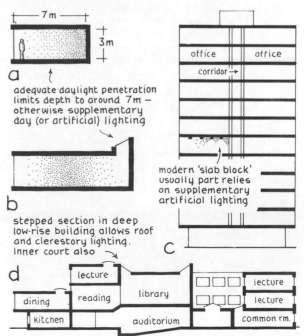

a

adequate daylight penetration limits depth to around 7 m – otherwise supplementary day (or artificial) lighting

b

stepped section in deep low-rise building allows roof and clerestory lighting. Inner court also

c

modern 'slab block' usually part relies on supplementary artificial lighting

d

5.4 Daylight penetration through side windows obviously reduces with plan depth, with consequent implications for design

Windows are the common daylighters not only because most rooms happen not to be directly under a roof but also, of course, for view and because they are well placed as ventilators. The constraint is limited light penetration.

Roof-lights can achieve a virtually even light distribution and, quantitatively, are more efficient daylighters, first, because they see the whole sky hemisphere whereas a window, at best, sees only half of it; and second, because roof-lighting arrives more vertically and hence more intensely on the horizontal working plane. But the light-and-shade modelling tends to be duller, and a thermal point is that roof-lights make a building particularly vulnerable to overheating by the high summer sun.

Daylight design – the variables

The daylight reaching a given point inside is the sum of the direct sky component, the externally reflected component from adjacent buildings and other surfaces outside, and the internally reflected component from the surfaces of the room (5.5)

$$\text{Daylight (lux)} = \text{SC} + \text{ERC} + \text{IRC}$$

The sky component is usually the most significant, but not always. The externally reflected component may be highly significant where a room looks onto a closely adjacent building or courtyard. And, in some cases, the internally reflected component can be contributing nearly half the illumination inside, so the reflectance of room surfaces can be highly significant.

The thing the designer usually needs to know is what geometry of room, and window or roof-light, will ensure a specified minimum illumination. The difficulty is the number of variables he has to contend with – the sky brightness which is unknowable, the degree of external obstruction, the comparative geometry of room and fenestration, and the reflectances of the room

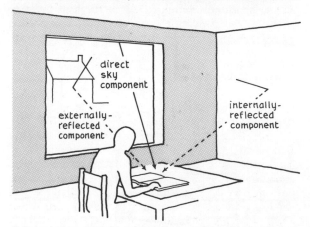

5.5 In design, daylight arriving at any point inside is taken as comprising direct sky component, and externally- and internally-reflected components

surfaces. Also, the daylight distribution across the room will vary – sometimes it is enough to find the average illuminance but more usually the illuminance needs to be found for a specified reference point, in which case, the position of the point needs to be considered – its distance back from the window, its height (usually taken as being at the height of the 'working plane') and its relative position in 'azimuth', the bearing in the horizontal plane, i.e. whether directly back from the window or offset to one side. A series of reference points may be required to build up the daylighting pattern.

Numerous calculation methods have evolved, their accuracy and complexity tending to increase as more of these variables are taken into account. They are amply covered in other lighting texts but, in illustrating the principles, they merit an outline here. They mostly evolved in the overcast maritime condition. Design under the clear tropical sky has, in part, adopted the maritime approach, albeit not always appropriately, and is often less concerned with light admission than with solar exclusion – again, see Chapter 2, *Enclosure*.

Agreeing a 'standard' sky – the CIE Sky

The first step is to agree a standard sky of average brightness or, rather, of a low brightness which the prevailing climate will equal or better for most of the time and which will, therefore, represent the critical design condition.

The UK originally agreed on a 'standard overcast sky', the uniform brightness of which produced an illuminance of 5000 lux on the unobstructed horizontal ground outside – in fact, pretty dull, and bettered for some 80 per cent of daylight hours overall. The real sky is not uniform, however, but increases in brightness from horizon to zenith overhead. An improved, and now more widely used, sky which acknowledges this is the *CIE Sky*, adopted by the Commission Internationale de l' Eclairage in the 1950s.

Measuring the daylight inside – lux and daylight factor

Then, there has to be a measurement yardstick for the daylight levels inside. Many countries use lux, of course, but an alternative measure developed in the UK is the *daylight factor*. The daylight factor is the illuminance at a given point in a room expressed as a percentage of the illuminance *simultaneously* prevailing on the unobstructed horizontal ground outside. *It is therefore a ratio, not a quantity*. Its value depends on the building design and, while it will vary across a room, it is independent of the actual sky brightness outside. If the sky happened to be producing 5000 lux, a 2 per cent daylight factor at a specified point inside would produce 100 lux there – a rise to 10 000 lux would produce 200 lux, the daylight factor staying the same. The concept has merits. It measures daylight in terms of the design, over which the architect has control, not the sky. It can be measured in an existing building, requiring only comparative light-meter readings through the open window and at the selected point inside. More subtly, since the apparent illumination inside is influenced by the eye's state of adaptation which, in turn, is influenced by the brightness of any window within the visual field, the indoor/outdoor ratio is helpful in going some way towards taking this adaptation influence into account.

When setting the target daylight level for a room, one can think of the daylight factor in two ways, either as the required average value over the whole room – commonly, over the whole floor area at the height of the working plane – or as the required value at a particular point, or series of points. The average daylight factor is convenient and, for most purposes, gives an adequate indication of how a room will appear. Broadly speaking, an average factor of 5 per cent will give a bright and cheerful appearance, while anything below about 2 per cent will appear gloomy and will need to be supplemented by artificial lighting. A point-by-point evaluation gives a more detailed picture but is more laborious, albeit one yardstick over the years has been simply to design for the worst likely case, namely, to achieve a minimum factor of, say, 2 per cent at the back of the room furthest from the window.

Formula methods
As a quick rule-of-thumb, the total area (m²) of window, or windows, needed to give a prescribed average daylight factor can be found from

$$W = \frac{4A\ DF\%}{t\ \theta}$$

where A is the floor area (m²); t is the transmittance of the window glass, about 0.85 for reasonably clean single-glazed; and θ is the angle in the vertical plane subtended by the portion of the illuminating sky 'visible' to the window itself – i.e. the angle between two lines projected outwards from the window centre, one going near vertically to the upper limit of the visible sky, usually the window head or roof overhang, and the other going to the lower limit, the distant skyline or external obstruction nearby.

There are more accurate formulae, for example:

$$W = \frac{2(DF\%)A(1-R)}{t\theta}$$

A, here, being the area of all the room surfaces, including that of the window itself, and R the area-weighted average surface reflectance – being, in fact, about 0.6 for a fairly light-surfaced room and about 0.4 for a fairly dark one. Clearly, this is taking account of most of the relevant factors – the room surface-area and relatedly the room size, the window transmittance, the surface reflectance and, hence, relatedly, the internally reflected component; and in θ, the extent and decli-

nation of the visible sky and, hence, the relative proportions of the sky and externally reflected components. Where windows are on different walls facing different degrees of external obstruction, the daylight factors can be separately found and simply added. The formula also works for rooflights – the higher θ value from the wider sky view, up to 180°, accords with roof-lighting's quantitative daylighting efficiency.

Apart from the average illuminance criterion, there can be further formula checks to ensure that the illumination fall-off away from the window is not too sharp, both from the point of view of the lighting appearance and the minimum illumination at the back of the room – we will come on to the question of illuminance 'uniformity'. But, in fact, if the room depth is no more than 6 m, there should be no problem.

The US approach inclines towards the use of formulae, but employing extensive tables allowing correction factors on window placements, room shape, internal reflectances and so on, tailoring the answer to the particular circumstances.

Daylight protractors
There are various 'on the drawing board' methods for finding the illuminance at a given point. Their limitation is that they need to know the room and window geometry as their starting point and, therefore, are more useful in checking rather than generating design. But they are accurate, especially since a point-by-point analysis can establish the complete illuminance contours across a room. The daylight protractor method of the UK Building Research Establishment (BRE) is one of the better known. There are protractors for vertical, 60°, 30° and horizontal glazing, and for unglazed apertures, and the more recent edition is calibrated for the CIE Sky. Actually, there are two protractors in each case, together comprising a circular instrument (5.6). As shown, the primary protractor is placed over the room section to find what the direct sky component would be at the reference point if the window were infinitely wide, and the auxiliary protractor is then placed over the room plan to correct the value for the actual window width and reference point position. The point's height and azimuth in relation to the window are inherently allowed for, and, also, the daylight's angle of incidence with the working plane. The externally-reflected component is found by repeating the process for the obstructed area the reference point sees through the window' but dividing the answer by (say) 10 to allow for the obstruction's lesser brightness. A very irregular obstruction outline may have to be approximated to an equivalent straight horizon on the drawing board for easier measurement. If there is a window on another wall, or a roof-light, their direct and externally-reflected components are similarly found and added in. If there are several windows in the same wall, only the plan correction factors need be

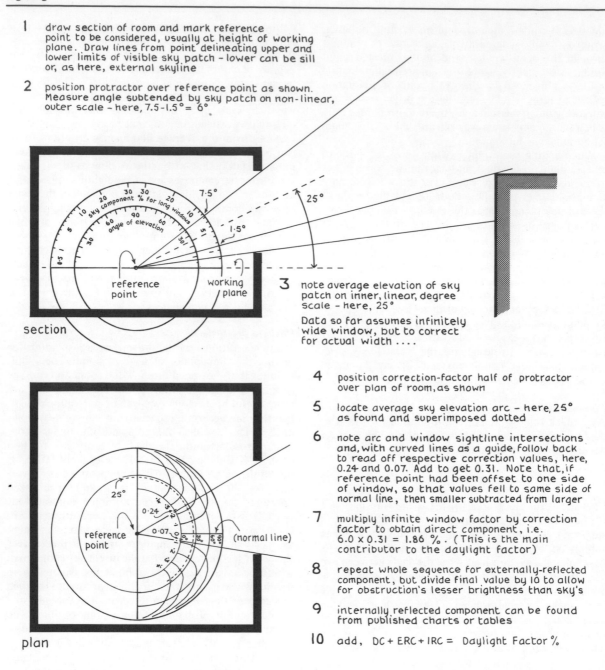

1 draw section of room and mark reference point to be considered, usually at height of working plane. Draw lines from point delineating upper and lower limits of visible sky patch – lower can be sill or, as here, external skyline

2 position protractor over reference point as shown. Measure angle subtended by sky patch on non-linear, outer scale – here, 7.5-1.5° = 6°

3 note average elevation of sky patch on inner, linear, degree scale – here, 25°
Data so far assumes infinitely wide window, but to correct for actual width ....

4 position correction-factor half of protractor over plan of room, as shown

5 locate average sky elevation arc – here, 25° as found and superimposed dotted

6 note arc and window sightline intersections and, with curved lines as a guide, follow back to read off respective correction values, here, 0.24 and 0.07. Add to get 0.31. Note that, if reference point had been offset to one side of window, so that values fell to same side of normal line, then smaller subtracted from larger

7 multiply infinite window factor by correction factor to obtain direct component, i.e. 6.0 x 0.31 = 1.86 %. (This is the main contributor to the daylight factor)

8 repeat whole sequence for externally-reflected component, but divide final value by 10 to allow for obstruction's lesser brightness than sky's

9 internally reflected component can be found from published charts or tables

10 add, DC + ERC + IRC = Daylight Factor %

5.6 Estimating the daylight factor at a specified point in a room by the BRE daylight protractor method

added, the total then being multiplied with the common section value already found.

The internally-reflected component can be found from tables. BRE tables here plot a range of percent-age values for various glazing area to floor area ratios, and wall reflectances – ceiling and floor having assumed set reflectances.

Equal-area methods – e.g. the Waldram Diagram
Equal-area methods are another approach, worth an explanation here for the neatness with which they help

convey the geometric parameters that control daylight intensity. They plot the areas of sky patch, and of any external obstructions seen through the window (or roof-light) from the reference point, not in their actual proportions *but in their light-contributing proportions*, which can then be measured directly. The Waldram Diagram is one of the better known (5.7). The sky vault, that is to say, the half hemisphere of sky the window wall faces, is represented by a grid (top right) so scaled in azimuth and elevation that vertical features like the sides of the window, and horizontal features like the head and sill, can be distorted into the light-representing area as shown. There are varieties of grid for different angles of glazing and types of 'standard sky'. The area of the resulting outline can then be compared to the area of the whole diagram to find the percentage sky component arriving at the reference point. Adding in the internally-reflected component, found from tables, then gives the percentage daylight factor.

An advantage of the Waldram Diagram is that it copes rather conveniently with the outlines of irregular external obstructions. For example, it has long been a useful tool in legal questions over rights of daylight. There is a version of the Waldram Diagram for evaluating direct sunlight penetration (particular relevance being to solar shading in hot climates) and there are many other sky-vault methods, each with their own applications and advantages.

Computer analysis

The fact of there being so many variables in lighting design will already have suggested the idea of computers to many readers. There is now a wide variety of programmes for daylight, sunlight and artificial lighting design. For the most part, they are used by lighting engineers rather than architects, and in large projects rather than small. However, even in large projects, the ordinary methods just described remain an essential guide to the architect in the early design stages.

Use of scale models

Lastly, there is the quite different approach of making a scaled-down model of the interior. This is viewed either under the real sky or, better, under a more stable and controllable artificial sky of diffused lighting panels and side mirrors in the laboratory. This is in no way a crude approach. Provided the model is an accurate simulation, the properties of light are such that the photometric result inside is an exact analogue of the real thing and, in fact, will take account of minutiae like glazing bars, surface reflectances and furniture, more than calculations ever can. Small photocells, placed in the model, can make the quantitative assessment of the daylight-factor distribution.

But it is in the *qualitative* assessment that models have perhaps their most telling advantage – in fact, leading us rather sweetly into our next topic. They

allow the designer to make a valuable subjective assessment of glare, the subtle effects of contrast, form, colour, texture, constancy – in short, the interior's whole visual character.

Summary

- Daylight design practice broadly divides into overcast maritime and clear non-maritime.
- Daylighting needs influence e.g. orientation, plan depth, sectional shape; and pose choices, e.g. between single-storey roof-lit and multi-storey side-lit.
- Windows are, obviously, the common daylighters. They also afford view. Light penetration is a planning limitation. Roof-lights are efficient quantitatively but summer overheating can be a problem.
- The daylight level inside is the sum of the direct component, the externally-reflected component and the internally-reflected component.
- The CIE Sky is now a widely adopted basic standard source for calculating daylighting.
- The daylight factor per cent compares simultaneous outdoor and indoor illuminance levels.
- The daylight factor is influenced by external obstruction, window and room geometry, and internal reflectances.
- Calculation methods include formula, graphical, computer, model.

DESIGNING FOR DAYLIGHT QUALITY

Bad lighting in any building is far more commonly due to inadequate quality than to inadequate quantity of light – there is hardly a lighting text or lighting engineer who would dispute this. The distribution, direction and character of the lighting, its diffusion, sparkle and potential for causing glare, the reflectance of room surfaces, all these play a part. *'How?' is at least as important as 'How much?'*.

Avoiding glare

Avoiding glare is the single most crucial factor in good lighting quality. Lighting design distinguishes between *disability glare* and *discomfort glare*, though they are really just different manifestations of bad adaptation conditions. They occur where windows (or artificial lights) appear too bright compared with the average brightness of the interior. When we are in a room, our eyes settle down to a general state of adaptation in response to the various brightnesses prevailing in the visual field. Suppose a room is lit through translucent blinds on a sunny day so that there is a pleasant, diffused illumination, adequate for working everywhere. If the blind is raised and direct sunlight streams in, the

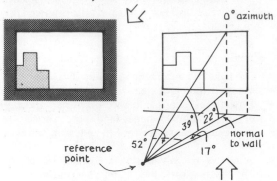

1   reference point 'sees' only a portion of the half hemisphere of sky vault the window faces

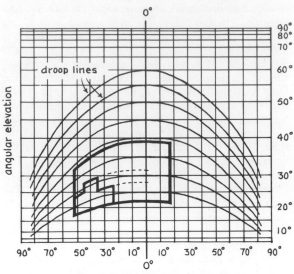

2   draw line from point normal to window wall – this would be on plan and section but the perspective shows the geometry conveniently here. Find angular bearing-off of vertical features in azimuth – here window sides are 52° and 17°. Similarly obstruction verticals. Find angular elevation of horizontal features on 0° azimuth vertical, i.e. 'dead-ahead' of point – head and sill are 39° and 22°. For obstruction horizontals, project sideways to cut 0° vertical

3   overlay published diagram with tracing paper. Transfer verticals using azimuth scale at base. Mark horizontals' elevations on 0° vertical and fade lines back to complete the image, using 'droop lines' as a guide – these allow for the horizontals' reducing elevation angle away from the 0° azimuth and save laborious plotting. The diagram grid proportions are such that the now distorted sky and obstruction images have areas equivalent to their daylight contributions, i.e. as comparable with the whole half-hemisphere sky area represented by the whole diagram area.

4   for easy measurement, lay distorted image over diagram sized grid subdivided into 500 squares. With total sky offering 100% component, and half hemisphere (represented by grid) offering 50%, then each sky square must offer 0.1% direct component and (dividing by 10 for lesser brightness) each obstruction square must offer 0.01% externally-reflected component. Add squares for total, here approximately $(43 \times 0.1) + (7 \times 0.01) = 4.37\%$

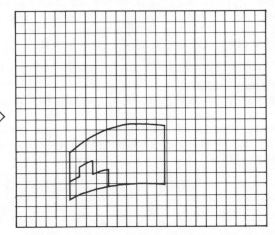

5   find smaller, internally reflected component as earlier described, e.g. from tables, and add in to get total Daylight Factor% at reference point

5.7 Estimating daylight level by the Waldram Diagram – one of the better known 'equal-area' methods

illuminance will rise several-fold, no question. But, in adapting to the window's increased brightness, the eye may actually see *less* in the room than formerly. The window wall itself may appear impenetrably gloomy – more light, less vision. *Increasing the brightness of the source can reduce the vision overall if the source is then overbright and awkwardly sited.*

There are various methods of glare prediction in buildings. Broadly speaking, the more precise assessments apply to artificial lighting, as we shall see. Glare control in daylighting is more by good practice and common sense.

Disability glare

This is the case where the luminance contrast is enough actually to impair vision, as when we are momentarily blinded by car headlights. In fact, its achievement in buildings requires a fair degree of idiocy on the part of the designer – the bright window as the only source of illumination smack at the end of a corridor, or the window right behind the lecturer's rostrum. Where such positionings are unavoidable, mitigating measures have to be taken. The area and/or brightness of the sky outside can be reduced by shades outside, tinted or reflective glazing, or translucent blinds inside. Also, the brightness inside can be increased by using high surface reflectances or supplementary light sources, such as other windows, clerestoreys, roof-lights – or artificial lighting.

Discomfort glare

This is a less dramatic upset but much more common. It is where overbright or ill-placed sources cause a nagging build-up of discomfort and distraction. A window wall adopts rather standard precautions against it, as follows.

Window surrounds. The window wall has relatively low illuminance but this effect will be reduced somewhat if it is light-painted, reducing its contrast with the bright window aperture (5.8). The reveals around the window should be light-painted for the same reason, in fact, their intermediate brightness helps visually by forming a 'buffer zone', grading the contrast between sky and interior. Suffice it to say, this grading is easier on the eye – the visual mechanisms are quite complex. Many older buildings, like the Georgian or Regency terraced house, have thick enough walls for reveals to be splayed. This is an effective device, since the increased area not only increases light reflection into the room quantitatively but, also, increases the buffer area qualitatively. Window frames and glazing bars are best light-painted to reduce the sharpness of their silhouette.

Window shape. Tall windows are efficient quantitatively, since they show the interior more of the sky source, and the brighter part of the sky at that. But, qualitatively, they increase the risk of unacceptable glare. It is better if the required lighting levels can be achieved without resort to very great window heights, especially in climates where skies are bright.

Window shading. External shading is mostly associated with solar control, as described in Chapter 2 *Enclosure* but, in masking off the upper, brighter part of the sky, it can have a glare-control function also. Sometimes, in fact, it will be specifically provided for

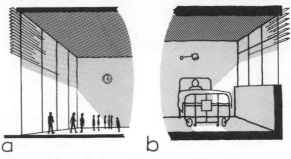

5.9 External shading against sky glare

glare control – 5.9a is one case, the high glazed wall of an airline terminal or other public concourse.

Solar glasses, also, have a related glare-control effect, and even the ordinarily glazed wall, say in an office building, can have the upper band tinted against glare, above the level of the horizontal view out.

The designer also has to take account of the activity in a space, some activities being more than averagely glare-sensitive. Visually, artistic activities are obviously so, whether creating in the studio or viewing in the gallery, though it is fair to say that, in such spaces, control against sky glare is only one aspect of the general need for diffuse, even lighting, associated with the use of translucent blinds and screens, and considered window placement.

The hospital ward (5.9b) is, perhaps, a more particular example. Patients in bed have a low eye-level and, if lying back, will actually be looking up – glare control is a critical care for beds near the window. The old, high-windowed 'Florence Nightingale' ward must have been especially unfortunate in this respect. Internal blinds are a possibility but have the limitation that they are not always used. People can be visually uncomfortable without actually knowing why – strange, but so it is – and, anyway, a patient may not be mobile enough to work a blind and staff are not always at hand to help. Also, internal blinds are scant protection against unwanted solar-heat gain. External shading over the window is again the best answer, blanking off the upper part of the sky view.

But total diffusion can be boring. There are limits to useful glare control, however. A single, large window appears relatively brighter than several equivalent, smaller ones, and more windows tend to mean greater diffusion – broadly speaking, glare and diffusion are inversely related. But, just to confuse the issue, while the daylight inside a marquee or translucent sports bubble might be wholly diffuse, making for optimum conditions of adaptation, most people would eventually find the bland modelling and the very *lack* of contrast psychologically boring. Although specular reflections and bright surfaces disfavour adaptation, they can vitally enhance the daylighting scheme. Take sunlight, for example.

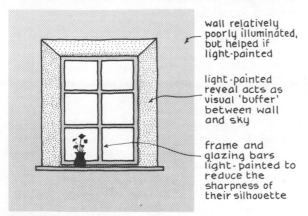

wall relatively poorly illuminated, but helped if light-painted

light-painted reveal acts as visual 'buffer' between wall and sky

frame and glazing bars light-painted to reduce the sharpness of their silhouette

5.8 Window design can reduce apparent sky brightness and, hence, glare

Our attitude to sunlight

We have a mixed attitude to direct sunlight in buildings. In hot climates, it is excluded, no question. In more temperate climates, it would still be excluded on strictly visual grounds and, sometimes, on thermal, but not so urgently. Psychologically, it has undeniable magic, enlivening and varying the character of a room for part of the long day. The desire for moderate amounts of sunlight admission seems to apply particularly to our homes. UK domestic codes suggest a minimum of one hour in each of the principal rooms, with the house preferably orientated for morning sun in bedrooms and kitchen, and later sun in other living areas.

Summary

- Disability glare actually impairs vision, discomfort glare may fatigue.
- Disability glare is avoided by sensible planning. Added measures are needed to reduce contrast between window and interior brightness – e.g. shading to windows and additional lighting to interior.
- Discomfort glare is reduced by limiting window height, splaying and light-painting the reveals as 'buffers', keeping the window wall light, and having shading outside or inside.
- Conversely, total diffusion risks being psychologically boring. For example, modest sunlight penetration is desirable, at least in the overcast maritime climate.

Direction and modelling

Strength of flow

The light 'flow' in a room causes contrasts in light and shade, sharpening our awareness of the shape of the room and the objects within it, their surface modelling and texture. The directional flow is strong with only one source such as a single window, and stronger still if the direct window component is very much greater than the internally-reflected component, i.e. where the reflectance of the room surfaces is low – think of the single window to a gloomy dungeon. Such arrangements would produce dramatic modelling, but too much so for ordinary preference, and would certainly cause glare. This is the converse of the total diffusion just described. There has to be a sensible balance.

Direction of flow

Strong down-lighting tends to model the human face rather curiously. Horizontal lighting models form rather curiously, too, and disfavours comfortable adaptation by bringing the source down more obviously into the visual field. The lighting most people prefer – or are accustomed to – is slightly downward and from one side. This gives agreeble modelling (5.10) and allows

5.10 Strength and direction of flow. Usual lighting preference for pleasant modelling of objects and visual comfort is as shown on the right, oblique and with helpful but not excessive contrast

people to organise themselves so that they are not working in their own shadows. Happily, it is just the sort of lighting that ordinary side windows tend to bring.

So, although quantitatively speaking, it is enough to know the *planar* illuminance, the light flux falling on the horizontal working plane, qualitative measurement needs to be subtler. We can just note in passing that there are other criteria for measuring illuminance, including *scalar* illuminance, the intensity of light from all directions falling onto the surface of a very small, notional sphere at a given point, and illumination *vector*, the net directional component of the light at a given point.

Illuminance distribution across a room – uniformity ratio

The lighting distribution, the varying daylight factor across a room, is a further aspect of diffusion. It is quantitative but, also, has qualitative effects. To within limits, the eye's adaptation copes with reducing luminance away from the source, and constancy allows us to accept the consequent reducing brightness of surfaces, and reducing value and apparent chroma of colour. In a deep room, though, it helps to make the back wall light and reflective, not only because it then adds to the actual light level there but, also, because it compensates by making the wall *appear* more bright.

Also, where individual windows are repeated along the wall of a large room, or where there are repeating roof-lights, their spacing must be close enough to give reasonable lighting uniformity. The *uniformity ratio* – the ratio of minimum to maximum illuminance – should be at least 0.7 for comfort.

Then again, the distribution can be varied for special purposes. Higher illuminance on the work surface can aid vision and help focus concentration, always provided the strength and direction of the light flow is not such as to produce reflected glare off the task. Selected highlights can articulate the visual progress through a building – the skylight over a corridor turning or capping a stairwell, the shaft of light on a church altar. This embraces aesthetics but, of course, so does all lighting in truth. And for all that the rules so far are valid so, too, is the intuition and spirit of the designer. We will return to these rather delicate points at the end of the chapter!

Summary

- Light flow is strongest with a single source and low interior reflectances – an undesirable, glaring extreme.
- Preferred is mild flow, slightly downwards and from one side – as from a window.
- Illuminance distribution across the working plane should normally achieve uniformity ratio (minimum to maximum level) of at least 0.7.
- Conversely, select highlighting can focus attention, and can display and enhance.

ARTIFICIAL LIGHTING

When you think about it, electric lighting in freely extending the working and social day has profoundly affected the way we organise our lives. Earlier times had torches, candles and oil lamps, but not cheaply, and most ordinary folk simply went to bed soon after it was dark. Gas lighting was fairly widespread during the last century, but the advent of relatively cheap, efficient electric light – Edison's system for New Jersey's Menlo Park in 1879 is accepted as the major milestone – heralded a social revolution, no less.

Basic source

Edison's bulb comprised an incandescing carbon resistance filament, suspended in a vacuum glass globe to stop its burning up. The modern bulb is not so different, usually having a coiled tungsten filament in inert gas (5.11). It has a warm pleasant light and is the principal source in the home. The fluorescent tube was initially a French development in the late 1930s. AC is applied across the electrodes at either end of a glass tube containing argon or krypton and mercury vapour, giving rise to electromagnetic radiation which excites the fluorescent coated walls and causes them to emit light. In other words, the otherwise useless infra-red and ultra-violet components, emitted by all lamps, are here converted to visible light – hence the greater efficiency. It is a rather 'cooler', perhaps harsher, light than tungsten, though much less so than formerly – but is at least four times as efficient in lumens put out for watts put in. This efficiency is highly relevant in today's energy context but, even before that, figured in the evolution of the large modern building, for example, the wide-bay factory and the deep-plan office block.

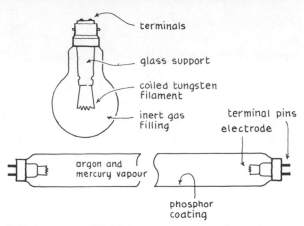

5.11 Common artificial light sources – the ordinary tungsten filament and the fluorescent tube

There are more efficient sources still, like sodium and mercury discharge lamps. To date, their unnatural colour rendering has tended to restrict them to outdoor or limited industrial uses, but types of high-intensity discharge lamp are now being developed, suitable for general uses.

Fitting the source to the building

Domestic fittings need no description. We are all perfectly familiar with ceiling and wall lights, table lamps and so on. In larger buildings, both tungsten, and fluorescent fittings now tend to be incorporated in the building fabric as a whole. True, individual tungsten fittings may be independently hung and so, too, fluorescents where utility is the watchword. But, as already illustrated in Chapter 3 *Climate Services*, the overall suspended ceiling, with integrated lighting and ventilation, is a commonplace in modern design. Dimensionally, the suspended ceiling will usually have a module, co-ordinating the fittings and panels within the larger structural grid. We shall see how, within this discipline, the spacing of fittings is restricted by the need for lighting uniformity over the working plane. There has to be easy access for the periodic cleaning of tubes and reflectors – important for efficiency – and for tubes to be replaced. There may have to be ventilation to avoid overheating, especially since the efficiency and colour appearance of fluorescents suffer if they run too hot. Simple convection slots in the fitting housing may be enough, but some positive mechanical extract is better, certainly over levels of 500 lux or so.

DESIGNING FOR ARTIFICIAL LIGHTING QUANTITY

Essentially, there are three categories of system – *general lighting*, usually from ceiling lights and intended

basic illuminance in lux	low task reflectance or contrast?	will errors seriously matter?	is task's duration short?	is space without window?	modified illuminance in lux
300	no / yes	no / yes	no / yes	no / yes	300
			yes		
500	no / yes	no / yes	no		500
			yes		
750	no / yes	no / yes	no		750
			yes		
1000	no / yes	no / yes	no		1000

5.12 Principle of the flow chart for modifying basic service illuminance for a particular activity (see Fig. 5.2) to allow for other helpful or unhelpful factors. 'Yes' answer gets dotted arrow

either for whole illumination over the working plane or for background amenity only, as in a foyer; *localised task lighting* e.g. over desks; and *special lighting* for architectural effect or display. Any or all of these can occur in combination. But first, lighting quantity.

Deciding on the illuminance required – the flow chart

Assuming there is a task, and there usually is, the first thing obviously is to establish the illuminance required to perform it. Tables like 5.2 (p. 141) are the main indication, but the values there are often further modified by reference to a *flow chart* which adjusts the basic illuminance according to the particular circumstances in which the task is being performed (5.12).

Satisfying the requirement

Then there are the sums. As with daylighting, so here – there are many methods for calculating artificial illuminance, their complexity generally increasing with their accuracy. The following examples are intended only to show the main factors at play.

Illuminance from a single source

If a single source (5.13a) is to provide a specified illuminance at a point, i.e. on a single task, the required source intensity I, in candelas, luminance units, can be found from:

$$\text{illuminance (lux)} = \frac{I \cos \theta}{D^2}$$

Where θ is the angular offset between the source direction from the task and the normal from the task (a task directly under the source gives cos θ its maximum value

of one) and D is the distance in metres – the inverse square law again. The appropriate intensity fitting is then found from manufacturer's tables.

Alternatively, manufacturer's tables may grade fittings by the lux (lumens/m^2) they achieve on a working plane, normal to their emission direction and a specified distance away.

Average illuminance from multiple sources

The average illuminance over the whole interior, like an office, is less of a private matter between source and task – room proportions and reflectances play a part, too. As shown (5.13b):

$$\text{average illuminance (lux)} = \frac{NFMU}{A}$$

This considers N the number of fittings and F the light

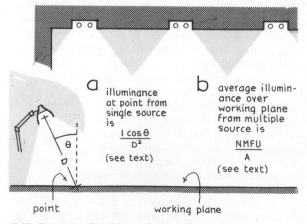

a illuminance at point from single source is

$$\frac{I \cos \theta}{D^2}$$

(see text)

b average illuminance over working plane from multiple source is

$$\frac{NMFU}{A}$$

(see text)

point   working plane

5.13 Estimating illuminance from artificial lighting

output in lumens of each lamp; M the maintenance factor (usually 0.8, acknowledging that both lamps and reflective room surfaces get dirtier in time); U the utilisation factor – all divided by square metre area A. The utilisation factor is the ratio between the lumens emitted and the useful lumens received. Found from lighting organisations' or manufacturers' published tables, it is there to take account of light losses in the fittings themselves owing to reflectors, shading louvres, etc., the losses dependent on the reflectance of principal room surfaces, and the effect of the room proportions and fitting height – known as the 'room index'. In the common case of a room with a fairly bright, 50 per cent or so, average reflectance and with fittings designed to distribute light generally rather than exclusively down on the working plane, the utilisation factor is about 0.4.

Usually the required illuminance is the known quantity, the unknown being the output from the lighting system (i.e. F × N) needed to achieve it. This FN value is easily found by substitution and, within it, the individual values of F and N can be juggled to suit – so many fittings of such and such an output.

Uniformity ratio
The number of fittings F is further governed by the need for reasonably uniform illuminance over the working plane. Obviously, there cannot just be one fitting with mammoth, glaring output. In practice, the minimum to maximum illuminance ratio should, again, be at least 0.7. As a rule-of-thumb, this is achieved if the fittings' spacing across the ceiling is no greater than $1\frac{1}{2}$ times their height above the working plane, or $1\frac{1}{4}$ times if they are louvred and predominantly downward directional. But we are verging on the qualitative.

Refining the method
Essentially, the above factors, the room and light fitting properties, must figure in any design method, though they are considered with varying degrees of accuracy. For instance, the term *effective reflectance* implies a closer consideration of the effect of carpets, furniture, wall coverings and such on the reflectance of surfaces. There is *ceiling cavity reflectance*, the effective reflectance of the ceiling and upper wall surfaces above the plane of a suspended system of fittings, and *floor cavity reflectance*, the effective reflectance of all surfaces, including furniture, below the working plane. Light-fitting output and distribution patterns are classified in various ways, as we shall see.

Summary
- Usual sources are tungsten or fluorescent.
- Usual systems are general lighting, localised or task lighting, special lighting.
- Basic illuminance required for task is modified to circumstances by flow chart.

- Factors affecting lux from single source are intensity, offset angle, distance.
- Factors affecting average illuminance over space include number and output of fittings, maintenance factor, utilisation factor acknowledging losses in fittings and at room surfaces, and room area.
- Minimum uniformity ratio is normally 0.7.

DESIGNING FOR ARTIFICIAL LIGHTING QUALITY

Source again – colour appearance and rendering
Artificial light sources vary in *colour appearance*, how the emitted light looks, and *colour rendering*, how it makes surrounding colours look.

Tungsten
Tungsten lighting appears comfortably 'warm' and, like all incandescent sources, it emits wavelengths right through the spectrum, giving a reasonable rendering of all the colours around. There may be some distortion, for example a white card will appear rather warmer than it would in daylight, but there the eye's colour constancy will cope and read 'white'.

Fluorescent
With fluorescent tubes, the colour appearance and rendering are more diverse. Their former clinical coolness has been much improved upon, and 'cool', 'intermediate' and 'warm' types are now available. *Colour correlated temperature* (CCT) grades appearance more precisely. It relates the apparent colour temperature to that of a heated 'black body' in a range from around 2500–6500 Kelvin – a black poker heats through red and white towards blue, so the higher the temperature, the *cooler* the apparent colour, the opposite of what one might expect. As to colour rendering, fluorescent emits unevenly through the spectrum, differing tube types favouring different colours.

The difficulty is that no single tube type can combine optimum efficiency, warmth and colour rendering all at once. So, depending on the activity to be lit, the designer has to compromise or favour one property at the others' expense. Sometimes efficiency will be the only criterion, as in a warehouse but, usually, there is some need for pleasant warmth, especially where lighting levels are below about 300 lux, when cooler fluorescents tend to produce a particularly miserable effect. Good colour rendering helps visual clarity generally, but many activities have a particular need for good colour rendering, like a textile design studio, a display showroom, or a hospital where patients' facial colouring can be important in effective diagnosis – medical staff may miss spotting jaundice if there is an absence of yellow in the incident light. Lighting intended to supplement daylight in an office needs daylight appearance for there to be an unobtrusive match, the diffuse overcast sky being the usual case to match. The CIE colour-rendering index grades lights in a comprehensive, colour-rendering code, with a maximum score of 100.

Glare

Whatever the source, it must avoid causing glare. Now, as with windows, direct disability glare is unlikely. Certainly, it could be caused by a bright unshaded source smack in the visual field but the remedies would then be terribly obvious.

Reflected glare

Reflected glare off the task is itself a risk, disfavouring adaptation and maybe causing *veiling* or *specular* reflections – the former would come off a matt page, the latter off a glossy one. It is a matter of geometry between source, task and eye and, as can be imagined, is easily controllable with adjustable task lighting, much less so with fixed ceiling lighting.

Discomfort glare – The IES Index

Discomfort glare is the other risk. As with windows so here, the source must not appear overbright, either in absolute terms or in relation to the whole visual field. For example, the absolute comfortable limit for an overall luminous ceiling is accepted as being around 1500 asb.

But relative source brightness is the more common concern. Here, good room surface reflectance is the first measure, doubly effective in that the surfaces then contrast less with the source and, also, enable a less intensely bright source to achieve the desired illuminance.

The position and design of the source are both important (5.14a,b). A ceiling source some way off is actually more likely to cause glare than one close overhead, since the former's lower angular elevation brings it more into the ordinary field of view – glare is more of a risk in long rooms. As to the allied question of fitting design, this can be roughly classified in a range from 'indirect', emitting the flux predominantly upwards to reflect from the ceiling, to 'direct'. Indirect fittings are inherently unlikely to cause glare. With direct fittings – the usual ceiling system away from domestic scale – the accepted rule-of-thumb is that glare will be unlikely anywhere provided the shading or underside louvres achieve a lighting cut-off angle within 35° of the vertical. The illumination should seek the task not the eye.

But there are more accurate methods for predicting and, therefore, avoiding, lighting glare. These include mathematical, by calibrated calculator, by computer,

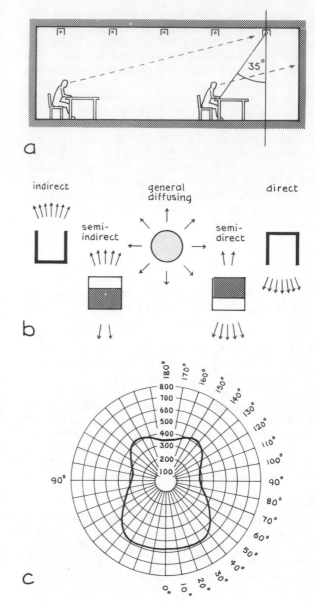

5.14 (a) Glare is more likely with distant, i.e. apparently 'lower', source in long room – but risk minimal if all fittings shaded to give 35° cut-off (b) General classification of lighting fittings (c) More specific 'polar curve' classification, plotting in vertical section the distribution intensity in candelas

of course, and by tables. In the US, the common way is by tables, used to establish the *visual comfort probability*, and there is the UK table method, the *IES Glare Index* (of the Illuminating Engineering Society). We will not go into glare indices in detail – suffice it to say the IES index takes into account the following: the number of fittings; their brightness; their area; their position in the visual field; and the reflectances of surrounding room surfaces. The index value derived for a proposed layout should then be less than the accepted, standard index for the activity in question. An office is around 19, some factories can be as high as 28. If the lighting proposals exceed the index, then the system design will have to be reviewed, for example, having lighting fittings more downward directional (and, therefore, probably more numerous and closely spaced for illuminance uniformity over the working plane), and/or having room surface reflectances higher.

Polar curve classification of light fittings

Coupled with the use of glare indices are the more accurate *polar curve* classifications of light fittings, plotting in vertical section the light output intensity round a particular fitting in candelas. 5.14c is a curve for a semi-direct fitting. British Zonal (BZ) polar curves are a variety which, relevantly enough from the glare point of view, concern themselves with only the light distribution *below* the fitting. They rate fittings from BZ 1, predominantly downwards, to BZ 10, predominantly towards the horizontal. The current European system uses lighting-fitting luminance as the sole basis for evaluating glare. This is adequate for many purposes, but the method may soon be extended to take account of the room context as well.

Brightness of task to surroundings

The well-lit task tends to be inherently brighter than its surroundings. As a simple example, take the reading lamp, directed on a book but also softly lighting the room around – a good balance. As well as allowing vision directly, highlighting draws the eye to the task and focuses attention. But how much brighter should the task be? It must not be so much as to contrast too sharply with its surroundings – the bright TV screen in the unlit room. In fact, progressive grading asks for a task about three times brighter. More specifically, an ideal ratio between say a book, desk top and room surfaces around is thought to be of the rough order 9:3:1, allowing, though, that people's subjective preferences, not to mention lighting researchers' opinions, differ. But this is based on strictly visual criteria and, in practice, the need to achieve good overall lighting with reasonable economy in, say, an office floor demands rather higher surface brightnesses – i.e. reflectances – than the ratio suggests. In the UK the IES code allows a ceiling luminance up to 0.9 that of the luminance of the immediate task surroundings. The walls, more directly in the visual field, have the rather tighter constraint of 0.5 to 0.8.

With overall ceiling lighting, reliance has to be placed on the task's being more reflective than the surrounding working plane, and on the fact of the task and working plane's being at right angles to – and hence better lit than the walls which are incident to – the mainly downward lighting flux. Ideally, a desk surface should neither be so dark as to contrast too greatly with the task nor so light as to vie with it. The old black desk top has been discarded in favour of mid-value green or grey, and is matt to avoid causing reflected glare from the source. Obviously, preferential task lighting increases the designer's control.

Relative surface brightnesses are further controllable by their reflectance, and by the intensity and direction of the lighting flux. If the ceiling risks being too dark, it can be made more reflective – and a reflective floor, also, helps by redirecting more of the lighting flux back upwards. Wall brightness can also be controlled by reflectance and by the direction of the flux around the room perimeter. Where a regularly spaced lighting layout happens not to come close to the walls, or where the ceiling lighting is, in any case, modest, perhaps where a scheme is part reliant on local task lighting near the working plane, the walls may be left too dark. Additional fittings may then be used around the ceiling perimeter, directed down the walls – a technique known as 'wall washing'.

Summary

- Choice of source depends on the relative need for accurate colour appearance and rendering, 'warmth', and/or efficiency.
- Reflected glare is avoided by source/task geometry.
- Discomfort glare is mainly avoided by fitting positions and design, and by good average room reflectance.
- IES Glare Index classifies maximum tolerable glare for various activities.
- BZ scale classifies fittings according to their distribution pattern.
- Visual ideal for brightness ratio task: desk: room is of the order 9 : 3 : 1, but variation reduces, in practice, as described. Controls are mainly by system choice (e.g. whether there are task lights or not), fitting design and surface reflectances.

... and so two choices of system compared

As an interim illustration of some of these points, suppose an office floor were to be lit by either a general ceiling lighting system only, or by general and local task lights combined. There are merits either way.

General lighting only. Clearly, ceiling lighting alone is simplest to install. The generality of its light coverage facilitates freely changing work patterns below. It lends itself easily to centralised and, possibly, automatic control. On the other hand, the very generality of its light makes it harder for the person seated at a desk to escape reflected glare off the task, and the ceiling's required higher brightness may increase the chance of its producing discomfort glare. True, the lighting output can be mainly downward, i.e. in the lower BZ values, but this can, arguably, lead to curious modelling. In fact, the even illuminance over the working plane may appear rather lifeless and become psychologically boring after a time – all right for 20 minutes but what about the whole working day? This last may be subjective point but it is one that modern design has come to consider. Residual sparkle from the ceiling system may add zest but then, again, it may distract from the task.

Additional task lighting. Assuming there *are* local tasks, additional task lighting, i.e. desk lights, will help focus attention. The lights can be directed to avoid reflected glare and the lower, general ceiling illumination now required must reduce the risk of discomfort glare there – albeit thoughtlessly directed task lights can trouble other occupants. The ratio of task to surrounding brightness is likely to be closer to the ideal and, undeniably, the varied lighting pattern across the space will enliven the mood. The user has the convenience and dignity of personal control.

The above is but a simple illustrative case. Each building type and interior has its own problem for the designer and good solutions will be intuitively, as well as empirically, conceived. But, for instance, in a drawing office, the higher illumination level will suggest additional task lighting, especially since the ceiling alone could cause unavoidable reflected glare off drawing boards. A library will call for general lighting to book shelves, but additional reading lights at desks. A factory will also, usually, call for both general and task lighting, i.e. having additional task lighting over machinery or zoned lighting over assembly lines. In general office spaces, the decision may be less clear-cut. Certainly, any need for flexible interaction between personnel across the floor will call for the more uniform coverage of a general system. Similarly, department stores, concourses and other public spaces will call for general lighting, albeit with occasional supplements for display or at any fixed staff work positions. The mobile activities in a sports hall will call for general lighting, without question.

Summary

- General lighting centrally controllable and favoured where activities are interactive and mobile.
- Task lighting added where fixed tasks need high illuminance. Adds personal control and, arguably, visual interest.

Other things

The effect of room proportions
Room proportions can modify many of the points above. In a small high room, the walls have very much

more area than the ceiling and are more in the visual field. They are, therefore, more significant in terms of their reflectance's effect on lighting quantity and their appearance's effect on quality – which is, perhaps, another way of saying you can paint the ceiling any colour you like.

Conversely, in a very large open-plan office, the ceiling is visually much more significant than the walls although, admittedly, this also depends on how close to the wall one happens to be. So, the ceiling fittings in a large office have a greater potential for glare. The rest of the ceiling surface should be light-painted to reduce contrast with the fittings and also, because the ceiling is now such an important reflector, i.e. if it is reflective, the lighting output can be lower.

Relative illuminance between adjacent areas
The proper lighting levels in a space may be influenced by the levels in adjacent spaces. In a hospital internal corridor, it is not enough simply to install the minimum safe illumination for a person walking along it. By day, the brightness must be enough to avoid an appearance of gloom in relation to the daylit adjacent wards. By night, the brightness must be capable of reduction to avoid intrusion into the then darkened wards and to avoid glare for the dark-adapted person leaving the ward – in fact, the corridor can be a valuable intermediate 'buffer' between a dark ward and a bright treatment area, allowing staff's eyes more time to adapt. Similar requirements, ideally, apply to the artificially lit entrance foyer, the preferred lighting level of which depends on whether it is day or night outside.

Window areas at night
The effect of exposed window areas at night is quite significant. Curtains and blinds have obvious thermal benefits in cold weather but they have a role in lighting performance, too. Glass as a room surface is a fairly poor reflector – it is, after all, a good transmitter – and light-coloured screening usefully increases the reflected component back into the room. Moreover, for all that glass *is* a poor reflector, dark windows can still cause extremely distracting images of the interior, especially of bright lights. Screening avoids this. It also avoids the uncomfortable contrast of dark windows with bright surrounding walls, the reverse, as it were, of the daylight glare risk. Sometimes, windows will be left uncovered for design reasons, for the importance of the night-time appearance to the passer-by outside but, as well as possible thermal loss, there must then be lighting loss also, the significance of which will depend on the night-time illumination levels in use.

Summary

- Walls are significant quantitatively and qualitatively

in a small, high room. Ceiling is more significant in long, low room.

● Brightness contrast between areas must not cause discomfort in darker area nor be enough to outstrip the adaptation capability of passers-through.

● Windows at night are screened not only against heat loss but to avoid light loss, specular reflections and reverse contrast.

Integrated day-time lighting

Obviously, artificial lighting is available to help out daylighting towards evening or when the weather darkens. But integrated day-time lighting is the more regularised approach in modern buildings, applicable as the plan gets too deep for natural lighting and as glazed areas are reduced, these days, for thermal reasons. 'PSALI', 'Permanent Supplementary Artificial Lighting Installation', is one of the better known design formats, introduced in the early 1960s but, in common with the other early approaches, it was, perhaps, more concerned with the strictly visual criteria and less with the energy-saving criteria than is becoming the practice today.

The good lighting points already made will still apply in the integrated day-time scheme and many of the buildings already quoted would have such schemes, often incorporated with the ordinary night-time fittings. But the thing to realise is that good integrated lighting means more than just switching on part of the straight night-time installation by day – to be sure, much of the packaged lighting design in commercial buildings is content with just that but, though such installations are cheap to buy and may appear adequate in use, they involve real losses in lighting quality and economy.

Ideally, the lighting design should achieve an unobtrusive synthesis, the artificial supplement blending with the daylight rather than rivalling it. So, for a start, the fluorescents need to be of the 'cool' daylight-matching type and, within this requirement, there may be finer adjustments, depending on whether the prevailing climate is predominantly clear or overcast, and on how far the glazing is north- or south-facing.

The illuminance of the mix must reduce gradually away from the windows, enough to simulate the daylight fall-off into the room but not so much that remote parts are inadequately lit (5.15). This requires adequate apparent brightness as well as adequate quantity of light, since adaptation is affected by the whole visual scene. Attention has again to be paid to the flux on, and reflectance of, the back walls, if they are not to appear gloomy in relation to the adaptation prevailing. The PSALI approach could even argue a supplement with a higher illuminance than that of the artificial lighting on its own at night, the brightness of the daytime interior being partly judged in relation to the occupant's view of the bright windows and view outdoors. But, again, this would be an extreme concern with visual parameters and, in practice, there needs to be a balancing concern about the energy a high daytime supplement consumes – in any case, the required added brightness is not as much as might be expected, since properly conceived integrated lighting will allow smaller windows and, hence, lower adaptation levels than did the indiscriminately glazed curtain walls in the past.

Really, one is talking about a graded balance with quantitative and qualitative parameters. The illumination appropriate to the room function must be a starting point but, beyond this, integrated lighting can be taken as being needed anywhere the daylight factor on the

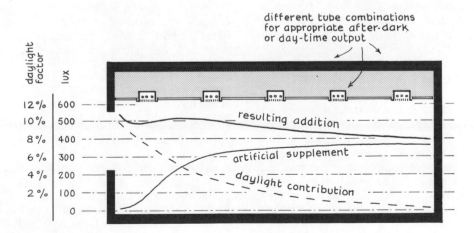

5.15 Integrated artificial lighting reinforcing daylight. The resulting illuminance gradient simulates daylight fall-off away from the window, only is much moderated. The lux levels shown are fairly typical for an office space – the equivalent daylight-factor values are assuming a standard sky producing 10 000 lux on the horizontal plane outside

working plane falls below 2 per cent or, if preferred, where the average daylight factor falls below 5 per cent. Also, excepting the very bright zone immediately adjacent to the windows, the mix should be so graded that the fall-off across the room results in an illuminance diversity no greater than 3:1 – effectively, a uniformity ratio of at least 0.3. In a very deep plan, the windows may lose significance as light contributors, but will still affect adaptation.

The gradation of the mix across the room will further depend on the planning, asking for a logical fall-off away from side windows, a centre dip between opposite wall windows, and so on. A very simple installation might have discreet supplementary fittings, in addition to the ordinary fittings, for after dark. The more usual case of the suspended ceiling with incorporated lighting grid may achieve the supplement by day-time use of *selected* fittings, or tubes, within that grid. For example, the grid might comprise three systems A, B and C, alternating in rows across the room, allowing an AB combination for after dark and a BC combination for the integrated mix.

Whatever the system, it must be under staff or automatic control rather than at the whim of the occupants. There are now sophisticated photoelectric controls, imperceptibly adapting the installation in response to the changing light conditions outside.

Summary

- Integrated day-time lighting can be taken as needed anywhere daylight factor falls below 2 per cent, or the average factor falls below 5 per cent.
- It may be fluorescent daylight-matching.
- Illuminance level of the mix depends on particular fenestration. Reduces away from windows, but not so much that remote areas are underlit or illuminance diversity excessive.

THERMAL IMPLICATIONS OF LIGHTING

We can elaborate for a moment on the thermal implications for lighting design, picking up the energy theme of Chapter 2 *Enclosure* and Chapter 3 *Climate services*. On the one hand, it would be an oversimplification to say that daylighting was free. Windows increase heat loss in winter and, possibly, unwanted heat gain in summer. On the other hand, unless the plan is very deep, daylighting will be a valuable and perhaps the dominant lighting contribution to the day-time operation of a building. This is the more significant remembering the

double energy penalty artificial lighting can bring, directly consuming electricity – which is very expensive – and indirectly adding to the heat load the climate services have to dissipate. So there has to be a balance. As a broad observation, artificial lighting levels have tended to reduce in recent years, partly because illuminance levels were originally set unnecessarily high – up to 1000 lux was the fashion for an ordinary office – and partly because more weight has come to be attached to the value of the daylight contribution in the reasonably shallow plan. All this is notwithstanding the increasing efficiency and coolness – or rather reduced warmth – of modern fluorescent types, and the improved methods of useful heat recovery from artificial lighting installations. Of course, the strategy for the deep-plan building, the factory or bürolandschaft office, has to be very different. There, the artificial lighting prevails, with windows primarily for psychological need and comprising as little as 6 per cent of the floor area.

Achieving optimum vision with minimum wattage in artificial lighting is mainly a matter of applying the standards described earlier. Where appropriate, fluorescent (or e.g. discharge) sources will be chosen rather than incandescent, and they will be the most efficient fluorescents consistent with the required colour appearance and rendering. The utilisation factor will be improved by emphasising the downwards (BZ) flux (which also helps glare-avoidance) and by having optimum surface reflectances, room index and so on. Localised lighting tends to be less efficient in terms of lumens out/watts put in, than the cool fluorescents in a general illuminating ceiling, especially if the former uses short fluorescents (less efficient than long) or incandescents. But it can bring savings where high task illuminance is needed over selected areas only, over remote office work spaces or over the long assembly line in a factory. It all depends on the application.

Control strategy is a factor, perhaps allowing banks of lights to be switched off when not needed, or having timed automatic cut-outs, for example of office lights at night, or having photoelectric adjustment of integrated lighting.

Summary

- In the reasonably shallow plan, the relative energy advantages of daylight versus day-time artificial lighting have to be balanced.
- Many modern factories and deep-plan offices are virtually windowless from the daylight quantity and thermal standpoints.
- More efficient artificial lighting means less energy used – factors are lighting type, utilisation factor, system type, control.

LIGHTING FOR 'ARCHITECTURAL EFFECT'

Little has been said on lighting for special architectural effect – it scarcely lends itself to neat analysis. We are aware that high contrast and strong directional flow can be used to reveal, enhance, select, exaggerate, even distort. The range is infinitely diverse.

At the one extreme, there is the cathedral arcade thrown into contrasting silhouette by light played on the vaulted ceiling of the side aisle behind, and with a spotlighting focus on the altar. At the other extreme, is the bürolandschaft office with sparingly splashed highlighting and patches of colour to enliven the otherwise bland scene. The lighting system in an art gallery or museum must reveal the interior, probably an interior with a highly conscious aesthetic, and also reveal the exhibits. Such display lighting must pay close attention to intensity, direction as regards modelling, colour rendering and the avoidance of specular reflections off the exhibit or its glass enclosure. The display lighting in a shop window is a very different case but it has to attend to similar things, as must theatre stage lighting and exterior floodlighting of buildings.

Colour can be relatively easily classified in appearance by the Munsell System but, again, it would be unrealistic to attempt a pat analysis of its effects in building. There can only be broad observations. Very generally speaking, warm, exciting colours will better suit a room that is orientated away from the sun, and which is large, with smooth surface textures, and where the occupancy is short-term, low-noise and involving only light exertion – perhaps a large restaurant? The converse tends to apply for cool, calm colours – a sun-facing music room? Pure chromas similarly tend to suit short-term occupancy and they are, also, often suggested for where the task involves low responsibility and where taste and smell are unimportant. But people's responses being subjective and different, the argument gets tenuous, and the designer's colour sense is, in any case, intuitive and complex – which brings us to our closing point.

GOOD LIGHTING DESIGN – EMPIRICAL AND INTUITIVE?

That rules, sums and standards must couple with an innate sense of 'design' to produce good lighting, has been a recurring theme of this chapter. Clearly, it has emerged that the factors contributing to a good lighting scheme are many and diverse, and that people's responses will, in any case, be subjective and slightly different. So lighting design *cannot* be wholly empirical, there needs to be intuition, experience, call it 'flair'. Clearly, the lighting that allows people to see what they are doing, and with comfort, must subtly interweave with the building's aesthetics and spirit. The balanced view, perhaps, is to accept that there have to be *minimum* standards, and rules for their application and that, while those rules are not enough in themselves, the overlying sense of good design has a common source with them and, not surprisingly, is the better for understanding the principles behind them?

6

ACOUSTICS

THE NATURE OF SOUND

The circular ripples from a stone splashed into a pond carry the energy of the disturbance through the water surface to the banks. Sound behaves in an analogous way in air except, of course, that the spread is three-dimensional. When the man shown shouts, the energy from the vibration of his vocal chords is causing sound waves to radiate in the air around him as a series of successive, concentric spheres growing in size (6.1). These waves vibrate the ear drum of the listener, producing the subjective sensation we call sound.

WAVE THEORY

In truth, wave theory is not all that simple but some grasp is essential to a working understanding of acoustics. If you have no particular physics background, take courage.

6.1 Sound energy waves can be pictured radiating in air as a series of concentric rings – or, in fact, spheres – growing in size.

Imagine a simple tuning fork. Normally, the air particles around it are at rest but, when the prongs are struck and vibrate at their designated frequency (this depends on their stiffness and mass), they impart rapid pushes to the surrounding layer of air, the particles of which are, therefore, compressed together at one instant and decompressed the next. These pushes or pressure fluctuations are then passed on to the next layer of air and so on, so that a wave motion is set up, carrying the prong's energy outwards. The air particles are momentarily displaced from their normal position as the wave passes, but do not themselves radiate, any more than does the water on the pond's surface. They merely oscillate within their own area of influence, each 'layer' compressing and decompressing like a sponge (6.2a).

The classroom analogy to sound waves, where a string is fixed at one end and shaken at the other, causing waves to pass along the length, is actually rather misleading. Waves so produced are *transverse*, meaning that the up and down oscillations are at right angles to the wave travel. Sound waves are *longitudinal*, the air particle oscillations occurring parallel to the wave travel. Think of a horizontal, coiled spring fixed at one end and given a series of horizontal shoves at the other, so that compression waves, momentarily closing the coils, appear as darker bands chasing each other along the length. And again, of course, it is the waves that travel, not the wave-transmitting coils.

Amplitude, wavelength and frequency

Any wave form, whether in water or of sound or light, can be described by its amplitude, wavelength and frequency. In the pond, the amplitude – the particles' maximum displacement from their mean position – is half the height from wave trough to crest. The wavelength is the waves' distance apart and the frequency is the number of waves passing through a fixed point in a given space of time.

In sound, the amplitude of the particle oscillations affects loudness, which we will come on to. Frequency

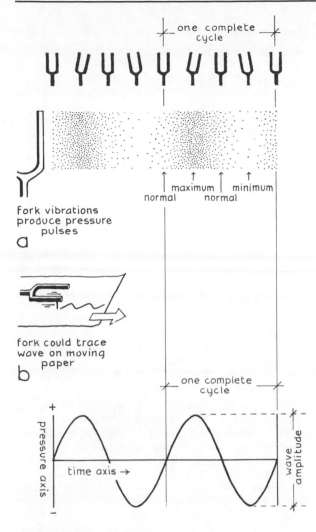

one complete cycle

fork vibrations produce pressure pulses

maximum minimum
normal normal

a

fork could trace wave on moving paper

b

one complete cycle

pressure axis

time axis →

wave amplitude

+

−

... similarly, graphical 'trace' of fork's tone plots pressure against time

c

6.2 Wave form of a pure tone. (a) Can be viewed in terms of space or time, either picturing at a frozen instant the pressure pulses radiating from the fork, or picturing over a period the pressure variations in a thin 'slice' of air – if the slice is an exact wavelength multiple distant from the fork, the variations will correspond to the fork oscillations shown above the wave. (b) The fork could physically trace a wave. (c) The graphical representation of what the air slice experiences

is the speed of the oscillations, the time taken for one complete compression-decompression cycle, and is measured in cycles per second (cps), or hertz (Hz) as it is now more commonly known. It is not hard to see that wavelength × frequency = speed. (You go faster if you take lots of big strides quickly.) The speed of sound is a constant at a given air density, at sea level, about 340 m/sec. And, given the speed, i.e. wavelength

× frequency, is a constant, then high frequencies must have short wavelengths and low frequencies, long wavelengths. Wavelength does not, in itself, affect what we hear but, of course, frequency does. High frequencies are high notes and low frequencies, low notes. Middle C, musically, is 256 Hz and the range of human hearing is from about 20 Hz to 20 000 Hz, the upper limit generally decreasing as the ear drum gets less sensitive with age.

Sound waves shown graphically

The undulating trace of a heart rhythm or sound wave on the screen of an oscilloscope is really only mathematical shorthand for describing wave form. Similarly, a sound wave can be expressed graphically. In fact, at a very basic level, it could be graphically transcribed by the prong of a tuning fork (6.2b), the point of the pencil recording a sound trace on the paper being speedily withdrawn from under it. A graphical trace can similarly describe the amplitude, wavelength and frequency of the pressure fluctuations in any 'slice' of air near the fork. The peaks and troughs correspond to the maximum and minimum pressures the slice experiences in a cycle.

Thus, thinking in terms of *time* (6.2c), if the cycle in our slice of air happens to start at the instant of normal atmospheric pressure then, one half-cycle later, we find that the pressure has increased to maximum and returned to normal. A further half-cycle and it has reduced to minimum and returned to normal. So the full cycle brings us back to where we were. Thinking in terms of *place*, the pressure fluctuations in a slice of air situated one wavelength distance nearer to, or further from, the source, would occur exactly synchronically. And, incidentally, thinking in terms of time and place gives us the familiar wave pattern in motion, the waves appearing to chase each other along an axis – or across an oscilloscope screen.

Loudness

As the spherical wave front radiates from the sound source, its size must increase and its energy must become more dissipated in consequence. The waves' amplitude becomes less. Sound intensity, like light intensity, reduces in proportion to the square of the distance from the source – the inverse square law again.

The metric unit of sound pressure is the Pascal (Pa), which is 1 dyne/cm^2. However, the loudness we hear is not directly proportional to sound pressure but to the ear's subjective response to sound pressure, a distinction that has to do with the enormous pressure range the ear has to cope with. The only way it can be sensitive to tiny pressure fluctuations at the threshold of audibility, around 0.000 02 Pa at 3000 Hz, and yet remain undamaged by noise like that of jet aircraft which, at anything up to 30 Pa, is over a million times greater, is for *its sensitivity to decrease as pressure level increases*. The practical unit of loudness is the *decibel* (dB). The decibel scale (6.3) is, effectively, a set of equal steps of perceptible loudness increase. It corresponds closely to what we hear and is much less unwieldy than the cumbersome units of a pressure scale would be. It is linear but the corresponding pressure increase is not. For definition, the number of decibels between two pressure intensities, Pa_1 and Pa_2, is 10 log Pa_1/Pa_2 but, as an adequate guide, a 10 dB increase requires 10 *times* the pressure increase.

Incidentally, the energy in sound, the pressure variations that make it, are relatively very small – compare even the extremely loud 30 Pa with the 100 000 Pa of normal atmospheric pressure!

Directionality

Our ability to tell where a sound has come from is called *auditory localisation*. This is achieved at higher frequencies by the ability of the ear most turned away from the sound source to detect a slight reduction in pressure in the sound shadow cast by the head and, at lower frequencies, by the brain's actually noting the slight time-lag between the sound reaching one ear and then the other. Localisation is rather better horizontally than for sounds coming from above or below – we shall see how this affects the way sound is electrically amplified.

Summary

- Sound waves in air radiate spherically away from the source as alternate pulses of slight compression and rarefaction.
- Waveform is described by its amplitude, wavelength and frequency.
- Frequency is measured in cycles/second (Hertz).
- Perceived loudness is measured in decibels.

ROOM ACOUSTICS

Sound occurring in an enclosed space will be partly transmitted through the boundaries, partly absorbed and partly reflected (6.4). Transmission we can safely leave until discussing noise control – it is mainly absorption and reflection that modify the sounds we hear indoors.

No material can be completely sound-reflective. Some of the energy in a pressure wave hitting a surface will be absorbed in causing it to vibrate and by conversion to heat owing to friction – and the more this absorption, the less the reflection. In general, hard

type of sound	pressure rise	dB
near threshold of pain	10 000 000 000 000	130
under airport exit path	1 000 000 000 000	120
	100 000 000 000	110
noisy traffic pneumatic drill	10 000 000 000	100
loud passage from orchestra	1 000 000 000	90
	100 000 000	80
noisy office	10 000 000	70
medium traffic	1 000 000	60
busy restaurant	100 000	50
ordinary conversation	10 000	40
quiet domestic interior	1 000	30
	100	20
rustle of leaves human breathing	10	10
threshold of hearing	1	0

6.3 The decibel scale – some approximate values

rigid surfaces are reflective and soft, porous ones, absorptive.

At the risk of overworking the analogy, absorption and reflection modify the pond's waves, which hit the banks and are partly absorbed and then partly reflected as smaller waves which, in turn, interfere with one another and re-reflect in an increasingly complex pattern, until the original energy from the stone's splash is lost and the surface is still. The rather notional diagram shows a single sound 'ray' (a further shorthand) being directly reflected but, in reality, of course, the waves will repeatedly reflect from all the boundaries in an infinitely complex, interactive pattern of disturbance, which may persist for some seconds after the original sound source is silenced.

The nature of a sound is inextricably linked with the

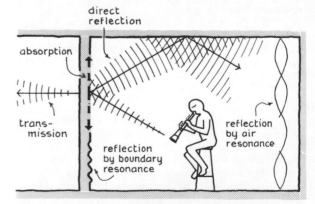

6.4 The behaviour of sound waves in a room. They are partly transmitted, partly absorbed and partly reflected. Reflections are the main cause of reverberation or 'echo'. Also, the air path between boundaries, and the boundaries themselves, can respond by resonating

nature of the space in which it occurs. A room is an instrument. A handclap in one will sound different from a handclap in another – the claps are the same but the responses are not. In a furnished living room, the response will be soft and muted, 'dead' acoustics, or 'dry' as musicians would say. In the hard, highly-reflective surfaces of a bathroom, it will be longer lasting and louder, giving 'live' or 'reverberant' acoustics – the kind of mushy response enjoyed by the bathroom singer. For the designer, the ideal acoustics depend entirely on the use for which the particular space is intended: the needs of speaker, actor, musician or orchestra are different.

The principal phenomena at play are *reverberation* which, loosely speaking, is how long the sound 'echo' lasts after the source has stopped, and *resonance* which is how a room's shape and surfaces contribute to the sound quality, rather as the sound box of a stringed instrument affects its tone. The two are interdependent, to an extent, but we will take the more important one, reverberation, first.

REVERBERATION – THE EFFECT OF ROOM SHAPE AND ABSORPTION

As you will have noticed, the walls and ceilings in spaces like music rooms and auditoria are often selectively treated with sound-absorptive material. This is to control reverberation. The amount of reverberation depends on the rate at which the sound decays. The decay is faster when the volume of the enclosing space is small, since the sound waves darting about at constant speed must then suffer more absorbing encounters at the boundaries in a given period of time. It is also faster where a space's boundaries and contents are more absorptive. *Reverberation increases with volume and decreases with absorption.*

But, of course, the louder a sound is, the longer its greater energy will take to dissipate and so, to provide a standard basis for comparing the reverberance of different spaces that is independent of loudness, we have the concept of *reverberation time*. The RT in a space is accepted as the time taken for a 60 dB transient sound to die away to silence – or for a transient sound to fade by 60 dB. Incidentally, the decay gradient will never be wholly regular but, for general purposes, it can be assumed to be so.

6.5 shows the reverberation time limits appropriate to various activities. The optimum RT for speech, in a lecture room or classroom, is short, at around 0.75 sec. There, the overriding aim is for clarity, for syllable articulation. When the speaker projects a syllable, it is the direct sound, the sound travelling along the direct path to the listener, that contributes most to

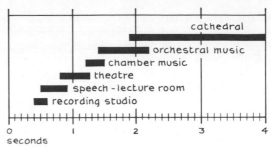

6.5 Approximate, desirable reverberation times

intelligibility. First-reflected sound may also help if it arrives hard enough on the heels of the direct sound for the brain to accept it as the same, within about 35 milliseconds, in fact. Later reflections tend to blur the syllable and intrude upon following syllables, but this cannot happen if a fast decay keeps the reverberant level safely low. Of course, if the listener is close to the speaker, the direct sound level will be safely above the background reverberant level. If the listener is remote, at the back of a larger hall, the fall-off in the direct sound intensity makes excessive background reverberation more significantly intrusive.

As shown, drama prefers a slightly longer RT. Clarity is still important but, then, the actor's voice is trained to be clear and to project to the remoter parts of the theatre and, for reasons that are frankly not fully understood, a certain amount of reverberation seems to help convey mood and the finer suggestions in meaning. Music asks for a further increase, since reverberation imparts power and richness of tone and, also, in comprising reflections from all around, it seems to help the audience's sense of total envelopment. How much of an increase is desirable depends on the kind of music. The notes of chamber music or a piano recital still call for definition, rather more so than orchestral – compare Chopin's works with a Beethoven symphony. The cathedral is the extreme. There, the large volume and the highly-reflective masonry and glass surfaces both contrive to give times up to 5 sec or more, blurring individual sounds but allowing choral and organ music to mass in power and majesty.

Design to control reverberation time

The Sabine formula

The relationship between reverberation time, volume and total absorption was first established by the Harvard physicist, W. C. Sabine, whose researches early this century pioneered so much of modern acoustic theory. Apparently, Sabine was obliged to become nocturnal during the period of his reverberation researches, in order to have the necessary quiet, labouring nightly with a starting pistol and stop-watch to measure the rate of sound decay. Today, there are various formulae for predicting the RT in a space, from rules-of-thumb to the quite complex and accurate. But Sabine's original formula remains a corner-stone in room acoustics, giving an adequate estimate for most design purposes. In metric terms, it is

$$RT_{60} \text{ in sec} = 0.16 \times \frac{\text{volume}}{\text{total absorption}}$$

where the volume is in m^3 and the absorption is in *square metre absorption units*.

Quantifying the absorption

Actually, Sabine's original unit, later named the *Sabin*, is equivalent to 1 ft² of totally absorptive material – a strictly theoretical idea which, in practice, would only be achieved by an open window. The metric version, the 'metric Sabin', is the square metre absorption unit (m² unit) in the formula.

The *absorption coefficient* of a material is its absorption compared with a totally absorptive surface of the same area. Multiplying a material's absorption coefficient by its surface area in m², gives its contribution to the absorption in a space in m² units. The table illustrated (6.6) gives a brief sample of the absorption data available to the designer. Estimating the total absorption in an intended space requires the coefficients and respective areas of the proposed surface materials to be multiplied and the results totalled – $A_1S_1 + A_2S_2$ etc. The considerable effect of people and furnishings must also be added in, the m² units of each audience member and his seat, and of the estimated proportion of remaining seats unoccupied, and of the performers themselves – an orchestra will absorb a fair amount of its own sound. Notice that absorption varies with frequency, and that different materials vary differently with frequency, some absorbing better at high frequencies and others at lower. And the absorption of a material can further depend on how it is applied,

item	unit	125 Hz	500 Hz	2000 Hz
concrete, stone	per m²	0.02	0.02	0.04
plaster on solid wall	per m²	0.02	0.03	0.04
brickwork	per m²	0.05	0.02	0.05
wood-block, plastic floor	per m²	0.02	0.05	0.1
plaster on plasterboard battened out from wall	per m²	0.25	0.1	0.04
t&g timber on solid wall	per m²	0.3	0.1	0.1
heavy curtains	per m²	0.1	0.4	0.5
polyurethane 'acoustic' tiles on solid backing	per m²	0.15	0.8	0.8
person in padded seat	m²	0.15	0.45	0.45
padded seat unoccupied	m²	0.15	0.4	0.45

6.6 Some absorption values – for the total absorption in a space, the coefficients (per m²) of the materials are multiplied by their respective areas, and the results added. The m² units of people, seats etc. are added in directly

whether as a continuous surface or (more effective) in patches, whether directly on a solid backing like a wall, or mounted slightly out from the backing – we will come on to all this in a moment. At high frequencies, the slight absorption provided by the air is a factor.

... and so achieving the required RT

Clearly, there are various calculation routes with the Sabine formula. A common case is where the desired RT for an auditorium is decided and, hence, substituted in the formula in order to find the combination of volume and total absorption that will achieve it.

Now, it is sometimes suggested that the volume of an auditorium is controlled by the particular activity and the size of the audience, and that it is, therefore, the absorption of the surfaces that has to be varied to control the reverberation. But this is something of a misconception, especially with the slightly longer RTs required for music. The function and audience size must, certainly, influence the plan but the ceiling height and, hence, volume can be altered – and to good effect. A large part of the absorption in an auditorium, often the greater part, is contributed by the audience and, with too low a ceiling and, hence, volume, one could well be *irretrievably* landed with too short a reverberation time, even though the enclosing boundaries were highly reflective. Looking at it another way, for a given purpose in an auditorium and assuming helpfully reflective boundaries maximising the sound energy to the listeners – and that is important – there is an optimum volume per person which will give approximately the right reverberation time. For interest, it works out at around 3 m³ to 4 m³ for speech and 6 m³ to 7 m³ for music.

A simpler case is where the intended volume and surfaces of a theatre, or lecture or assembly hall or, say, a school gymnasium to be used occasionally for assembly, are known. There, if anything, the RT may turn out too long. For example, if the RT at 500 Hz – around the middle of the working range – came to 2 sec in an auditorium ideally requiring 1 sec, then there would be an immediate indication for added absorption. In practice, the RTs would be further calculated for lower and higher frequencies in the range (the latter taking into account the air's absorption contribution) and, if either diverged significantly from the 500 Hz value, then materials could be included with appropriate low- or high-frequency absorption characteristics.

Adding absorption – what kind and where?

The type and positioning of absorptive treatment to modify reverberation is usually a matter for the acoustical engineer but there are points of interest. Essentially, there are three types of absorber.

The main category are the *porous absorbers* that dissipate the sound within their thickness – carpets, heavy fabrics and soft furnishings, and the array of materials specifically designed to absorb sound, such as spray-on foams and the familiar, perforated 'acoustic' tilings.

There are *panel absorbers*. Any panel of non-porous material, mounted so as to create a narrow air gap between it and a solid backing wall or other part of the solid room enclosure, will tend to resonate, together with the air behind and, in doing so, will absorb sound energy. In design, it may be a case of allowing for the panel-absorber effect of construction installed for other than acoustic reasons, for example a plasterboard false ceiling; or it may be a case of deliberately installing the panels for a specific absorptive effect. Panel absorbers' absorption coefficients tend to increase towards the lower frequencies.

Finally, there are *cavity absorbers*, sometimes called 'Helmholtz resonators'. A cavity resonator can be likened to an empty bottle with narrow neck and larger air-space behind. In response to a select forcing frequency from the sound waves outside, the air in the cavity resonates and, in so doing, absorbs sound energy at that frequency. If one blows across the open top of an empty bottle, it resonantly responds at a frequency that depends on its shape. In fact, certain types of perforated tile are, to an extent, employing cavity resonance at small scale. At larger scale, are panels with holes drilled in, mounted proud of a solid backing to create a resonating air cavity behind. Larger still are perforated concrete or honeycomb brick screen walls, built slightly forward from the main enclosure wall. This last device is occasionally found in churches or other formal public spaces, where it offers a hard finish but still some absorption. If the screen is slightly angled to the backing wall, the frequency range absorbed can be widened. But, generally speaking, cavity resonators are, by their nature, best suited to selectively correcting where a particular frequency in a space is found to be reverberating too long (we will see later how this can happen), rather than for correcting the RT as a whole.

As to the positioning of absorptive treatment in a hall, the usual order of preference is first on the back walls, then on the side walls and ceiling margins, and on the centre of the ceiling last of all. The reasons for this will also be seen later.

Unavoidable variation in reverberation time

There is a limit to the accuracy worthwhile when predicting reverberation time however, since it must, in any case, vary as audience numbers vary from one performance to the next. Sensibly, the design calculations will be based on an average anticipated attendance rather than a full attendance. And heavily padded seats help. They bring more than comfort, since their absorption empty can compensate by almost matching their absorption filled – bad enough to have a poor attendance without ruined acoustics as well!

Intended variation in reverberation time

Another problem occurs when an auditorium is needed for a variety of uses, perhaps for lecturing, drama and even music, all asking for different reverberation times. What is to be done? Well, at a very simple level, there could be heavy curtains drawing back to expose a more reflective wall behind, albeit the curtains would offer little absorption and, hence, little change from the bare wall, at the lower frequencies. More elaborately – and this has often been done – there can be some sort of reversible wall panel arrangement, enabling either the reflective or the absorptive side to be exposed. The trouble is, it takes an awful lot of absorption to affect the reverberation significantly, especially when the bulk of the absorption is being provided by the audience, anyway.

A more effective and sophisticated development today is *electrically assisted resonance*, where a battery of microphones, amplifier channels and loudspeakers covers the frequency range, amplifying slightly and effectively increasing reverberation time as the type of performance requires. For instance, the system might be turned on for musical performances but left off for drama, offering two types of hall for only slightly more than the price of one.

Summary

- Sound waves hitting room boundary are partly transmitted, partly absorbed and partly reflected.
- Reflected waves form an interactive pattern of disturbance, decaying to silence.
- Reverberation time in a space is the time for a 60 dB decay. It increases with volume and reduces with total absorption.
- A short RT favours clarity and, hence, speech; a longer RT adds 'richness', especially for music.
- The Sabine formula (and others) relates total absorption, room volume and RT.
- A material's absorption in m^2 units (metric sabins) is found by multiplying its surface area by its absorption coefficient; the total absorption in a space is found by simple summation.
- In an auditorium for music, the required RT is mainly achieved by ensuring an appropriate volume
- In a hall for assembly or drama, a suitably short RT might call for the addition of a calculated amount of absorptive treatment.

DIFFUSION, RESONANCE AND OTHER EFFECTS OF ROOM SHAPE

The splays and buckles of the principal surfaces of auditoria, or the simple slope of a music-room ceiling, will tell anyone that shape must be acoustically very important. The main reason for this contouring is to try and ensure that the response in a space is even to all frequencies of the source, wherever the listener may be. A 'lumpy' response makes for perverse listening conditions, with some frequencies sounding louder and longer than others and with reverberation time impossible to predict realistically. We expect a hi-fi to respond evenly and it is reasonable to expect an auditorium to do the same. The whole question of shape and response has to do with *diffusion* of sound in general and *resonance* in particular.

Diffusion

Broadly speaking, diffusion is the even scattering of the sound reflections in a space, a blending of the sound pattern, if you like. We shall see that the things that help diffusion help even the response. Mark you, total diffusion would be a mixed blessing since, with the sound pressure everywhere the same, the harassed listener would be theoretically incapable of hearing where the sound had come from but, in practice, there should always be enough fall-off away from the source for auditory localisation to tie up with what the listener sees.

The desirable scatter is produced partly by sound absorption, notably by absorptive treatment at the boundaries, especially if it is in strips or patches. Also and, perhaps more obviously, the scatter is produced by uneven surfaces – think of the effect of an irregularly-shaped bank reflecting the pond wave. In the case of sound waves, the unevennesses must be bold in scale, the dimension across any individual surface contour being at least as great as the length of the wave it is intended to disperse, otherwise the wave will reflect as if the surface were flat. For reference, middle C has a wavelength of 1.33 m.

Conversely, the main detractors from diffusion are flat, reflective surfaces. This is especially so when they are opposite and parallel – room resonance will tell us why.

Resonance

Uneven resonance wrecks diffusion. Just to confirm what we mean by resonance, a simple case is when a child is pushed on a garden swing; only if the frequency of the pushes, the *forcing frequency*, coincides with the inherent *natural frequency* of the swing pendulum, will the motion grow. There are dangerous parallels in wind-gusting frequency on building structure. In acoustics, the forcing frequency is that of the sound source and the many natural frequencies which respond are the inherent property of the room, both through the resonance of the air it contains, which depends on the room shape, and through the resonance of the boundaries.

Air resonance

In the midst of the complicated pattern of direct and reflected sound in a room, there will always be *stationary waves*, or 'standing' waves, waves which oscillate on a fixed path between the fixed boundaries without actually getting anywhere. A simple example of a stationary wave is a plucked violin string vibrating between its fixed ends – except that, again, the vibrations are transverse, whereas a sound stationary wave in air oscillates longitudinally, i.e. parallel to its path. It is as if the coiled spring had an end fixed to each opposite boundary: the compression bands would no longer be chasing each other but would remain stationary as the individual portions of the spring oscillated in space. In acoustics, what is happening is that the frequency of the particular sound and, hence, its wavelength, is such that an exact number of wavelengths can fit into the total path length, *so that the reflected wave exactly coincides with the incoming wave, reinforcing it and making the sound louder*. In the same way that the resonant, 'harmonic' frequencies of an organ pipe depend on the length of the air column inside, so the particular *resonant or so-called 'natural' frequencies of a room depend on the various path lengths of the stationary waves its shape allows*.

The graphical representation will help here (6.7). It shows the first three possibilities of a whole series of stationary waves possible in one path length across a room. In each case, there are points where the air is still and without pressure fluctuation. These points are called *stationary nodes*. The simplest case shown at the top has nodes only at the walls – it is easy to imagine that the standing-wave air at the point in contact with the walls is still – making for a half-wavelength standing wave. Then the addition of a node in the middle makes for a complete wavelength. Then, comes one with two nodes in the middle and so on, in an infinitely expanding series of natural frequencies. At each step, the wavelength decreases and the frequency increases –

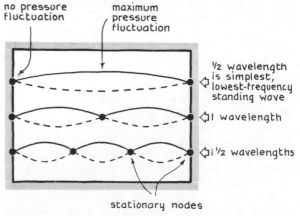

6.7 Standing waves – the first three resonant possibilities, i.e. natural frequencies, across just one dimension of a room

moreover, the frequency jump becomes progressively less, so that the natural frequencies become progressively more closely grouped, the higher they are in the frequency range.

So we can see that any room can have a multitude of natural frequencies waiting to be excited – wall to opposite wall or floor to ceiling. There are also countless more complex standing-wave paths, incorporating more than two boundaries and, possibly, the angles between boundaries as well, albeit their effect is usually less significant since the greater boundary contact tends to speed their decay. As the source's forcing frequency approaches a natural frequency or, as is more often the case, a concentration of natural frequencies, the room's response will be louder. Curious effects can result. A church organ may play the notes of the scale at the same sound pressure but some of them may sound louder than others to the congregation. Many a mediaeval chant was pitched to exploit the resonance characteristics of a particular church. But uneven response in a modern building for music or drama would be a serious defect.

The problem is mainly apparent with smaller spaces and lower frequencies – with dimensions less than about 10 m and frequencies below about 350 Hz. Smaller spaces have fewer, more widely spaced, natural, resonant frequency modes – especially at the lower frequencies – and lower frequencies are harder to randomly disperse. All this tends to make the resonant response a more lumpy, hit-and-miss affair. At higher frequency, the closer grouping and the presence of the numerous, complex waves ensure that the resonance peaks satisfactorily overlap.

Anything that increases the resonance possibilities helps even the response. The designer will avoid having the principal room dimensions simply related arithmetically, since that would only duplicate the available path lengths and reduce the number of natural frequencies on offer. Sound-absorptive treatment helps because it tends to broaden the resonance peaks so that they further overlap. Boldly contoured surfaces similarly help. So, to some extent, may out-of-parallel surfaces like the traditional sloping music-room ceiling though, in fairness, many acousticians have come to doubt this, arguing that they only make the troublesome frequencies harder to predict.

Other effects of room shape

Leaving aside resonance and the special case of stationary waves, there are other, simpler ways in which room shape affects reflection and response – some harmful, some helpful.

Faults
Echoes. Sound reflected to a listener from an adjacent boundary must obviously have travelled further than sound received directly from the source and, in a reflective room when these path lengths differ by more than about 30 m, a distracting echo will be heard, and even 15 m difference will cause blurring. A common case in auditoria is the rear wall or the front of the balcony, echoing back to the front rows of seats – again, the prevention is broken and/or absorptive-treated surfaces.

Another echo problem is *flutter*, where a sharp impulse like a handclap rebounds between opposite, parallel, reflective surfaces, to be heard as a repeating echo dying away to inaudibility. With the surfaces 20 m apart, it would sound as separate impulses, clap . . . clap . . . clap. Between the closer flanking walls of a stage recess, it would be a buzz. Flutter problems are unrelated to those from standing waves, despite their similar prevention – by absorptive treatment or angling one of the surfaces slightly.

Concave surfaces – foci and dead spots. Convex surfaces are really only enormous bumps and so tend to scatter the sound they reflect and usefully promote diffusion. But concave surfaces, such as curved walls and domes or barrel ceilings, may cause *sound foci*. The effect of these foci ranges from the merely curious, as where a domed ceiling transmits snatches of very private conversation across a crowded room, to the acoustically chaotic, as in an auditorium, where they could cause undesirable increases in sound level in some parts (possibly heard as echoes) and corresponding reductions in other parts, which then become dead spots with poor audibility (6.8). Worse still, if the source is moving, where an actor crosses the stage or different sections of an orchestra play, the sound intensity to some parts of the audience can vary from one moment to the next – making acoustic nonsense. The designer has to ensure that the radii are of such length that no foci occur at audience level, or that concave surfaces are absorptive, or avoided altogether. Acoustically, as well as structur-

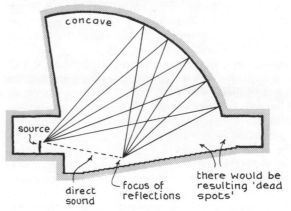

6.8 Concave ceiling causing discernible echo and preferential foci – any resemblance to an existing opera house is, of course, purely coincidental

ally, Sydney Opera House posed its functional problems. Clearly, the curved shells would have been an acoustic disaster. In the event, a profiled, suspended ceiling of moulded plywood was used for the acoustic boundary above the audience.

Additional measures for audibility

In larger spaces, there arises the problem of getting an adequate quantity, as well as quality, of sound to all listeners, especially those most removed from the source.

The auditorium plan

The fan-shaped plan (6.9a) has the immediate advantage of bringing the audience closer to the source than would the deeper rectangle of equivalent area. The narrower width at the stage end usefully intensifies the sound reflections there in a way that can both help projection to the audience and, an often forgotten point, help the performers to hear *each other*. Also, speech intensity is fairly directional and the fan shape helps keep the front rows of the audience within a reasonable angle of a forward-facing speaker. But there are disadvantages. From the overall planning point of view, the fan can be a rather awkward shape to reconcile with the other accommodation in a building. Acoustically, it can tend to funnel sound from the stage towards the rear corners of the auditorium in an unbalanced way and is more likely to generate concave surfaces in the back wall, calling for absorptive treatment there to avoid foci and preserve diffusion. And the more rectangular plan (6.9b) has the merit that its side walls tend to increase the lateral reflections to the audience which, as explained, can be helpful to the appreciation of music – opposed side walls may increase the risk of uneven resonance and flutter, but only slightly, since the source does not, itself, lie between them.

Very broadly speaking, the factors tend to favour the shallower plan for lecturing and drama, and the deeper, rectangular plan for music. But, as instanced (6.9c), there are finer compromises possible between these alternatives, contoured to retain some of the merits of both and to promote diffusion.

Balconies, also, help get people closer to the source in a large auditorium, but their seating area, i.e. the extent of their projection forward from the back wall, is limited, lest they cast shadows – dead spots – on the audience seated under them.

The auditorium section

6.9d is a very simple auditorium section, the essential features of which are a raised stage and a progressive rise in the audience seating. These would be needed, in any case, to provide good sight-lines, but it is important to realise that good sight-lines make for good direct sound-lines as well. Sound passing at 'grazing

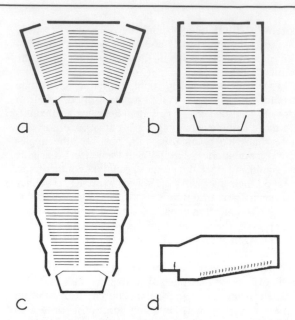

6.9 Auditorium plans. (a) The shallow (lecture theatre) fan tending to favour speech (b) The deeper rectangle tending to favour music. In practice, there are compromises between these extremes (c). (d) Simple auditorium section

incidence', close over the audience's heads, is strongly absorbed, increasing the rate of fall-off away from the source. To avoid this, the arrangements should be such that the sound-line from any row is at least 100 mm above that from the row below – we show a regular seating rise but, by geometry, the sight-/sound-line requirement may call for a progressively steeper rise towards the back of the hall.

Use of beneficial reflections

As well as designing to get the people near the sound, one can design to get the sound near the people – by use of beneficially reflecting surfaces. A pure example is the parabolic reflector behind the orchestra outdoors, beaming the sound over the audience. The inclined timber panel behind a church pulpit is doing much the same thing at smaller scale. As to the room surfaces themselves, take, first, the simple case of a classroom (6.10a). Amidst all the reflected pattern within the boundaries, some reflections will be more specifically helpful in reinforcing the sound of the teacher's voice for the pupils – notably those from the ceiling because that is the surface with the best uninterrupted view of the whole class. Clarity, in a large classroom, can be helped by absorptive treatment to control reverberation time but, equally for clarity, such treatment should be withheld from the ceiling. There is no conflict here, it is just a question of shaping the sound in the optimum way.

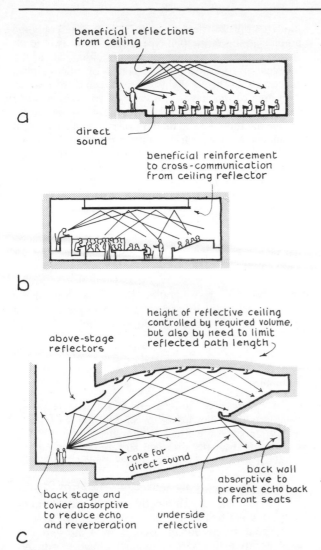

a

beneficial reflections
from ceiling

direct
sound

beneficial reinforcement
to cross-communication
from ceiling reflector

b

height of reflective ceiling
controlled by required volume,
but also by need to limit
reflected path length

above-stage
reflectors

rake for
direct sound

back stage and
tower absorptive
to reduce echo
and reverberation

underside
reflective

back wall
absorptive to
prevent echo back
to front seats

c

6.10 Beneficial reflections from (a) the ordinary, reflective ceiling in a classroom or lecture hall, (b) the ceiling reflector in a courtroom and (c) The profiled ceiling in a theatre auditorium

In a conference room or courtroom (6.10b), there needs to be clear communication between all participants to-and-fro across the space. Hence, the installed ceiling reflector. This will often be in timber, whose even, resonant response across the frequencies is a further bonus.

As we might expect, the most forcefully expressive use of reflective surfaces is in auditoria. We have seen, generally, how an auditorium can be made reasonably reflective and sound-reinforcing, provided the optimum volume per person is chosen to give the desired reverberation time- or at least get close to it without therefore requiring excessive and, hence, sound-losing absorptive treatment on the one hand, or having con-

ditions irretrievably dead, on the other. But design for reflection is taken further. As said, if the reflectors are ill-shaped, they can harmfully unbalance the acoustic response but, by the same token, they can be used in a judicious, controlled way to reinforce the sound intensity and, indeed, preferentially to reinforce it to the farther seats (6.10c). The exact surface profile is the result of highly specialised design but here are some pointers. The boundary must be close enough to the direct sound path to prevent the path difference and, hence, lag between direct and reflected sound from being great enough to cause blurring. The contours – which, incidentally, may be exploited to accommodate ventilation extract and lighting – must be bold enough to reflect the low frequencies of long wavelength. Further, their shape may have a convex ingredient, recognising the additional need for diffusion – again, the needs for reasonable diffusion and yet preferential reflection should be seen as mutual contributors to the optimum shape, rather than as conflicting. Going a little bit further, one can say that, in recent years, auditorium design specifically for speech and drama has come to regard the ceiling as the most important reflector – it covers the whole area and, also, auditory localisation is less sensitive to and, hence, less confused by, reinforcing sound coming from above. So diffusion and absorption there must come more from the walls and carpeted floors and, of course, people. For music, however, where intensity reinforcement is less critical, the ceiling is seen more as a diffuser, with walls adding lateral reflections for envelopment. This ties up with the earlier planning comments and it is about as far as this text can reasonably go.

Electrical amplification

Where numbers get much above 500 people, shaping alone will not guarantee audibility, certainly not for speech. Electrical amplification is needed. Amplification need not distort sound, in fact, it can restore its quality as well as intelligibility, and even improve on otherwise defective acoustics. The test of a really good system is that the audience be unaware of it – in a sense, the system must lie.

Single-channel systems

Amplification is easiest when the source is stationary, like the speaker's voice in a lecture hall. At the very simplest, a single microphone is used with one or two high-mounted speakers 'overlooking' the audience. Obviously, the directional illusion will be helped if the speakers are close to or behind the visible source, but they must not be too close, otherwise they may cause *feedback*, where the sound from the speakers is picked up by the microphone again, re-amplified, picked up yet again, and so on, in a vicious circle, rapidly building up to cacophony.

Microphones are normally designed directionally selective, so that they favour the wanted source over any any background noise. Also, the feedback risk is reduced.

A design problem with loudspeakers for music is their tendency to beam directionally the higher frequencies more than the low ones, meaning that listeners removed to either side of the speaker axis receive a distorted, high-frequency-starved sound. It happens, though, that smaller speakers are inherently less directional and this fact can be exploited to reduce the frequency directional effect. A loudspeaker unit will usually comprise a large speaker, or speakers, for the low frequencies (large, because the lower frequencies need more power for the same loudness) and smaller speakers for the higher frequencies. Logically enough, if they are mounted above each other ('woofer' above 'tweeter'), making the familiar vertical loudspeaker unit, people's less sensitive vertical auditory localisation allows the source position to seem static as the individual speakers sound.

Of course, conditions will often call for more than a single loudspeaker near the source. Take an extreme and awkward case – a cathedral, which is large, irregularly shaped and with columns and other obstructions blocking the direct sound path from pulpit or lectern to the congregation. Ordinary loudspeakers at one end, beaming down the nave, would never be adequate. Turning up their volume would be no great help for, in such reverberant surroundings, that would only cause a confused build-up of sound, reducing syllable articulation to the point of incomprehensibility and, possibly, causing feedback into the bargain. A common technique nowadays is actually to exploit the extreme directionality possible with modern 'column' speakers, mounting them above the source and beaming them at the most effective absorber, namely, the audience. This gives adequate received volume without over-reverberance. In fact, the large space may have several such speakers, separately located to cover sections of the audience progressively farther away from the source. Here, lest the sound seem to be coming from the masonry itself, some means is again needed to perpetrate the lie that it is coming from the source.

An electronic delay built into the system allows the direct sound to reach the listener fractionally before the reinforcing sound from the loudspeakers – in terms of auditory localisation, it is this first received sound, albeit faint, that counts. The built-in delay can be quite sophisticated, allowing the farthest listeners for example, to hear the direct sound first, an intermediate loudspeaker between them and the source fractionally after, and their local speaker supplying the main reinforcement, fractionally after that.

Stereophonic systems

A moving sound source, such as in theatre and many kinds of musical performance, calls for stereophonic amplification. A simple type might have two loudspeakers facing the audience, one on either side of the proscenium opening, and a series of microphones spaced along the footlights, each directed towards its own area of the stage. The voice of an actor crossing from one side to the other is preferentially picked up by one microphone and then the next, and the electrical arrangements are such that this transfer results in one loudspeaker, increasingly more than the other, contributing to the total volume of the reproduced sound. This gives the audience the auditory illusion that the sound has moved in confirmation of what they see.

Clearly, the combined needs for progressive reinforcement towards the farther seats, and for stereo, lead to quite complicated systems. But the technology has responded to the need and, also, to people's higher expectations of sound quality these days. Systems for amplification and, possibly, for assisting the control of resonance and reverberation time, are increasingly seen as part of the modern acoustic package in a large auditorium. The trend is likely to continue.

Summary

- In diffuse conditions, there is an even response throughout the space to all frequencies of the source. Measures for diffusion are:
- – contouring the principal surfaces and/or adding absorptive treatment to scatter the sound reflections;
- – slightly angling the principal surfaces to maximise the number of path lengths and, hence, standing waves and hence, natural frequencies, available to respond to any forcing frequency of the source;
- – ensuring that surface materials have reasonably even boundary resonance characteristics.
- Echoes are heard when the reflected path from the source to the listener is more than 30 m greater than the direct path. Remedy is again by broken or absorptive surfaces.
- Flutter is a repeatedly rebounding echo between opposite, parallel, reflective surfaces – avoided by slightly angling or absorptive-treating one of the surfaces.
- Concave, reflective surfaces bring a risk of preferential foci and dead spots.
- Audibility helped by planning for people to be close to the source. But there are other general planning and sound-quality considerations – compare the fan-shape plan, favouring speech, and the more rectangular plan, favouring music.
- Amplification is usual in larger spaces – loudspeakers are usually placed high, with single channel for a stationary source and stereo for a moving one.

NOISE CONTROL

'Noise' is unwanted sound and it is a very real form of pollution today, eroding the quality of shelter our buildings give us. And whether it comes from outside, such as from heavy traffic of aircraft, or from inside, transmitted between rooms, the lightweight nature of much modern construction has, inevitably, tended to make the problem worse.

The decibel scale is an approach to our subjective assessment of sound intensity and, as such, is the starting point for measuring noise. But there are added factors. For one thing, apparent loudness varies slightly with frequency, tending to peak in the middle range from around 1000 to 5000 Hz. This has led to various modified forms of measurement. One is the *phon* scale. Phons, like decibels, measure apparent loudness but they take account of frequency. A given level in phons requires a higher actual pressure level towards the low and high frequencies than in the middle range. Noise-level meters can be designed to screen out the lower and higher frequencies and/or have specially weighted scales to allow for the frequency influence. The *A-weighting network* is one of the most commonly used, readings being designated dBA as distinct from the simple sound-pressure level dB (SPL).

Another factor – and this one is very subjective – is that the disturbance potential of any given sound depends also on its nature. Sounds tend to bother us more if they are intermittent (like a lift next to a hotel bedroom), or interesting and therefore attention-seeking (like conversation or rhythmical music), or at night. In Europe, Noise Criteria (NC) curves are one standard, plotting acceptable sound-pressure level against frequency for various activities. A US standard is the Perceived Noise Criterion (PNC) curve. But do not worry *too* much about all this – it is useful to be aware of these measurement distinctions but their application is really for the acoustical engineer.

Noise transmission can, obviously, be *airborne*. It can, also, be *solid-borne*, transmitted through the building elements, such as the impact noise of a footfall in the room above or vibration from mechanical plant. In either case, the key control is preventing the sound vibrations from travelling.

DESIGN AGAINST AIRBORNE NOISE

First steps – initial planning

Airborne noise is the most common disturbance in buildings. In domestic buildings – in the small house

– it is enough to ensure that the ordinary, good construction of the elements that enclose and internally divide the building meet the designated, minimum standards for sound insulation. For example, the ordinary domestic enclosure will usually serve tolerably to exclude noise outdoors, that is, provided town planning has not been especially delinquent in its residential-area zoning. But, in most larger building types, there should be thought for noise control from the early design stages. Ambient noise levels outside may be such as to affect the enclosure design, and indeed high levels outside and/or building functions asking for low criteria inside (perhaps a hospital, hotel, or library on a busy urban site), may call for sealed windows and full air-conditioning in consequence. Often, noise-meter surveys for a proposed site, taken over, say, a 24-hour period, will be a valuable guide. *In the internal planning, sensible zoning can be absolutely critical.* Again, take a hospital, with wards and treatment areas requiring quiet and yet with much noisy plant and servicing – *planning these conflict areas well apart is a first step in noise control, before any thought for the insulation performance of the internal walls and floors.*

Sound reduction through walls and floors

But, of course, the principal protection against noise is the building's walls, floors, and other divisions – generically, we will call them partitions. In a nutshell, when sound waves hit the surface of a single-leaf partition, they cause it to vibrate so that the far surface becomes a secondary sound source for the receiving space beyond. Some of the incident sound energy goes to making those vibrations and, in consequence, there is a sound-level drop. The particular insulation performance in decibels of a partition is called its *Sound Reduction Index* (SRIdB), or to use the simple, synonymous term more common in the USA, the *Transmission Loss* (TLdB).

The actual sound insulation achieved, i.e. the sound reduction into the receiving space, is mainly governed by the transmission loss but not solely. There are two other factors. First, the total absorption in the receiving space affects – reduces – the sound level the transmitting partition achieves there. Cars are quieter inside with passengers. Second, and much more important, the designer has to beware of sound's fickle knack of finding the path of least resistance. A gap in an ill-fitting window will reduce sound insulation out of all proportion to its own small area – there is a limit to the useful deadening in the bulkhead behind a car engine when sound can still come up the gear shift. Imagine sound, leaking, like water. In buildings, *flanking transmission*, where sound bypasses a partition by routing round via a corridor of ceiling void, is part of the same problem. But, returning to direct transmission for the moment – as a first idea, table 6.11 compares some

type of construction	TL dB
single-glazed window, 6mm glass	26
double-glazed window, 4 and 6mm glass with 100mm separation, absorptive reveal	34
lightweight partition – 100x50mm timber studs, plasterboard and plaster both sides	35
50mm wood-wool slab partition, plastered both sides	35
75mm blockwork, plastered both sides	41
100mm brick wall, plastered both sides	45
200mm brick wall, plastered both sides	50
175mm dense concrete	50
cavity brick wall with 100mm leaves, inner leaf plastered, wire ties	54

6.11 Approximate transmission losses for airborne sound through some common constructions. Values averaged from 100 to 3000 Hz

common constructions' ability to insulate against airborne sound. The values can only be approximate, as will become apparent.

Insulation performance – partition mass and stiffness, and the effect of frequency

Logic suggests that the thicker and denser the partition, i.e. the more the inertia, the greater the transmission loss will be. For most practical applications, this is, indeed, the case and the transmission loss is said to be 'mass-controlled'. Thus, *sound-insulating* materials tend to be distinct from *sound-absorptive* materials, the latter being generally light and porous. Averaged over the frequency range, a 100 mm (i.e. single-leaf) plastered brick wall achieves a transmission loss of about 45 dB. Doubling the thickness and, hence, mass, a 200 mm brick wall achieves about 50 dB, a 5 dB increase. As a rule-of-thumb, each thickness doubling of a homogeneous partition brings a transmission loss increase in the region of 5 dB.

But, as we may have come to expect from acoustics, we would be lucky were it that simple! Again, there are other factors. As just implied, transmission loss varies with frequency, being generally poorest at the lower frequencies and improving towards the middle and higher frequencies. This is why traffic noise persists as a low rumble heard indoors and why the base notes of the hi-fi next door are harder to exclude than the high ones or speech. Further, at the lower frequencies especially, the constructional stiffness of a partition comes significantly into play in its insulation performance, except that, to complicate matter further, certain lower frequencies may resonate with the natural vibration frequency of a partition – for example, window panes to traffic. And, as well as these primary resonances, there is the 'coincidence effect' where, at the

higher frequencies, the wavelength of the incident sound happens to coincide with the natural flexural (bending) patterns of the partition, again dipping the transmission loss below the mass law value. So values for various constructions must either be listed at intervals in the audible frequency range or, as a rougher indication, must be averaged over the range. In the UK, the Sound Reduction Index of a partition is taken as rated under laboratory conditions and is frequency-related. And there are Mean Sound Reduction Indices which relate to averaged values in ranges appropriate to traffic, music, speech and so on. US practice further rates partitions by their Sound Transmission Class, a single value that can be adjusted to frequency and other circumstances via standard curves and tables – analogous, perhaps, to the US tabular approach to the estimation of indoor daylighting.

The value of discontinuity

If doubling the thickness of each partition adds a 5 dB or so transmission loss, there is clearly a reducing benefit the further we go. The requirement for high transmission losses would involve uneconomically heavy construction. A 55 dB loss might easily be needed on a busy street or near an airport but it would take a rather idiotic 460 mm brick thickness to supply it. *Instead, the partition design must seek some form of discontinuity to sever the direct sound-transmission path.* If, instead of doubling the thickness of our 100 mm wall, and improving from around 45 dB to 50 dB, we double-up by having two *separate* 100 mm leaves then, theoretically, there should be an adding-up effect, giving 45 dB + 45 dB, i.e. 90 dB, for the same weight of brick. In practice, we would never get near it. The walls can never be wholly 'discontinuous', since they will probably spring off the same foundation or floor slab and be linked elastically, to some extent, by the air-gap between them. But a 10 dB improvement to 55 dB is now possible (6.11a). (This is assuming that piers, or some other means of stability, have been found to replace wall ties directly connecting the leaves.)

The same principle applies to the stud partition (6.11b), where each plasterboard face has its own support studding. Incidentally, the insulation-type quilting shown would only have a marginal effect, and mainly at the higher frequencies. Similar improvements could be expected in the all-timber floor, where the floor boarding above and ceiling plasterboarding below are each supported on separate joisting.

Perhaps the most familiar discontinuous construction is double glazing (6.11c), usually installed for thermal benefit and adding extra noise-exclusion as a bonus but, sometimes, specifically installed for noise-exclusion. There is a point here, though. Thermally, double glazing performs best with an air-gap around 10 to 15 mm, and there are the thin, factory-made sandwich

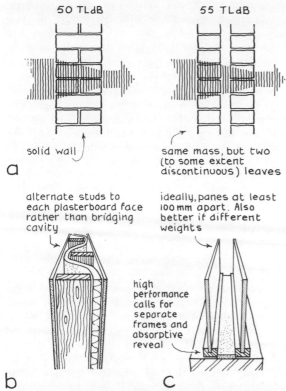

50 TLdB   55 TLdB

solid wall

a

same mass, but two
(to some extent
discontinuous) leaves

alternate studs to
each plasterboard face
rather than bridging
cavity

ideally, panes at least
100 mm apart. Also
better if different
weights

high
performance
calls for
separate
frames and
absorptive
reveal

b   c

6.12 In addition to mass and stiffness, discontinuity valuably
adds to transmission loss – (a) Cavity wall (b) Timber stud
partition (c) Double-glazed window

How much sound insulation?

Often, the required transmission loss (or equivalent
sound-reduction index or sound-transmission class) of
a building's dividing partitions is controlled by minimum legal standards, varying with building type, occupancy or the nature of adjacent occupancies, e.g.
bedroom to bedroom, bedroom to stairwell, and so
on. Alternatively, the appropriate insulation level will
be found by calculation, and a brief outline of one
approach will help summarise many of the points so far.

The first thing for the designer is to set the criteria,
to know the likely nature of the potential noise – outdoors or in the adjacent room – and the acceptable
level in the receiving room. For example, determining
the likely level outdoors may be a matter of reasonable
assessment (6.13a) or 24-hour metering. The acceptable
level for the activity in the receiving room may be
found by reference to Noise Criteria curves across the
frequencies but the suggested values at 1000 Hz (b),
again, give a general idea. The difference is then the
target, except that 5 dB safety increments may be
added where the perceived noise is increased by being
e.g. intermittent, 'interesting' or at night or, conversely, there may be relaxation, e.g. where peaks are
only by day.

The actual drop into the receiving room must match
this target and, providing there is no significant flanking transmission, the

$$\text{Actual drop} = \text{partition TL} + 10 \log A/S$$

location	dB	
	night	day
quiet rural area	25	35
suburban residential area	35	45
town road, light traffic	40	50
busy city street	55	60

a

activity	dB
auditorium	20
conference room	25
bedroom, hospital ward, cinema	
living room, lounge	30
library, private office	35
general office space, restaurant	40
cafeteria	45
noisy computer room, games hall	50
typing office	55
workshop,	60
factory (depending on process)	65

b

6.13 Suggested likely noise levels outdoors (a) and acceptable noise levels indoors (b). Values are very approximate and
assumed at 1000 Hz.

units which can be glazed like a single pane directly into
the frame. But windows like these provide only a marginal improvement in noise-exclusion, because the narrow air-gap tends elastically to connect the panes.
Noise-wise, the panes want to be at least 100 mm apart
and, preferably, more. The performance will also
improve if the frames are discontinuous, i.e. each pane
having its own frame. And it is better if one of the
panes has less thickness, less weight, than the other,
since then their vulnerable primary resonance and
coincidence frequencies will be different, ensuring that
at least one pane is providing full integrity at every
point in the range. Absorptive lining in the reveal will
help by reducing sound build-up within and, hence,
transmission across the cavity – it is, after all, only a
small 'room' in there.

But windows are only insulators if they remain
closed, obviously, and it is not much use if having quiet
means being swelteringly uncomfortable. An intermediate measure is fresh-air ventilators, sound-proofed by
having the air route and, hence, sound path zig-zag
through a grid of absorptive baffles – but permanently
shut windows will call for mechanical ventilation and,
probably, cooling. Buildings near airports have been
fitted with sound detectors which automatically close
the windows when the sound pressure rises.

where A is the absorption of the receiving room in m²
units and S the partition area in m². So, estimating the
required TL is a matter of easy substitution. The
amount of absorption can be estimated (surface areas
× their coefficients, as explained earlier) or, in an
existing building, can be found by taking reverberation-
time readings and substituting in Sabine's RT formula.
Note, from the above formula, that more absorption
will increase the drop and more transmitting partition
area will reduce it, which is what we would expect.
Anyway, knowing the required TL, the appropriate
partition construction can be chosen, using standard
tables or manufacturers' specifications.

Of course, a partition may comprise more than one
element, each having its own TL, for example, a wall
with a window. From the 'weakest link' tendency, it is
obvious that one cannot simply multiply the respective
areas and their TLs to obtain the average for the whole,
as one can U values in thermal insulation. The graph
(6.14) is one way of assessing the aggregate result. Take
the case of a wall the area of which consists of 4/5 brickwork, offering 50 dB loss, and 1/5 window, offering
22 dB loss. The area ratio 1 : 4 is found on the vertical
scale and this is carried horizontally across to intersect
the appropriate dB difference curve, in this case 50–22
= 28 dB (the 28 dB position can be estimated). The
intersection is found to correspond to a 21 dB required
reduction on the bottom scale, i.e. indicating a combined TL of 50–21 = 29 TLdB. As expected, the window has disproportionally reduced the TL and, if its

area were increased to half the wall area (ratio 1 : 1),
the combined value would virtually reduce to that of
the window itself.

The likely transmission loss through a wall of three
or more elements can be assessed by calculation, essentially, by averaging the products of the component
areas multiplied by their *transmission coefficients* and
divided by the total area – the transmission coefficient
t being the fraction of the incident sound energy transmitted through a partition and found from TL = 10 log
1/t dB – but here we are getting outside the architect's
scope and, in truth, close calculation cannot predict
performance absolutely since, in practice, the individual components will perform differently with frequency. And so much will depend on the construction
integrity, how well the glass is fitted into a window
frame, for example, and how well-made are the vulnerable joints, as where brickwork meets the side wall
or the slab above – good site supervision will count for
a lot.

In short, where there is a need for high insulation
levels, there will be little point in up-grading, say, a
partition wall without up-grading component windows
and doors as well, i.e. having windows double-glazed
at least, and doors heavy and well-fitted and, indeed,
preferably in the form of a two-door lobby to give the
'adding-up' effect – the lobby between hotel bedroom
and corridor is a good example. Reasonable absorption
in the receiving space helps. Establishing the cost-
effective balance between respectively up-grading the
wall, the other components or the absorption, would be
a matter of fine calculation and, in practice, the designer needs to attend to all parts of the package. And
then there is flanking transmission, as follows:

Flanking transmission

Flanking transmission is the indirect transmission of
sound between spaces, as distinct from the direct transmission through the dividing partitions, and its significance will depend on the degree of sound insulation
the partitions themselves are intended to supply. At
ordinary insulation levels between spaces, say 35 to
45 dB, it takes specific design blunders for flanking
transmission to annoy. 6.15 shows two familiar ones,
the dividing partition wall stopping short of the floor
slab above, and adjacent doors leading onto a common,
over-reverberant corridor. Adjacent windows can offer
a flanking path through the outside air, especially
where hinged windows are so hung that they reflect and
reinforce the path. Speaking-tube effects through
ducts can be a problem – spurs off a common branch
leading to ventilation grills in different rooms must be
so arranged that the sound path is not direct, and they
may also have absorptive inner lining. Thoughtlessly
sited service trunking can form troublesome, resonant
paths between the spaces they serve.

Where high insulation levels are required between

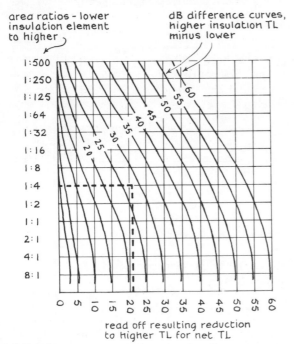

area ratios - lower
insulation element
to higher

dB difference curves,
higher insulation TL
minus lower

read off resulting reduction
to higher TL for net TL

6.14 Graph for finding net insulation provided by a partition
that comprises two elements with different transmission
losses. For method see text.

6.15 Sound can bypass a partition by flanking transmission

spaces, perhaps between adjacent music practice rooms or recording studios, the inevitable, residual flanking transmission through the structural elements themselves becomes that much more significant and is hard to avoid. For example, a transmitting partition wall will always tend to impart slight vibration to the floor slabs and side walls it abuts which will, in turn, form flanking routes to the spaces beyond. We will discuss solid-borne transmission in a moment.

Ventilation-system noise
Air movement through ducts and grills can be noisy if the flow is too fast and turbulent. As explained in *Climate services*, this places minimum limits on duct and grill sizes and asks for gentle bends in the system. There has to be careful attention to the design of grills and nozzles in auditoria, where the noise criteria are particularly low and yet where remoter parts of large volumes may call for a fair 'throw' of air to serve them.

Ways have to be found to prevent the noise of the air-handling plant itself, particularly the fans, from being transmitted through the ducts and around the building. Absorptive inner lining is a common precaution, absorbing the sound as it travels along the duct. Where a duct happens to pass through a very noisy space – notably, before it leaves its own plant room – external lagging will avoid noise 'breaking-in' and so being transmitted to other spaces. Noise *attenuators* are effective. These are silencers built into the duct travel and consist of a battery of absorptive-lined baffles over which the airflow passes.

Masking sound – some background sound can actually help
For all that has been said, very low levels of background sound are not necessarily a disadvantage. Someone asleep in a hotel bedroom will probably be less disturbed by an occasional noise from the room next door or from the carpark outside, if it occurs against the gentle hum of the ventilation system. The perceived noise will be relatively less.

Background sound can help privacy. A medical con-

sulting area sited off an active, bustling ward needs less acoustic insulation than it would if sited off a quiet waiting area. An amorous intrigue can be pursued with confidence in a crowded bar.

In the large open-plan office, the double need to exclude perceived disturbance from adjacent workspaces and to maintain reasonable privacy may be partly answered by having *masking sound*, sometimes rather unfairly called masking 'noise'. It is now an accepted treatment, albeit one that is very hard to quantify. The modest background sound of the ventilation system may suffice, or steady recorded sound, 'white noise' may be used – so called because it comprises a mix of the frequencies analogous to white light.

Summary
- Phons, like decibels, measure apparent loudness but allow that loudness is slightly greater towards middle frequencies.
- Noisiness is slightly different from loudness; measurement yardsticks include the weighted dBA scale, the perceived noise (PNdB) scale and noise criteria (NC) curves.
- Sensible planning is the first step in noise control – proper urban zoning in town planning and, in building planning, proper zoning of noisy areas away from low-noise areas.
- Airborne sound transmission loss (TLdB) through a partition is primarily mass-controlled but there are other frequency and resonance/coincidence effects; performance measured by e.g. Sound Reduction Index (SRI) and Sound Transmission Class (STC).
- High insulation requirement suggests discontinuity for 'adding-up' effect
- Reduction into receiving space also depends on absorption there and on avoidance of flanking transmission
- Different TLs of multiple-element partitions cannot be simply averaged (sound tending to find weakest link). Low TL elements, like windows, disproportionally reduce the insulation value.
- Some masking sound can reduce disturbance and help privacy.

DESIGN AGAINST SOLID-BORNE NOISE

The term solid-borne noise is, perhaps, self explanatory – the impact noise of a footfall, carrying down through the floor structure overhead, or the vibration of ventilating plant, carrying through the structure or through the ventilation ducts themselves. Attenuation of sound vibration through the building structure is very slight and very little affected by mass – think of how the

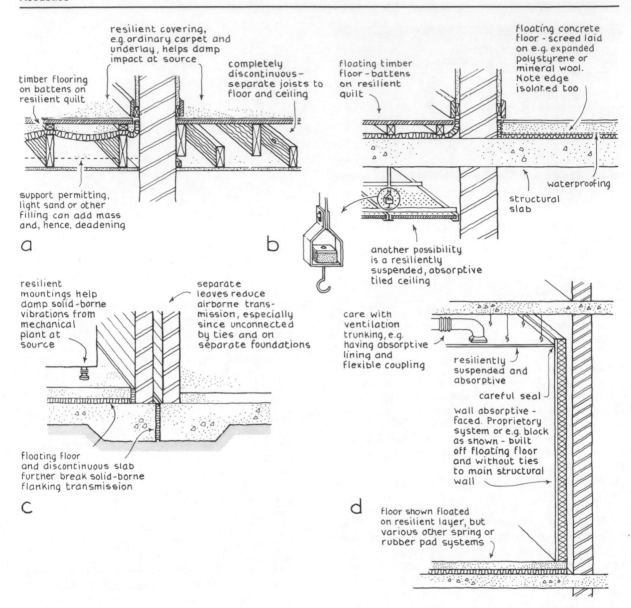

6.16 Construction against solid-borne as well as airborne noise – (a) Discontinuous timber floors (b) Floating timber and concrete floors and resiliently suspended ceiling (c) Structural isolation of a noisy space, like a plant room and (d) The 'box within a box' construction of a sound-recording studio

sound of a train precedes it far along the rail. Discontinuity, valuable against airborne noise, is vital against solid-borne.

Of course, airborne noise is solid-borne, as it passes through a partition, and solid-borne is ultimately airborne to the ear. And, furthermore, structural discontinuity to prevent solid-borne transmission, e.g. through a floor construction, must also reduce airborne as well. But the airborne/solid-borne distinction is useful from the design point of view.

Consider the diagrams (6.16). In (a), notice that the first measure to prevent impact sounds of footfalls, moving furniture and so on, from carrying to the space below – or indeed from flanking via the floor structure to the space adjacent – is to damp them at source by use of resilient floor coverings. Further upgradings are then achieved by built-in structural discontinuities, as shown. (6.16b) shows two examples of high-insulation *floating floors*, where discontinuity is achieved by floating the timber or screeded top floor on a resilient underlay rather than directly on the main slab. The resiliently suspended ceiling is an alternative, the void and absorptive tiles reducing the airborne transmission and the resilient hangers specifically designed against the solid-borne. For very high insulation levels, one might well have a floating floor *and* a resiliently suspended ceiling. But then, of course, something would have to be done to break the now significant solid-borne flanking path through the walls and slab, such as building the wall above off the floating screed rather than off the slab directly and/or having a flexible insert where the underside of the slab meets the wall head below. The construction (6.16c) is isolating noisy basement plant. There is three-stage isolation through the floor construction – the plant's resilient mountings to damp vibrations at source, the floating floor and the slab's split in its travel under the wall. Moreover, the cavity wall will be the more effective against airborne transmission for having no ties (the stability would need to come from piers or articulations of the wall-line on plan) and, also, since the separate leaves are built off effectively separate slabs. In fact, we see the distinction between design against airborne as opposed to solid-borne starting to get rather fuzzy. The room section (6.16d) is taking the discontinuity principle to its logical conclusion. This is the 'box within a box'

construction typical of spaces like recording studios, for which very low noise criteria call for high insulation levels from the buildings around them. The inner skin is, as far as possible, acoustically separated from the main structure. Care has to be taken lest ventilation or other trunkings solidly bridge between the boxes – hence, the flexible coupling. Doors are another critical point. The usual thing is to have a two-door sequence, with the doors heavy, absorptive and well fitted in their frames, and with the lobby enclosure between them absorptive-lined, and, itself, incorporating some form of discontinuity to maintain the inner box's isolation.

In both airborne and solid-borne transmission, design to achieve ordinary insulation levels is not difficult as long as the effects of mass, stiffness, discontinuity, absorption and flanking are appreciated. The same principles obtain in an extended way for high insulation levels, but it is fair to say that their really effective application at the drawing board and then on site is, again, usually a matter for the acoustical engineer.

7

FIRE SAFETY

Most of us, mercifully, have no experience of what it would be like in a building on fire. The day-to-day normality in a building gives not the slightest hint of what would be the sudden transformation and horror – the choking, vomit-inducing, blinding, disorientating, trapping menace – that a developing fire can bring.

But need lack of experience mean lack of awareness? The annual death toll, grim in some countries and, certainly, no cause for complacency in any country, tells of an inadequate appreciation of the dangers and how to reduce them. A leaking roof is soon noticed and can be mended. The trouble is that fire-safety lacks may never be noticed except in the rare event that a fire reveals them, by which time it is too late and often tragic. Hardly surprisingly, good fire safety in design comes as much from the observation of enforced controls as from intuitive provision and if, of all the building controls, these are the ones the designer finds the most awkward; the most intrusive from the planning and the aesthetic point of view, it can only be remembered that *every piece of fire legislation ever enacted has its origin in a repeated, grim statistic*.

And if it is hard for the designer to foresee all the risks, how much harder is it for the owner or manager who must safely operate a building, or for the ordinary user who could help spot safety lacks and, in the widest sense, help strengthen the political will towards achieving better legislation in countries where it is needed. Undeniably, there is widespread misinformation about what constitutes good fire safety. One only has to look at the media response when a serious fire costs lives, so often focusing on peripheral factors like how long it took the fire service to arrive, or how long their ladders were. *When people die in fires, the most architecturally significant factor is that the building did not provide for their escape*. It may help to discount some of the more common fallacies.

SOME FALLACIES

- Flames are the immediate danger – untrue. Smoke is the primary threat to life. Choking, blinding and, very possibly, lethally hot or poisonous, it can quickly disorientate and kill far from the seat of a fire. Its natural buoyancy and the convection induced by the fire will spread it rapidly in an unprotected building. Sleeping people in residential buildings may never live to wake up.
- Buildings with non-combustible construction and surface finishes need less precaution owing to the lower risk – untrue. A building's contents are the most significant threat to life safety in the early stages of a fire. By the time the structure itself is on fire, the occupants are normally out or dead. In one case last century, some 2000 people died in a South American cathedral when festive decorations caught fire. A building of low fire-load may reduce the severity of a fire but a serious fire can still develop.
- Extinguishers can be relied upon to save people – untrue. Their use can be crucial, or too late and wholly irrelevant.
- The fire service can be relied upon to save people – untrue. This is in no way to discount the fire services' vital work but simply to say that, if a building has to rely on outside help to save its occupants, then it has failed. A Boston night-club fire in 1942 killed 500 people, although the fire service (responding to a slightly earlier false alarm) were on the scene within a minute of the outbreak.
- '. . . in event of fire' notices instructing building occupants to avoid panic, to crawl under smoke, to wedge wet towels around doors, are helpful. They are not. They are philosophies of despair, tacit admissions that basic safety precautions are absent.

This chapter divides into three sections – *escape, compartmentation*, i.e. the subdivision of the building by fire- and smoke-resisting construction, and *structural protection*, i.e. preventing collapse. The topics are often interrelated, especially escape-route provision and compartmentation.

ESCAPE

BASIC PLANNING PRINCIPLES

The first and chief need, and the one with most impact on building design, is for escape. Measures to prevent fire are important, of course – the safe installation and routine checking of electrical systems, the provision and maintenance of fire-fighting equipment – but it has to be accepted that the fire-immune building is an unattainable aim.

A fire in a building produces an environment which is changing and increasingly threatening to life. If no precautions were taken, escaping would be a simple – or not so simple – matter of beating the clock. The life-safety approach is to buy time by providing routes to enable people to get out of the building more *quickly*, and providing compartmentation to contain the fire so that the threat develops more *slowly*. Escape provision and compartmentation are related in that they attack the time problem from both ends, helping the people

and hindering the fire. They are also constructionally related, in that the enclosing walls and doors designed to protect an escape route are obviously a form of compartmentation, in addition to the other compartment divisions that divide up the building as a whole.

The 'alternatives' principle – the conventional wisdom

The conventional wisdom, the most common approach, is to provide alternative routes so that if one route is cut off, an alternative remains. In other words, the escape planning – and this largely includes the everyday circulation planning – must ensure that *no matter where the fire and no matter where the occupant, there will remain a route for that occupant to safety*.

In the building (7.1a), there are no alternative routes from any room. The corridors are 'dead-ends' in which only one exit direction is possible, and possibly into danger. A fire on a lower floor could soon trap everyone above. If it started near the stairway, the trapping would be immediate. Even if it were some distance from the stairway along a corridor, people could be trapped in that corridor and, unavoidably, smoke would spread to the only stairway as escapers opened doors onto it. The building is unsafe.

The building (7.1b) is safer, because it offers alternative, protected routes from virtually all points. There are two stairways and they are positioned towards the plan extremities to avoid long corridor dead-ends beyond them. *Crucially*, they are enclosed by fire- and smoke-resisting construction, with self-closing access doors at each level. This isolates them from the rest of the building and, even if the isolation fails for one of them, leaves the other one isolated. Actually, stair

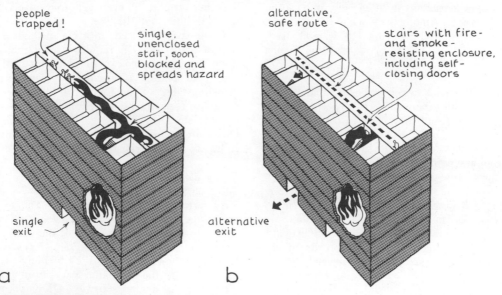

7.1 Escape provision by having compartmented alternatives. Building (a) is unsafe, (b) is much safer

enclosure is the prime example where escape provision and compartmentation are the same – an open stair is a potentially disastrous fire path, flueing the smoke, and then the fire itself, rapidly up the building.

The translation of this into legislation (let alone the question of enforcement) varies from one country to the next. But a fire is a fire anywhere and the *basics* of any credible code are similar, though their quantities – requirements on maximum travel distances, partition fire-resistance times and so on – may vary in detail. The UK codes are comprehensive and we will touch on them at points but only to illustrate the theme and not to imply they are most 'right'.

Applying the 'alternatives' principle

Escape to the final, safe exit can call for travel in rooms, travel along corridors and, if the building is multi-storey, travel in stairways – usually, though not necessarily, in that order.

Escape from rooms

Small rooms. Hotel bedrooms, small offices and so on, often have only one exit and, hence, no alternative route. But this is obviously all right so long as the room really is 'small'. UK hotel codes put the maximum travel from the remotest point to the exit at 9 m.

Large rooms and spaces. We can take 'large rooms' to refer not only to larger component spaces in big buildings, like the hotel ballroom and school assembly room, but also to any large-volume building – the factory, the theatre. (Theatres of all building types have, perhaps, done most to jolt fire-safety awareness historically, proper legislation so often only following in the wake of disaster. Among the worst in the black roll were the Lehman, St Petersburg (1836), 800 dead; Canton theatre, China (1845), 1670 dead; Ring, Vienna (1881), 450 dead; and Iroquois, Chicago (1903), 572 dead).

There has to be quick escape available in alternative directions (7.2). This means at least two exits (and more, depending on size) and they have to be far enough apart to constitute proper alternatives – for example, UK legislation asks for them to subtend at least 45° on plan from any point they serve.

Again, the maximum allowable travel from any point to an exit can be measured in terms of simple distance, for example, UK hotel codes have this at 18 m. In any building type, this can sensibly increase where there are favourable factors, like a low hazard (low-combustibility construction and contents), numerous exits or the presence of sprinklers (see later). Conversely, it can reduce where there is a particular hazard, for example, where there are highly-combustible contents and disorientating obstructions, as in an exhibition space.

In an alternative approach, codes can measure the

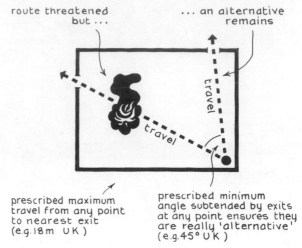

7.2 Ensuring alternative exit from 'large' rooms

maximum allowable travel in terms of *time*, requiring it to be such that all the occupants of a space be able to reach a safe place within a specified period of time – say, $2\frac{1}{2}$ minutes, or shorter or longer, depending on the nature of the likely hazard. The 'safe place' will usually be the outside but it could be a staircase, were that staircase fully protected, or it could be another part of the building, if that part were fully compartmented from the first. We will come on to this. The advantage of the time measure is that it acknowledges that escapers' mobility differs in different kinds of occupancy, i.e. allowing only short travels where people are likely to have restricted mobility, in hospitals, and homes for elderly or handicapped people and, at the other extreme, allowing generous travels in schools, sports facilities and the like.

There have to be rules ensuring adequate exit widths from a space, the UK requiring a 750 mm minimum when the occupancy exceeds 50, with further increments, depending on numbers. Many regulations, including those of the US, think in terms of 'stream widths' of escapers, i.e. depending on numbers, requiring at least a 23in. single stream or 46in. double stream. Often, fire codes control maximum occupancy – US fire codes are particularly strong on this.

Escape along corridors

There have to be sensible limits to 'dead-end' travels, corridors with exit in one direction only. They are potential traps, whether leading away from an *apparent* exit in a public space or forming part of the overall corridor circulation. The important diagram (7.3) shows this. It also shows how, in multi-storey buildings, there have to be limits to the maximum travel from any point to the nearest protected stairway. UK limits are 7.6 m (i.e. 25 ft) for dead-ends and 18 m (60 ft) for maximum travels. These are quite strict by other countries' stand-

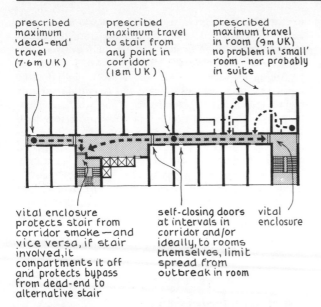

prescribed maximum 'dead-end' travel (7·6m UK)

prescribed maximum travel to stair from any point in corridor (18m UK)

prescribed maximum travel in room (9m UK) no problem in 'small' room - nor probably in suite

vital enclosure protects stair from corridor smoke—and vice versa, if stair involved, it compartments it off and protects bypass from dead-end to alternative stair

self-closing doors at intervals in corridor and/or ideally, to rooms themselves, limit spread from outbreak in room

vital enclosure

7.3 On (e.g. hotel) floor, compartmentation ensures that the two stairways are truly 'alternative' to escapers and that any outbreak is confined to its floor (and, it is to be hoped, even room) of origin

ards and, again, could arguably be exceeded where there were other helpful factors, like the presence of sprinklers or where routes are strongly protected against fire and smoke intrusion from adjoining rooms.

The cardinal point in multi-storey buildings is that any dead-end limit underlines the need for at least two stairways, and an overall travel limit may call for more than two stairways where the floor plan is large.

Now, of course, escape routes have to be 'protected', if they are to be reliable. We will come back to the concept of constructional fire-resistance but, for the moment, note the placement of the doors and screens in the diagram. A quite ordinary door will impede the spread of fire and smoke if it is closed, so it is immediately helpful if room doors have self-closers – they will then normally be closed and, important, will close behind people escaping from a room on fire. Alternatively, the hazard containment can be left until the corridors themselves – screens with self-closing doors at intervals of 20 m or so – can delay the smoke spread along the corridor *and keep the alternative directions open*.

Ideally, there would be containment both at room doors *and* at intervals in corridors. Total reliance on room doors, arguably, leaves the route undefended against a fire in the corridor itself. Granted a corridor outbreak is unlikely but, certainly, an extended corridor layout needs additional divisions at critical points. On the other hand, sole reliance on containment at intervals along the corridors leaves escape from rooms rather undefended against an outbreak in a room in

that same corridor length. There has to be an assessment of what is practicable, for safety, like everything else, has its price. Different countries do different things – some do nothing at all.

By similar token, any potential fire sources adjoining the route should be contained. For example, doors to storage cupboards or electrical distribution boards should be fire-resistant and carry signs instructing that they be locked shut when not in use.

Escape in stairways

Escape in stairways is normally downwards to a final exit but it can be up from a basement, or up from an upper floor to a safe refuge or roof-route to an adjoining building.

Single stairways. A single stairway can, in fact, be safe enough in a small building – (the UK definition of 'small' is here not over 4 storeys for hotels and similar residential buildings, but is more lenient for various other occupancy types). This is provided:

1. There are no rooms accessing onto the stairs that could produce smoke.
2. Corridor access at all levels is short and through two stages of self-closing doors, i.e. making smoke-checking lobbies.
3. The final route away from the stairway is short and through an area of virtually no hazard.

The fact of a single stairway may strengthen the argument for added precautions, for example, for having fire detectors in ground floors and basements, especially if these areas are likely to be unsupervised at night.

But, given that the acceptability of having only one stairway depends on the presence of mitigating factors as well as on building size and, notably, on the reliability of the stair compartmentation, we start to see the limitation in rigidly interpreting any one country's rules. Where the standard of local enforcement is in doubt, single stairways must seriously increase the risk, even in small buildings. On the other hand, some countries achieve adequate safety with single stairways even from high-rise buildings – we shall see how later.

Alternative stairways. Still, the conventional wisdom is, again, to have alternatives. Returning to the corridor plan (7.3), if stairways are to be truly alternative escapes, and never spread the fire or smoke, they must be properly enclosed. Forgive the repetition, but it is so important. Logically, the stairways must discharge via *separate* exits to the outside – one sometimes sees otherwise well-protected stairs discharging to a common final route through ground-floor areas and it takes little imagination to see what an outbreak there could do. Ideally, there should be no store cupboards or other potential fire sources within the stair enclosure. Nor should any habitable rooms lead into it directly.

An outbreak within the enclosure would not only knock out that stair in seconds but, worse, if there happened to be a defective enclosure door at another level, smoke could well reach the adjacent corridor there and block access to the alternative stair as well.

Other vertical escapes. External escape stairs occur most frequently as safety improvements to existing buildings. Ideally, they should be in fire-resistant enclosures (the enclosure also protecting them against the weather – an icy tread on a dark night can be as lethal as a fire). But, if unenclosed, they should be clear of windows or, at the very least, windows near them should be fire-resistant and fixed closed. They should be emergency-lit (we will talk about this shortly) and safely traversible. Metal ladders dropping down many storeys are extremely hazardous. They might be all very well for the agile in daylight but again, picture them ice-covered at night, and what of elderly or disabled people, or the hotel guest who is less than sober?

Other escapes include steel-rope ladders, harness and rope arrangements unwinding from inertia reels, and an innovatory French system where flexible fabric tubes, through which escapers can safely slide, break out down the outside of a building. These rather acrobatic devices can certainly save lives but, realistically, must be seen as part-remedies and not basic provision.

Ordinary lifts should *never* be used for escape. The electrical failure so likely in a fire can leave people trapped to die. And a lift shaft is a natural fire and smoke path.

The final exit

The final exit is the door to a permanently safe place, usually the outside, and it should be as directly accessible as planning can make it. If there has to be a route to it from a large public room, or from the discharge point of an escape stair, then that route should be protected. The doors must be immediately openable from the inside at all times. Probably the safest system is the push-bar release lock, or 'panic bar', a sharp push on which bursts the door open outwards, in the direction of escape. You will have seen other systems, exit catches behind breakable Perspex covers, or locks with the key hanging adjacent behind a breakable cover. These are all right provided they are emergency-lit and as long as the exit's location guarantees the escaper a moment's pause to use it safely, for example, at the end of a longish protected corridor. But they could cause a nightmare delay to a crush of people struggling to escape from a smoke-filled discothèque. In the 1970 St Laurent du Pont discothèque fire in France, in which 145 people died, there were three exits. One was a slow turnstile, one was locked, and one was hidden behind a screen and never found.

There is the argument that doors unlockable from the inside present a security problem, allowing thieves to exit from back premises, hotel guests to leave without paying bills and so on. Nearer the truth is that *the locked escape door is just a cheap and irresponsible solution to the security problem.* Apart from direct patrolling by security staff, there are other effective measures, such as locks connected to burglar alarms and closed-circuit TV monitoring – but, whatever form they take, they must not be allowed to interfere with the paramount need for escape. Sometimes, managements claim that locked escape doors can be opened in an emergency by staff carrying keys. This is not only patently unreliable owing to the human element it introduces but it is also patently absurd. If staff can travel from, say, reception areas to escape doors then by the same token the routes must be open for escapers from those doors back to reception and safety, and the doors would not be needed in the first place!

The physical nature of the escape route

Constructional fire-resistance

The fire-resistance of a 'partition' – wall, floor or screen – is the time for which it will resist the passage of fire and smoke but, of course, this measure is only significant if the severity of the fire is stipulated. Fires vary widely in practice, and not only in magnitude. The rate of heat release is a factor. The particular building construction and contents, and the available air supply, will result in a particular time-temperature curve – from a quick flare-up to a long, slow burn. And even though there is an obvious need to standardise furnace testing criteria, in order to compare the performance of various building constructions and components, different countries apply different criteria. So, although here we simply assume 'a fire' for convenience, it is with reservation!

Basically, to shield an escape route from a fire, from heat and smoke, a partition must: 1. Remain intact as a whole assembly. 2. Develop no cracks, for example, by warping. 3. Keep heat transmission down to safe limits. (*Stability, intergrity* and *insulation* are the respective terms used in UK furnace testing).

Walls and doors flanking a corridor are normally designated half-hour fire-resistant. Most ordinary constructions will supply this, even simple timber studding plasterboarded and plastered both sides, or lightweight wired-glass screens. We will look at doors in a moment. As to the stairway enclosure, for all that escapers may be in it for longer, and for all that it may have to endure for fire-fighting access, it is at least a stage further removed from any potential fire source. Half-hour designation is still usual from the point of view of escape. But where there are adjacent hazards, the need to compartment the stairway to prevent its spreading fire up the building is an added factor, and as much from that point of view, the designation may upgrade to one hour and, sometimes, two. 'It all depends.'

Doors are always potential weak points in the enclosure. Timber doors need around 25 mm solid thickness to satisfy half-hour criteria. Alternatively, new doors can be faced or interlayered with fire-resisting sheet material. Existing doors can be similarly upgraded – where appearance is important, they can be split up their centres and have the fire-resisting sheet inserted. 7.4 outlines most of the points in construction. Note that doors in circulation routes generally need vision panels if they are not to pose an accident threat in ordinary use. Inevitably, this brings the possibility of intense radiation coming through in a fire but, at least, the panels can be kept sufficiently small and high to allow escapers or firemen to crawl by, underneath the radiant hazard. There are radiation-reducing glasses now, but they are about three times as expensive as the wired-glass equivalent.

Fire-resisting steel doors are mostly appropriate to obvious major hazards, like fuel stores. At the other extreme, lightweight aluminium doors, often highly glazed, can be used where the need is for smoke-resistance alone, such as subdivision of the escape route itself, where the non-combustible corridor surfaces isolate the doors from the immediate threat of flames and heat. The low melting-point of aluminium and high transmission properties of glass make such doors unsuitable between escape corridors and any adjoining rooms in which fires could occur. In short, there is a distinction between enclosure of the route against adjoining hazard and subdivision in the route itself – assuming the combustibility there is low.

All these measures are to no purpose unless the doors can be kept shut – but how to ensure this? Self-closing mechanisms are the obvious starting point. The difficulty is to make people realise the critical importance of these doors, to realise that the broken self-closure, or worse, the innocent-looking wedge inserted by staff to hold a door open for everyday convenience, can mean the difference between life and death in a fire. Prominent 'Fire door, keep shut' notices are a help and so is patrolling by security staff but the problem recurs. Doors are sometimes permanently held open by fusible links, links which melt and release the door in a fire. This is valid for fire compartmentation but offers no protection against the initial, possibly lethal, spread of cool smoke. A modern device is the automatic-release door. Its use is likely to increase. Typically, it is held open by electromagnetic catches, possibly wired to a time clock, say for night release when supervision is at a minimum, and connected to the alarm or to adjacent smoke detectors.

Surface combustibility – rate of flame spread

The surface materials in escape routes should be non-combustible, both to reduce the hazard in the route itself, so that one can assume the fire will not actually start there, and further to prevent the route spreading fire from a space adjoining. Not only is the total calorific value and the rate of heat release of the materials important but so, also, is the rate of surface spread of flame, i.e. the speed at which fire will normally spread across it: it is important to control both, and the grading of various materials' fire properties is, again, the job of fire-testing laboratories. Controls in routes are mainly applied to ceilings and walls – tests have shown that floors play a lesser part in fire spread. Surfaces like masonry, concrete and plaster are, of course, inherently safe, and there are fire-retardant papers, tiles, fabrics and other decorative finishes. There are also fire-retardant treatments for upgrading otherwise combustible materials: timber linings can be factory-impregnated and most materials can be retardant-sprayed or intumescent-coated. But these treatments should be mostly seen as palliatives in existing buildings rather than as first design measures. They can undoubtedly help but it is often hard to be sure how effective some will be in practice or whether they will remain effective with time.

Shape

Ordinary planning needs *tend* to keep circulation routes wide enough for escape but most countries' codes specify minimum widths for purpose-made escapes – similar to the minimum room exit widths outlined earlier.

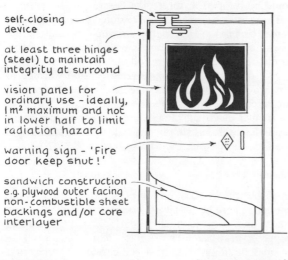

self-closing device

at least three hinges (steel) to maintain integrity at surround

vision panel for ordinary use - ideally, 1m² maximum and not in lower half to limit radiation hazard

warning sign - 'Fire door keep shut!'

sandwich construction e.g. plywood outer facing non-combustible sheet backings and/or core interlayer

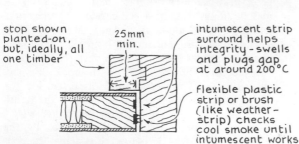

stop shown planted-on, but, ideally, all one timber

25mm min.

intumescent strip surround helps integrity - swells and plugs gap at around 200°C

flexible plastic strip or brush (like weather-strip) checks cool smoke until intumescent works

7.4 Typical timber fire- and smoke-resisting door

There should be no snagging hazards – no sudden projections from walls, single-step changes in level a corridor or irregular tread patterns in a flight of stairs. Intermediate access doors into routes should be recessed or otherwise positioned so that they cannot be opened into the face of escapers who have already entered the route earlier in its travel.

Keeping the route clear

But it is one thing for the architect to provide an escape that is safely usable and quite another to ensure that it stays that way during the building's use. Building management, possibly coupled with periodic fire-service inspection, must see that routes are kept clear. Seating in an assembly hall should, preferably, be fixed down in regular rows since, disarranged, it can fatally obstruct quick evacuation. Stacking seats can, at least, be linkable in use. Furnishings looking innocent enough on a corridor or stair landing can kill or injure should these ways ever be used for escape. And how often is the purpose-made and, hence, seldom-used escape stairway, or the service stairway, used for storage? Bedding, collected refuse and so on, can dangerously block escape – and can, themselves, be a serious fire-hazard.

Summary

- All rooms and spaces (unless 'small' as described) need exits in such numbers, positions and widths, as to give quick alternative escape.
- Room exits onto corridors should, preferably, be self-closing. Alternatively/additionally there can be fire- and smoke-resisting doors at intervals in corridors.
- Dead-ends in corridors should be 'short' and, except in 'small' buildings, corridors should offer alternative directions to separate exits or stairs. Extensive layouts may need three or more stairs.
- Stairs, to be safe alternatives and safely compartmented, must be enclosed. They must lead to separate, final exits. Single stairs in 'small' buildings need special precautions, including smoke lobbies with self-closing doors at each level.
- Final exits need to be quickly openable at all times.
- All routes need to be safely traversible, unobstructed and non-combustibly-surfaced.

ANCILLARY PROVISIONS

Fire detection

Buying time is a matter not only of escape-route design but, also, of knowing there is a fire and, hence, warning people and getting them onto the escape route early. Automatic detectors are by no means essential to fire

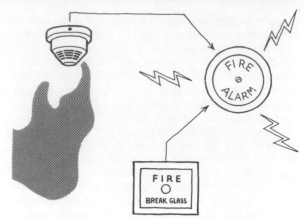

7.5 Alarm sounder activated at call point and possibly by ceiling-mounted smoke (or heat or flame) detector

safety in every building, but they are not particularly expensive within safety budgets and can be valuable where there is a chance of a fire's developing undetected, such as in an unpatrolled building at night – and, anyway, they are better and cheaper than having a person patrolling. More specifically, the argument for having them increases with the likelihood of outbreak of fire, for example, in unattended store-rooms, and with the seriousness of the consequences, for example, in residential buildings where large numbers of people could be trapped. Insurers like them.

There are various kinds of fire detector, sensing heat, flames or smoke. Heat detectors would be most effective in a kitchen, for instance, where smoke detectors would be giving false alarms all the time but, generally, smoke detectors, which give a more rapid warning of fire, are most common (7.5). Detectors, like all electrical fire-safety installations (alarms, emergency lights) have their own electrical circuit and battery or generator supply. The mains supply can fail in the first stages of the fire – may cause the fire, in fact.

Alarm

In all but the smallest buildings, shouting 'Fire!' is not enough. There has to be a proper alarm, audible everywhere and, in residential buildings, capable of *waking* people everywhere. The simplest alarm, again talking of small buildings, is to have self-contained mechanical or battery-powered sounders, and at least two of them, sited so that no fire can render both simultaneously inaccessible. Larger buildings must have the alarm on its own electrical circuit, usually with break-glass call points, sufficiently numerous and dispersed for the alarm to be quickly triggered in any part of the building. The alarm may be, additionally, triggered by detectors, by internal telephone link-ins or by the opening of escape exits. In elaborate systems, the bells or sirens may be supplemented by public-address speakers which allow more information to be given in an emergency, helping the escape and, possibly, reducing the chance of panic.

Fire-fighting

Extinguishing is the most important ancillary precaution and, certainly, it is the one that comes first to many people's minds when thinking of fire safety. But it is the odd-one-out here, in that it has no *direct* relation to escape.

'First-aid' – extinguishers and hoses

If the fire is early in its development, there is a good chance it can be controlled by extinguishers or, possibly, hoses (7.6). The provision of these in any area, or on any floor in a multi-storey building, must be in such numbers and positions as to ensure both quick *and* alternative availability – in practice, the 'alternatives' need tends to be automatically met if the extinguishing points are by the exits. Extinguishers are normally water-filled but special types include CO_2 for electrical risks, and foam or dry powder for inflammable liquid, in places like kitchen areas and boiler rooms. Hoses can be supplied off the mains direct, possibly pump-assisted to give adequate head. Alternatively, there can be storage tanking within the building, either located high enough for gravity feed or, again, pumped.

Sprinkler systems

Sprinkler systems were first developed in the 1870s in the USA for protecting high-hazard industrial buildings. They are increasingly used today, wherever the risk is enough to justify their considerable expense – by 'risk', we here mean risk to property as much as to life, insurers' requirements being prime movers in the decision to have them. They are reliable, highly effective and, in any unsupervised area, have the enormous advantage of coming on automatically. A ceiling-level pipe grid supplies delivery heads at around 3 m intervals on plan. As shown in 7.6, these heads incorporate valves held shut against the water pressure by temperature-sensitive, or rather heat-breakable, glass bulbs. Or there can be fusible metal links. The activation temperature and density of discharge depend on both the particular risk and the ceiling height. Sprinklers demand copious water supply – a good capacity mains supply may serve modest systems but, otherwise, there will be on-site tanking. Large, low-rise premises often have their own private reservoir nearby – we will come back to the particular needs of high-rise.

A point though, is that the effect of sprinklers is more to protect property than life and, though the two are related, they are not the same. Sprinklers will, undoubtedly, reduce the hazard to escapers in most circumstances, especially in large spaces with high fire-

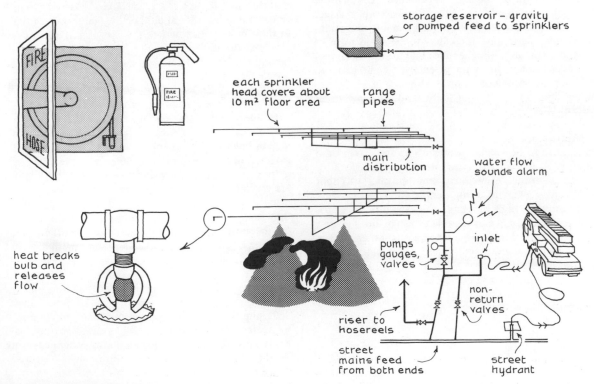

7.6 Fire fighting equipment – hosereel, portable extinguisher and automatic sprinkler system

load, like department stores and exhibitions. Their presence may allow relaxations in maximum travels to exits, fire-resistance times and so on. But quantities of lethal smoke can be emitted before sprinklers can control a fire and the basic need for protected escape remains.

High buildings

Supply to hoses and sprinklers in high buildings follows similar principles to those described for the ordinary water supply in Chapter 4 *Utility services*. Thus, a common thing is to have separate header tanks at stages up the building and, possibly, a large 'break-tank' reservoir near ground level. This keeps the pipe pressures manageable and avoids carrying large volumes of water high in the structure.

Rising mains. Rising mains, or 'stand-pipes' in the USA, are a further need in high buildings. They are pipes running the full height, with outlets on each floor, to which firemen can connect their hoses directly rather than having to drag them long distances from the street. It is often convenient to site these mains in the stairwells but, US practice argues, it is a mistake (whatever UK codes may say) to have the outlets there as well, since a hose run into the corridor will then wedge open the important stair-enclosure door. 'Dry' rising mains are empty of water, having a bottom inlet to which the fire appliance supply (in turn supplied by a street hydrant) is connected. 'Wet' rising mains are maintained full. Generally, there has to be some form of rising main in buildings over about 20 m and, more specifically, a wet rising main when the height gets too much for the appliance pumps – about 60 m – or where the building is very deep or there is no adequate street hydrant back-up.

Fire-service access

There have to be vehicle routes up to and around the building. And apart from the internal protected routes (note that protection will help maintain access to fight the fire as well as to allow people to escape), quick intervention in high buildings calls for lifts with protected power supply and capable of use by fire-fighters. Such a lift will have to open into an area on each floor that is protected from smoke and fire penetration, so that the fire-fighters have a reasonably secure bridgehead. This area also needs to be directly linked to a protected stairway, as a retreat if the lift and bridgehead failed.

Emergency lighting

Safe escape means being able to see where you are going. Ordinary mains lighting is quite likely to fail in the early stages of a fire and, without some back-up

system, the resulting black-out can be a nightmare addition to people's plight. Emergency lighting may be incorporated in ordinary lighting units or mounted independently. One system – the traditional one – has the lights on a separate, fire-protected circuit, supplied by batteries or generators which cut in if the mains fails. But perhaps simpler now, are 'self-contained' units, which are energised by the ordinary mains but still come on if the mains fails. Each unit has its own battery which the mains keeps charged. If the mains fails, the switching automatically cuts in the battery and light. If you tore out the mains wiring leading to the unit, the light would come on. In some cases, for example internal escape routes with little or no natural lighting, the emergency lighting is 'maintained', i.e. kept permanently on, but the back-up principles are similar. The illumination from emergency lighting is generally much lower than from ordinary lighting.

Escape direction signs

You must *know* where to go. There have to be signs marking the escape routes and exits. They have to be clear and may have to be capable of remaining visible in smoke and, therefore, have to be luminous, near emergency lights or, ideally, take the form of emergency lights themselves. Exits from large rooms need marking and, in high risk places like darkened cinemas or discothèques, they should be lit. Naturally, signs are at their most critical where the escape direction is not obvious, the indirect corridor route, or the stair that, having reached its final, safe exit, happens to descend on past it – escapers there could pass within inches of safety and not know it. But, even in straight corridors, clear signs are a reassuring focal objective.

Further, in ordinary circumstances when there is no emergency, signs and escape plans on walls implant the idea of fire security in people's minds. And, putting the cart before the horse rather, the requirement for signs helps to ensure that there *are* routes and that the routes are available. Building users and staff are more likely to notice and remedy blocked routes or locked exits if they are properly marked. Admittedly though, signs can be unsightly and institutionalising and, in many building interiors, their choice and placement may call for considerable thought.

Summary

So, using the routes needs:
- Detection, possibly.
- Accessible call points for an alarm system covering the whole building.
- (Extinguishers).
- Emergency lighting, to see.
- Signs, to direct.

DESIGN APPROACHES OTHER THAN ALTERNATIVE ROUTES

The alternative-routes philosophy is a regularised approach to escape in regular corridor and stair buildings. But, as already implied, there are other valid approaches to such buildings – and, for that matter, there are other buildings whose form could never be served by the alternatives approach.

Stair towers compartmented as refuges

Suppose we take a multi-storey building (7.7), whose floor plan has a central spine corridor of relatively modest length, say 30 m, with a single escape-stair half way along. There is no alternative stair. Many countries' codes would prevent such a building, but not all, for such an arrangement is not *necessarily* unsafe. If there are mitigating factors, such as good floor and wall compartmentation throughout the building, possibly sprinklers, self-closing doors from rooms onto the corridor and non-combustibility within the corridor, it is hard to see that people could be prevented from reaching the stair. As to the stair, if it is so constructed as to be, effectively, a separate entity from the fire point of view, with protected air supply and able to withstand a complete burn-out of the building around it, then it is hard to see that people could be threatened, once they had attained its refuge. The stair enclosure must be highly fire-resisting concrete or masonry – 'fully protected' happens to be the UK term, 'fire-proof', the

US. It must be non-combustibly lined and without rooms or cupboards of any sort directly off. The connecting routes into it must be lobbied and should, ideally, include a link that is permanently open to the outside so that, even in the unlikely event of smoke reaching that final link, it can never reach the stair itself.

Mind you, the idea of the stair tower constructed as a permanent refuge is in no way exclusive of the alternative stairs approach – it can be incorporated in it, especially where building height makes escape all the way down to the ground level impracticable.

Refuges within buildings

A related idea is to construct large buildings with internal refuges, compartments capable of riding out the fire. Escape can be into the refuge rather than to the outside. As we shall see, tall buildings, like offices, tend to be inherently compartmented by their concrete floors. This can be exploited for escape, the approach then being to upgrade the division, say, at every fourth floor or so, to give complete fire- and smoke-resistance. Measures like access from each floor to the stairs only through fire doors with vented lobbies, and automatic fire-stopping in vertical ducts, mean that escapers only have to cross the nearest division above or below to be safe. Further, large-floor-plan buildings can have fire separation within each floor allowing easy horizontal escape across the division – for example, this is particularly relevant to hospitals where an estimated 5 per cent of patients would die by the bare fact of evacuation to the street.

Large spaces – fire and smoke control

Large, single volumes, such as factories, auditoria and all kinds of large public space, do not have protected routes in the physical sense. There, the typical safety approach is to limit the fire hazard and get rid of the smoke – a double defence to slow or even prevent the threat's development. Take a building like the San Francisco Hyatt Hotel. Its large central atrium over the public spaces is flanked by banks of open corridors, all in the form of open galleries and, apparently, threatened. But the hazard is limited by the strict control of materials used in the atrium to limit combustibility and the chance of toxic gas emission. There is extensive sprinkler coverage. A fire would still produce smoke but there are large automatic extract vents in the roof to cope.

The Hyatt, of course, has the advantages of low fire-load and large volume, both reducing the potential density of the smoke. In a non-combustibly constructed sports building or large airline terminal, there might hardly be need for escape at all, except that there is always the chance of crowd panic, of real fire threats resulting from some temporary change of use or, in this day and age, of deliberate fire-raising.

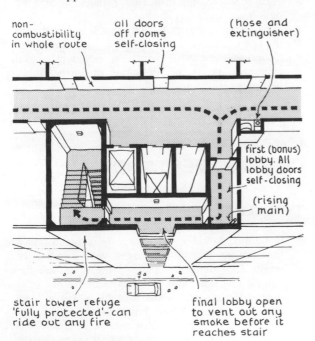

non-combustibility in whole route

all doors off rooms self-closing

(hose and extinguisher)

first (bonus) lobby. All lobby doors self-closing

(rising main)

stair tower refuge 'fully protected'- can ride out any fire

final lobby open to vent out any smoke before it reaches stair

7.7 Stair tower as fully protected refuge. Assuming other good factors, for example, short, protected corridor access, some codes allow this as the only stair

ventilation cowl has full open position for smoke venting. Manually operated or automatic and detector-linked

grid of smoke curtains drops to form collecting reservoirs under roof, impeding lateral smoke spread

7.8 Large, low spaces with high fire-load, like factories, need special measures for smoke venting

But, at the other extreme, occupancies like factories, exhibition spaces and interlinking shopping areas can have a high fire-load to volume ratio, sharply increasing the threat – a threat compounded where travels to exits are long and indirect, and where ceilings are low. *Sprinklers and smoke venting are virtually essential to life safety in the kind of rapidly developing fire possible in these places.* The venting can be by natural convection or mechanical extract and can be initiated manually or by detectors. Dropped screens, 'smoke curtains', are often used to form smoke-collecting reservoirs beneath the vents and to reduce ceiling level smoke spread (7.8).

There is a counter-argument that venting feeds a fire, increasing its development but, actually, one could equally point out that the sort of venting that occurs when a factory roof collapses can help by allowing a fire to burn out quickly within its compartment of origin, preventing horizontal spread and allowing the fire services to close in. The decision whether to operate vents or not will sometimes be left to the discretion of the fire services, but escape is highly dependent on smoke control, often by venting, and there is no merit in suffocating the fire if it means suffocating the people as well.

Pressurising and venting escape routes

Returning full circle, there is a place for active smoke control in conventional escape routes. A simple form is the venting of escape stairs, usually by a high-level ventilator, either manually operated, or motorised and, probably, smoke-detector linked. More elaborately, a buildings normal air-handling system can be linked to smoke control. Airflow directions dictated by the placing of inlet and extract grills can give corridors a small positive pressure over the rooms adjoining them, so that air tending to flow out of a corridor will inhibit the flow of smoke into it. There are installations where the entire air-handling system can be thrown into emergency 'purge' mode when a fire is detected – maximum fresh-air input to escape routes and maximum extract from all adjoining rooms, creating a smoke-purging regime to the outside.

Sophistication in ancillary systems

Apart from all the constructional measures towards fire safety, the installed ancillary systems in a large, modern building can be elaborate and highly sophisticated. There will be a central control room, a nerve centre, the functions of which include the monitoring of ordinary security, as well as fire safety – in fact, the two functions are nowadays related and tend to be seen as part of the security package as a whole. Fire detectors, telephone systems and alarm call points, and the operation of sprinklers, all can alert the control. There, security staff or automatic response will alert the building's fire-fighting team and the city fire service, issue a full or zoned alarm, initiate a full or staged evacuation and activate the appropriate air-handling mode and closure of compartmenting doors.

Summary

- Having alternative routes is the conventional wisdom, but there can be:
- Stair towers compartmented as permanent refuges.
- Escape into compartmented refuges within the building.
- Escape across horizontal or vertical compartment divisions.
- Other smoke-control measures, including vented ceiling reservoirs, pressurisation of escape routes and purge-mode air-handling.

COMPARTMENTATION

THE NEED

- Generally, to subdivide a large building so that a fire starting anywhere is caged within its cell of origin. In a high-rise building, the floors are the natural divisions; and in a school or factory, it is upgraded walls and doors forming bulkhead lines across the plan. Some buildings are inherently beter compartmented than others. A reinforced concrete building with non-combustible internal divisions has a fire advantage over one with timber divisions. A cellular hotel layout is less vulnerable than a large open-plan office.

- As far as possible, to avoid large volumes and interconnecting spaces. These are a hazard in themselves and not only because there is more to burn. They can make harder both escape from the fire and effective fire-fighting approach to the fire – the latter given that the throw from hoses is limited and that the chance of collapse keeps fire-fighters back. Large volumes may ask for sprinklers and smoke venting in compensation, particularly if there is a high fire-load.
- To contain specific hazards like boilers, fuel stores and store rooms, separating either a fire outbreak in them from the building, or a fire in the building from them. Also, to separate particular hazards from high-risk occupancies (a notable example being the theatre safety-curtain between stage and audience).
- To isolate escape routes, as described, both for their protection and, also, because they are potential fire paths.
- To separate one building from another.

THE METHOD WITHIN THE BUILDING

Compartment size

Obviously, the smaller the compartments, the less potential for damage there is, but there has to be a sensible compromise. Size limits vary between different codes but, generally, reduce as the fire-load and the risk of the particular occupancy increase. The presence of sprinklers is a significant mitigation, allowing larger compartment sizes – perhaps double.

Walls and floors

Major compartment divisions will normally have longer fire-resistance periods than the enclosure to escape routes. One hour is a common minimum, but resistances up to four hours or more are possible, demanded by fire authorities or by insurers anxious to minimise their risk.

Large wall and floor areas need expansion joints to minimise their distortion and consequent failure or disconnection from adjoining elements. A long wall needs small expansion gaps at intervals, non-combustibly packed. Compartment walls in industrial buildings tend to follow column lines, clear away from the possible mid-point collapse of roof trusses above (7.9). Extreme risks may call for double-leaf walls, making two buildings of one.

But compartment divisions are as strong as their weakest links and can be all but invalidated by unprotected doors or other openings and bypasses. You do not leave out planks on a boat hull just because they are hard to fix.

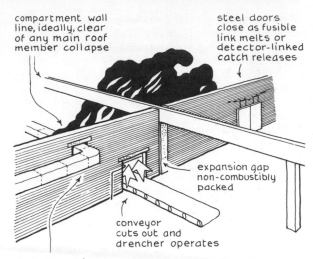

7.9 Some features of industrial-type compartment walling

Doors

Doors will, typically, be timber up to one hour, or steel. They can be self-closing and, possibly, locked. If they have to be open for ordinary circulation, as in a factory, they can be counterweighted or hung on inclined runners so that they shut on the melting of a fusible link or the release of an automatic catch. Doors on industrial scale often take the form of roller steel shutters.

Awkward openings, where a conveyor belt pierces the dividing wall, can be protected by a drencher system – a sprinkler-type water curtain – or by solid shutters. Escalator wells, in places like department stores, often break the compartmentation between floors. There, enclosure can include horizontal steel roller shutters activated by detectors.

Ducts and pipes

Service runs often have to pass through compartment divisions. Duct and pipe sections immediately adjacent to the compartment wall or floor are made fire-resisting, and the gap around them, where they pass through, is minimal and filled with a non-combustible or intumescent packing. The packing has to be flexible enough to allow for the ordinary vibration of the service run, and its expansion and distortion in a fire.

Fire spread within a duct is prevented by hinged damper plates inside, normally open, but fusible-link or sensor-connected to close in fire. Dampers can get fouled up by grease and dirt in time, so there have to be access covers for cleaning or removal. Intumescent honeycomb grills across ducts are another fire-stopping method.

Fire enclosure of lifts and duct shafts generally – and

189

of course stairs – is an essential prevention to rapid vertical fire spread. Sprinkler heads in vertical shafts are possible. Ceiling voids can be a disastrous horizontal fire route – witness the fatal CLASP building system fires in Fairfield, England (1975) and Paris (1973). Partition walls should carry up through a suspended ceiling, right to the underside of the floor or roof above.

Compartmentation bypassed by the external enclosure

Fire spread via the external walls and roof could make nonsense of the compartment divisions inside a building. 7.10 shows some of the more obvious remedies. Unless the roof is of non-combustible construction and fire-resistant surfaced inside and out, the internal compartment walls should carry up through it, rising, perhaps, half a metre above its upper surface. This applies at domestic scale, also, where the party walls are vital fire-stop, projecting through the roof, and, if their fire-resistance is inadequate, the walls too.

The multi-storey façade is a particular problem. Clearly, its external surface must be non-combustible but there remains the chance of flames breaking out through the windows of one storey and entering those above. Spandrel walls have been quite widely adopted as a simple defence against this, a precaution reflected in many countries' codes, including the US. But tests at the UK Fire Research Station in the early 1960s suggested that spandrels were little or no help, and the requirement for them was later dropped from the UK Building Regulations. Wired glass will help but is unsightly, and armoured glasses are expensive. Controlling the constructional combustibility around the perimeter area of each floor will help but then it is hard to prevent there being combustible contents and window hangings there once the building is in use. It is a tricky problem without easy solution.

Fire spread within double-skin wall cavities is

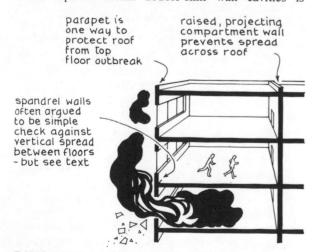

7.10 Some measures preventing fire from bypassing internal compartment/divisions via the enclosure

parapet is one way to protect roof from top floor outbreak

raised, projecting compartment wall prevents spread across roof

spandrel walls often argued to be simple check against vertical spread between floors – but see text

another risk. The void between, e.g., an external cladding and its solid backing must be non-combustible and fire-stopped at each floor, otherwise there could be a lightning and catastrophic upwards fire spread within it. This was a crucial factor at the Isle of Man Summerland leisure centre disaster in 1974. Similarly, there should be vertical fire stops to prevent horizontal spread past internal dividing walls.

The risk where low- and high-rise adjoin

A serious threat occurs where the roof of a low building adjoins the façade of a higher one. The lower roof must then be capable of containing any outbreak in the building it covers. It has to be of fully fire-resistant construction, and there are rigorous controls on the nature and placement of roof-lights.

COMPARTMENTATION BETWEEN NEIGHBOURING BUILDINGS

There is the risk that fire may spread from one building to another. Conflagrations like the Great Fire of London in 1666, or the great Chicago fire in 1871, are ultimate examples of what can happen without proper fire-spread controls – and many of the controls today stem from such disasters.

The threat can be by flames and hot gases directly. It can be by flying brands – these have been known to ignite buildings as much as half a mile from an outbreak. It can be by radiation – timber façades can be ignited at 50 m distance and, even if neighbouring buildings are not ignited directly, they can be heated to the point where ignition by flying brands is more likely.

Factors

The chance of fire spread between an exposing and an exposed building depends on:

- Their distance apart.
- The fire's severity, potentially depending on what, and how much, there is to burn, i.e. on the fire-load. This, in turn, depends on the exposing building's construction type and use. It also depends on its size obviously or, more exactly, the maximum size of any compartment within the building in which the fire occurs – logically, when you think about it, compartmentation *within* a building increases its effective compartmentation from its neighbours. Sprinklers are a mitigation. Clearly, the threat from an uncompartmented inflammable paint store (were it ever allowed) would far exceed that from a concrete-constructed, small-cell, sprinklered office.

- The fire resistance of the enclosures and, most immediately, the exposing enclosure, including the extent to which its area is perforated by doors, windows (and sometimes roof-lights). If the building is compartmented, then the area of openings within the 'enclosing rectangle' of wall fronting the largest compartment is considered in assessing the threat, not the area in the whole elevation.
- Other enclosure-related factors. For example there could be a solid division wall on the site boundary, or there might be a drencher system to discharge a protective, cooling curtain of water down a threatened (or threatening) wall.
- The combustibility of both buildings' external enclosure surfaces. Wood-shingle roofs in the US have repeatedly proved disastrous – whole neighbourhoods have been destroyed.

Applying the factors

Recent research has done much to quantify the relative importance of these factors. Typically, there are fairly simple, tabulated data and calculations to establish how far a building of a certain configuration should be from its site boundary or neighbour – or to establish what grade of enclosure is appropriate to the particular building separation and occupancy enclosed.

Incidentally though, the question whether to design on the basis of distance from boundary or distance from neighbour is an interesting one. If a building's walls have low fire-resistance, there will be an increased transmission risk, whether it is the building or its neighbour that is on fire. An architect is concerned with designing one building, not two, and fire legislation is simplified if it can be focused on one building, not two. So it is helpful to have a way of avoiding the question of what is on the other side of a boundary or, indeed, having to speculate as to what kind of building hazard might arrive across the boundary in the future. A 'mirror' approach is often adopted, in which the calculated enforcement standards assume that a building being designed is faced with an identical building, regardless of what is, or may later be, there. If a high risk neighbour arrives later, the mirror approach applied to *it* will ensure *its* adequate enclosure.

Summary

- Compartmentation within buildings is to subdivide the risk, isolate specific hazards, and protect escape routes.
- Divisions are, typically, of masonry or concrete and with up to 4 hours designated fire-resistance.
- Doors, shafts or other penetrations need automatic closure or e.g. drencher protection. Ducts need automatic dampers.
- Bypass by external enclosure is avoided by non-combustibility, and penetration of compartment division through to outside.
- Low-rise roof design is critical when it abuts a high-rise façade.
- Chance of fire spread between buildings depends on distance apart, fire severity (fire-load and compartment size), and fire-resistance and combustibility of enclosures.

PROTECTING THE STRUCTURE

The supporting structure must be fire-resistant. Failure to protect it can lead to irreparable damage and, indeed, total collapse.

Masonry

Masonry is generally fire-resistant. Some types, granites for example, can be damaged, but bricks and blocks will usually survive.

Timber

Timber is unlikely in supporting wall construction – away from domestic scale, that is. But columns, beams, laminated portal frames and so on require thought, particularly when they support floors above – fire-resistance is rather less critical in single-storey roof spans.

Timber is not, in fact, totally vulnerable in fire. It will burn initially but the charring this produces soon acts as an insulating shield, slowing the destruction. The denser the timber, the slower the charring eats into the section but the rate is generally taken at around 0.7 mm/minute. Knowing the charring rate for a particular timber, a designed fire life can be assumed by proportionally oversizing the members. Laminates with exposed steel connectors risk delamination, but developments in structural adhesives allow glued laminates to be regarded as if they were solid. Incidentally, impregnation with fire retardant does little to increase timber's fire-resistance, merely reducing its rate of surface flame spread. But non-combustible sheet materials, or expanded metal for plastering, are easily attached by screws or nails.

Concrete

Reinforced concrete is, in the main, fire-resistant, assuming there is enough cover to protect the vulnerable steel. But there are weaknesses associated with different aggregates. Some types may suffer surface spalling in a severe fire, requiring a light steel mesh under the surface to hold the cracked concrete cover in place. Some concretes are perceptibly weakened by changes in chemistry and moisture content with temperature, and fire-safe design has to allow for this. But lightweight aggregates, like pulverised fuel ash and clinker, are very stable – they are formed at high temperatures.

Steel

Protective wrappings

Chapter 1 *Structure* described fire performance as steel's Achilles' heel. The increasing loss of strength from around the 550° C mark means that unprotected steel would certainly collapse in a severe fire – the consequences of this in a high-rise steel frame building can be imagined. Brussels' disastrous Innovation department store fire in 1967 brought the building down: curiously, it was the lower two floors' steelwork that was inadequately protected – just where the stresses were greatest.

Columns, beams and floors have to be protected by a fire-resistant, insulating wrapping (7.11). Concrete is common for this, and can add to the ordinary structural properties as well. Concrete filling to hollow steel sections adds strength *and* thermal capacity to absorb heat, but heat transfer to the concrete is fairly slow, so there has still to be an outside wrapping. Other protections include asbestos and plaster, or fire-stable minerals like vermiculite. These can take the form of board encasements or, especially where sections are complicated, can be sprayed onto the section directly or onto an attached expanded-metal lath.

Water cooling

There are innovations. An interesting development is the fire-protection of hollow members by water filling. Heat is then rapidly conducted from the threatened steel to the water inside, the thermal capacity of which delays the temperature rise. There are vents to allow for expansion and steam escape, and the water can have additives to inhibit corrosion and, where appropriate, freezing. *Non-replenishment* systems have a rather limited fire-resistance. *Replenishment* systems, where there are hollow columns and beams connected to header tanks – think of gravity central-heating systems – can, theoretically, protect the steel as long as there is cool water to circulate and replace the hot. The heat-dissipating convection is normally induced by the fire itself.

Steel outside the façade

Finally, although this is not so much an innovation as a new way of arguing fire protection, tests have shown that steelwork a metre or so outside the building enclos-

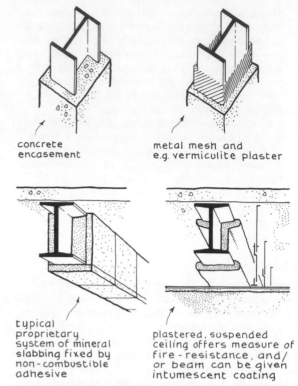

concrete encasement

metal mesh and e.g. vermiculite plaster

typical proprietary system of mineral slabbing fixed by non-combustible adhesive

plastered, suspended ceiling offers measure of fire-resistance, and/or beam can be given intumescent coating

7.11 Fire protection of structural steelwork

ure is less likely to approach the critical temperature than had been previously thought. The degree of internal compartmentation and type of enclosure again play a part here, but, generally speaking, unprotected external steel is now increasingly accepted.

Summary

- Masonry is generally fire-resistant.
- Timber is partially protected by charring – thickness and (usually) density are positive factors.
- Concrete is generally fire-resistant but needs thick enough cover to protect reinforcement and, possibly, secondary reinforcement against surface spalling.
- Steel needs protection by concrete or some other non-combustible, fire-resistant, insulating wrapping. Hollow sections can have concrete filling to help absorb heat, or water filling to absorb it and, ideally, circulate it away.

This chapter has touched on various countries' legislation. Inevitably, people's understanding of fire safety and its influence on the buildings around them – in terms of escape provision, compartmentation or anything else – tends to reflect the particular, local code which applies. Codes vary but it is worth remembering that this must owe as much to the extreme difficulty of writing a workable code as to any belief that one set of rules is indisputably best. And, anyway, it is more the details that vary: unsurprisingly, one finds the central features of all credible codes to be fairly similar – a fire, after all, is a fire anywhere. Again, the thing is to understand the underlying principles, allowing the rules to be interpreted flexibly and intelligently, rather than as a rigid doctrine.

8

THE FUTURE?

In the course of writing this book, I have talked quite widely with experts in the various departments of building design and technology; in the specialist professions, in manufacturing and in research; and often, as a parting shot, I have asked how they see the future in their particular fields – the immediate future, that is, what is or needs to be happening now and what the most likely trends are to be. The outline that emerges is an interesting one and I pass it on, together with a few thoughts of my own.

STRUCTURE

Structural innovation has so often been generated by the arrival of more capable building materials, witness the impact of steel in the last century and reinforced concrete in this. But steel and concrete have now, more or less, realised their potential in strength and versatility, and there are no higher-performance materials in immediate view, not, at any rate, at a price allowing them seriously to compete. There are select developments. There are carbon fibre and other improved reinforcements for concrete, but costs restrict these to specific uses, such as for high-strength pile caps. GRP and other plastics are also costly and mostly inappropriate for main supporting elements.

In any case, preoccupation with increasing materials' strength, to free-up the form of building structure or increase its scale, is rather redundant these days. Materials are now strong to the point where desired building form can generate structural form (at least theoretically, if not economically) in contrast to the historical case, where building form was overridingly limited by what could structurally be done. And scale increase, particularly in height, is constrained by other factors. There is the question of what is environmentally acceptable, and there are economic disincentives, not only in terms of the structure itself but also in servicing – lifts and the floor space they absorb are the

obvious case in point. The ultra-high-rise phenomenon in Manhattan owed to local liberal planning attitudes on height, and to economic factors which included the intense pressure on land space and the fortuitous presence of a stable rock bed to build on, enormously simplifying foundation construction.

Equally, in the immediate future, one has to be wary in predicting striking development in constructional technique. Recent history has shown how passing the trends can be. In the post-war period, it was barrel-vault roofs and ubiquitous prestressing of concrete building components. In the mid-1950s to mid-1960s, it was HP roofs. In the 1960s to early 1970s, it was industrial building systems and large-panel precast construction. These techniques often posed hidden problems for contractors on site, who, once caught out, subsequently tended to price them out of court on the next tender. They were not bad techniques but were, sometimes, applied without being properly understood or where they were inappropriate. The industry has become more cautious.

Exploitation of potential strength and innovatory technique further depends on the calculations engineers can make to test their hypotheses. Undeniably, computers have made huge strides in what can be attempted. For example, engineering practice is now familiar with concepts like 'finite element analysis', where a structural frame's performance is predicted in terms of its individual components, and 'global analysis' where the frame is analysed as an interactive whole. Certainly, computers have been invaluable in extending the real applicability of known theory – but it is only theory, for all that. The relevance of calculating to the Nth degree of accuracy is limited when, in practice, component interaction is incapable of finite prediction and when safety factors based on anticipated strength of materials are, inevitably, arbitrary to a degree.

Materials and methods will continue to evolve, and even today's materials and methods are capable of greater structural height and span, and of further exciting innovation in structural form. But engineers tend to see immediate development couched in realistically down-to-earth terms. The pioneering spirit is doubtless not dead, but structural technology has perhaps reached the point where, for the moment, the question is not so much what can theoretically be done as what is most practically and economically sensible to do.

ENCLOSURE

Clearly, the building enclosure has similarly been subject to rapid development in pursuit of lightness at high rise and erection economy generally, while maintaining, or improving upon, traditional performance standards. Here too, innovation in materials and constructional methods brought teething problems. We have touched on many of these earlier in the book, but to cite just one example, the massively adopted and rapidly evolving concrete technology post-war, particularly in industrialised systems and novel forms of pre-casting, proved almost perversely fickle at times. Deficiencies, or misapprehensions, in material quality, placement and detailing, brought all kinds of failures, in constructional strength and durability, in basement waterproofing, in the weather-tightness of pre-cast cladding, in the visual weathering of façades generally and, notoriously, in flat-roof constructions. In building construction generally, the post-war development boom unhappily coincided with, or perhaps indeed caused, a rapidly evolving technology that was understandably untried and, as it turned out, at points inadequate – an observation by no means exclusive to concrete. Now the technology is largely sorted out. At domestic scale the enclosure may no longer be 'traditional' in many of its components and sealants, but there is a detectable return to traditional forms – brick walls are after all convenient to build and pitched roofs do after all shed the rain. And at all scales of building the risks in using unproven materials and methods are better appreciated. The whole vocabulary of the large scale enclosure is far better understood and more reliably successful in performance and looks. But, as with building structure, so with the enclosure – there has arisen a sense of caution and, in the view of some practitioners, even a licking of wounds.

A remaining point on the enclosure's physical performance is on the face of it more mundane – but it is important. There is a growing awareness of the need to improve on durability and ease of maintenance. Claddings will doubtless continue to improve and gaskets and sealants are promising longer lives. But here at least manufacturers are speaking of more radical, eventual changes. Analogous to the motor industry there is talk of charts to building owners to guide on inspection times and vulnerable parts. Future claddings may be more easily detachable from the main structure for re-

placement or to be sent back to the factory for reconditioning. All this would call for a standardisation and technology not yet attained, but the economic stakes are very high considering that some 40 per cent of the building industry is engaged on non-productive maintenance. Whatever the approach, tax structures encouraging initial capital expenditure to save later running costs would be a considerable incentive.

ENCLOSURE, CLIMATE SERVICES AND ENERGY

Energy conservation has become, and will remain, a vital, critical factor in building design. Actually, it is rather tricky to order our story over the next few pages, because energy is multi-disciplinary. But, as the starting point, we can isolate for consideration the enclosure's performance as simple barrier against heat loss and unwanted solar gain. Consider the component products available for resistance insulation and solar reflectance. Their capability has improved radically in recent years in response to the urgent demand but it is hard to see the pace being maintained. For instance, double glazing is increasingly specified but, short of tripling or quadrupling the panes, one cannot foresee U values better than about 1.9. Resistance-insulation materials and methods have proliferated in blocks, screeds and all manner of external, internal and infill linings. High performance linings are already bettering U values of 0.8 in a 25 mm thickness but, though it would be nice to be proved wrong, no one is currently talking of any magical insulant on the way – after all insulation is mainly trapped air, so there is no obvious scope for anything new.

In solar exclusion, the methods of structural, fixed and movable external shading are well established and, what with further innovations, including micromesh external screening that is transparent normal to its surface but opaque to oblique insolation – and mirror glasses achieving around 80 per cent solar radiation reflectance – the rate of improvement must, again, be flattening out. Insulating/reflective shutters are increasingly being used.

The enclosure's other thermal capability is, of course, where it uses its mass as a thermal-capacity damper against the effects of short-term temperature fluctuation. On the basis that capacity is inherently mass-controlled, and accepting that high mass is irreconcilable with lightweight construction – notably for high-rise – any chance for improvement would seem to be thwarted. But, in fact, there is a glimpse of something coming here, namely, in 'phase-change' materials. These are, as yet, costly and with development problems to be solved, but they are worth knowing about not only for their enclosure application but also, as we shall see, for their potential use as heat-storage media in association with the building services. Briefly, the heat energy needed to raise the temperature of a material by a given amount is proportional to the material's mass. But, if a change of state is involved, such as a solid melting to liquid, then extra, 'latent', heat is needed for the actual phase-change to occur. Conversely, latent heat is released on resolidification. Phase-change materials are being developed with melting points near the comfort zone, for incorporation in building construction. Temperature fluctuations there will then require heat transfers many times greater than would be the case with conventional components – low mass but high capacity. Further, since the capacity at temperatures above or below the critical zone is as for an ordinary lightweight material, the building response to the heating or cooling services is better than with heavy materials. You get the capacity where you want it but not where you do not. Chemical compounds based on Glauber's salt, sodium sulphate decahydrate, are among the most promising.

Before looking at the energy design strategy as a whole, we must review the climate services side of the equation – clearly, enclosure, services and, for that matter, lighting, all interact in the thermal picture. Mechanical and practical developments in the climate services have been discussed earlier in the book, new ducting materials, computer programmes to obtain optimum system layouts and so on. But what are the energy aspects?

Considering the wealth and sophistication of systems now on the market, including accelerated, microbore and high-pressure water systems, and the whole variety of systems in mechanical ventilation and air-conditioning, the capability for moving water and air around to climatise buildings is already here in multiple measure. Theoretically, the telling developments should now be coming in system flexibility and response speed, in the ability to tailor economically to the varying demand around a building. Control systems are at the centre of this. However, services engineers are becoming rather divided as to how far climate services installations ought to have this kind of sophistication. Some seek it, looking forward to an ever greater multiplicity of room sensors, and system valves and dampers around the building – and, if there is indeed a market for such complexity, experience shows that system manufacturers will be more than willing to supply it! But most engineers are warning that servicing is already too complicated. Again, the innovatory trends have tended to be short-lived. Systems that worked perfectly well on paper have too often proved unreliable in practice, especially where 'cowboy' installers disclaimed later responsibility or where the complexity was beyond the reasonable competence of the building staff to operate and maintain. Often, theoretically efficient systems have been reduced to working in a distorted, half-

baked sort of a way, ironically wasting energy rather than saving it – what point in having a Ferrari, if you cannot keep it tuned? Effective flexibility and simplicity would seem to be the allied goals. For example, computerised control, with all the potential of microchip, may increasingly be able to guide on operation, fault-finding and maintenance.

'Distributed energy management' is a promising recent development. This is where the climate services of an extended range of buildings, such as schools throughout a region, factories in an industrial estate, or shops in a chain, are centrally controlled by coded signals over the ordinary telephone system. Usually there is a mix of central, skilled control, probably semi-automatic, and local caretaker control.

Then there are the more directly energy-orientated products. First solar panels – the very nature of which, obviously, places them between the enclosure and services discussions. Panel effectiveness clearly varies with climate and site and, certainly in cool-temperate climates, the investment pay-back period is proving longer than had been hoped, always remembering the panels' inherent drawback of having demand and supply curves out of phase – they collect least energy in winter when you need it most. But, as with all energy products, viability could improve if increased production reduces costs, especially in a context of rising fuel costs. And there are possibilities for further improvement in panel technology, for example, in 'moth-eye' glasses offering lower reflectance to incoming radiation, and in 'selective surface' coatings to the plate collectors, increasing the radiation-collection efficiency.

Solar cells, generating electrical energy, are currently too costly to have serious promise for buildings.

Heat pumps are another matter. These may have proved disappointing in their efficiency as scavengers of low-grade heat energy from outside, partly because they consume expensive electrical energy and partly because supply and demand are, again, out of phase. Certainly there was over-optimism in the domestic sector. But they are having growing impact in improving the efficiency of other energy systems, with solar panels and in the whole field of heat recovery and transfer within buildings.

In heat recovery and transfer, the services can positively contribute to energy conservation. For example, at domestic scale, assuming that resistance insulation is to a good standard, more than half the heat loss may be occurring through draughts and uncontrolled ventilation through the enclosure. Draught-sealing and controlled ventilation, possibly mechanical, are first remedies. But fairly simple air-handling systems are now on the market, which extract from kitchens and bathrooms, and intake to bedrooms and living areas, with a heat-pump-assisted heat-recovery link between the two. Then there is the question of washing-water waste, baths and dishwashers consuming energy which is shortly dumped in the public sewer. A simple recovery system here passes the waste through a 'catch-tank', a temporary holding tank of about $\frac{1}{2}$ m^3 in size and well insulated: a heat-pump circuit, with coils in the tank and in the hot-water cylinder, can then intercept some 70 per cent of the waste water energy.

Currently though, heat recovery is mostly associated with the mechanical ventilation systems in large buildings, not only in the basic philosophy of recirculating as much as possible of the ventilating air but also where heat is transferred from the sunny side to underheated parts, or where process heat is scavenged, notably from lighting fittings. Heat recovery is a developing and increasingly viable technology.

But, however efficient energy collection or recovery systems may be in themselves, one is still left with the fact that they are not necessarily releasing the heat when it is needed. In other words, there is an allied need to be able to *store* the heat, and there are important developments here, too. In the ordinary way, the domestic hot-water cylinder can be exploited as a heat store, and there can be hot-water storage reservoirs in larger buildings, but these have only limited efficiency and are certainly only adequate for the short term, a day, or a few days, at most. The possibility of having effective long-term storage, *seasonal* even, collecting in summer and using in winter, is a highly appealing one. Water is pretty unsuitable for this, owing to the large volumes required and the consequent problems in containment and weight. Denser materials, such as rock, are being used – at ground level, to avoid the weight penalty. But, again, phase-change materials are the most promising prospect. In the comfort temperature range, Glauber's salt has about 7 times the thermal capacity of rock and 12 times that of water. Other phase-change materials include waxes, hydrated salts which absorb heat as they dissolve in their water of crystallisation, and hygroscopic salts which absorb heat energy in drying and conveniently release it again when water vapour is passed in. The somewhat miscalled 'ice-heating' storage systems are another intriguing possibility. In a simple system for a house, heat is scavenged from a water and additive mixture contained in a reservoir sunk in the garden, causing the liquid to freeze to a slush. This then continues to soak up low-grade heat from the surrounding, warmer soil – and from any domestic recovery systems – and this heat is available to the house as the cycle repeats.

Concluding on products, even now there are pointers towards the chance of buildings eventually becoming energy self-sufficient. At the large scale, the common problem of having too much heat, too much energy, is currently resolved by importing more energy to remove it. This is a topsy-turvy state of affairs. An efficient means of converting surplus heat energy to useful power, possibly on the vapour-compression cycle principle, would really be the philosopher's stone. But at least there is already increasing attention to ways of effectively cooling by natural ventilation. At domestic scale, where underheating is the more pressing problem, there is continuing research into energy collection. In addition to all the measures just described, further opportunities include glazed Trombe-wall collectors with thermal shutters for cold periods, 'aerogenerators' for electrical power (we cannot call them windmills when there is nothing to mill, and waste digesters producing methane gas for short-term power uses, such as cooking. The energy 'autonomous' house is a theoretical possibility.

But looking at the immediate future of enclosure and services technology as a whole, and even allowing for the considerable chances for refinement on the services side, the major need for development now is in the thermal *strategy* in design. In other words, we now have enough product technology for our main thermal purposes: the pressing question is how to achieve its optimum application at the drawing board to the *particular* needs of *particular* buildings. The factors at play in a buildings thermal performance are so numerous – climate and site; building size, orientation and shape; enclosure type, including resistance- and capacity-insulation and the further effects of window arrangements on insolation and heat loss; pattern of use and processes inside; and the servicing possibilities in energy collection, recovery and storage. It is one thing for the designer to appreciate these factors and quite another to evaluate each one's *relative* importance in a *particular* case. Even a small house poses questions apparently simple but whose proper answer requires complex analysis. How much of the thermal budget is worth spending on resistance-insulation? How far is it worth enlarging the sun-side windows for useful solar gain, bearing in mind the insulation loss and possible structural cost penalties? Is it worth insulating the lowest floor slab? If so, is it better to insulate above the slab for fast response to the heating system, or below it, to contribute the slab to the total thermal capacity, possibly as a heat store adjacent to the sun-side windows? Is solar panel collection or heat recovery viable? And so it goes on. Many of the factors are far from fully researched, but data is improving as the basis for minimum standards and tables and, actually, there are already computer programmes appearing to advise the designer on likely, appropriate strategies for the particular building condition. Before long, access to such computer evaluation may be part of the ordinary practice's stock-in-trade.

Nor is that the end of the story. Apart from evaluating the factors within the thermal budget in *relative* terms, the thermal budget itself has to be assessed in absolute terms – what proportion it should be of the whole building budget. This question is, itself, bounded by other sliding economic and social factors. Theoretically, one can design a house to have virtually negligible heat loss. There is the widely publicised house in Upsala, Sweden, which is principally warmed by the mere presence of the family in it. It has 500 mm of insulation in the roof, 300 mm in the walls, quadruple glazing and a heat-recovery ventilation system. But Sweden is highly conservation-minded, importing over 70 per cent of its energy and with a long, hard heating season. Thermal budgeting has to acknowledge market costs to the building owner and, not necessarily quite the same thing, energy costs in producing energy components, i.e. 'total energy costing' in widest terms. The relationship between initial capital cost and resulting pay-back period has to be measured against estimated fuel-cost trends, and this raises supplementary points. For example, we again have to question tax structures tending to disfavour initial investment as compared with later running cost and – though this is very much widening the context of the discussion and tempting digression – there has to be the realisation that conservation only *slows* the consumption of finite resources and that, ultimately, we have to *produce* our way out of the problem by investing to develop alternative energy sources. The arguments pull all ways and, while thermal standards are generally too low right now, it is true, also, that they are in a bit of a muddle, with the targets ill-defined.

UTILITY SERVICES

Turning to the utility services and starting with water supply – the hydraulics here have been pretty well researched and the pipe systems and appliances continually refined. But practice still tends to lag and to vary widely between countries. In some cases, this is owing to restrictions placed on building systems by the limitations in existing street mains supplies. Direct mains supply to taps, which has long been possible on the hot side as well as the cold, avoiding the need for water storage in the building itself, is unacceptable where mains supply could not cope with the resulting peaks in demand. In other places, notably many US cities, the mains are awash with supply to the point where even the break-cistern reservoirs at the entry to large buildings can be discarded. But often resistance to innovation, to such as direct system and hot-side plastic piping, can owe simply to innate disinclination for change on the part of legislation and practice.

A concern voiced by many water authorities, which has ecological as well as economic implications, is the need to reduce water waste. In many older networks, London's for example, as much as 30 per cent of the water is simply leaking away. Much of this is underground and calling for wholesale piping renewal, but there is scope for economy in the way we use water in buildings. Apart from intelligent system design to avoid long, hot dead-legs and undetected overflows, there are potential appliance economies. Baths using about four times the water of showers is the usual comparison but then, many people like baths. However, some uses are crazy by any standards. Scandinavian design has shown that WC bowls can be so shaped that a 3-litre flush is perfectly adequate to clear the contents – the traditional European flush is about 6 litres and, in the USA, it is as much as 3 gallons! Other potential savers are use-related urinal flushes and (however irritating) limited-delivery basin taps.

In drainage, hydraulic theory and potential system innovation again tend to be well ahead of practice, in the lingering resistance to new materials, such as plastics, to the adoption of more economic pipe sizes and, particularly, to the adoption of the economic and now perfectly well-proven single, unvented stack. In the US, the resistance to single-stack is mostly coming from the construction unions.

In refuse-handling, the main changes in buildings are most likely to spring from external changes in refuse management at the large ecological scale. Increased material recycling – 'resource recovery' – is, in some areas, already calling for refuse to be sorted before it is collected from buildings, isolating glass and paper, for example. Authorities already incinerate refuse for energy recovery and, indeed, to combat spiralling costs in collection and in transport to land-fill dumping sites. But continuing fuel cost rises may further increase the benefit of local incineration. This might breathe life into the Garchey system, for instance.

As to electrical services, one has to remember that they have only limited impact on overall building form. And, in fact, electrical engineers are not anticipating significant changes though, doubtless, there may be refinements, such as improving switching and reducing the bulk of plant, like transformers. Of course, electrical systems are relatedly involved in other areas of innovation, in services control systems and in the whole developing field of tele-communications and security. And, without getting too much back into the energy question at the moment, there is the point that though electricity is currently expensive, electrical services may acquire increased significance as the convenient distributors of energy from 'alternative energy' sources, i.e. those replacing fossil fuels.

Lift technology is already pretty slick in terms of speed and computerised control. But considering that the shafts can absorb as much as 20 per cent of the high-rise building volume, there is a continuing, economic incentive for improvement. A limiting factor, at the moment, is that the conventional cable-and-pulley drive allows only one car per shaft and, also, over long travels, the whip and stretch in the cables tends to hinder floor centring and acceleration. One suggestion has been to run linear induction-motor lifts on rails, computer controlled and able to switch horizontally between shafts. But, ultimately, the margin for development is limited – other than putting people in pneumatic tubes or dematerialising them for instant transfer, one is still rather left with the fact of a comfortable, vertically travelling box.

LIGHTING

In daylighting, the distinction between empirical and intuitive design will always be a point for discussion. However, on the empirical side, and accepting there should be an empirical side, the separation between lighting as quantifiable engineering and lighting as qualitative design by the architect is now pretty well recognised in European standards. In the USA, it is currently rather less so. Refinements are still possible in the design methodology. For example, the 'standard sky' basis for calculation allows for window elevation on the sky view but not for window orientation. And, while the various calculation methods, both on the drawing board and by use of formulae, are now pretty well evolved, the varying factors influencing daylight levels in a room are clearly susceptible to computer evaluation.

Also, daylighting design must play an increasing part in energy conservation, i.e. as an input to the overall energy strategy described earlier. Again, the factors are numerous. If you have more windows, you get more daylight and need less electric light, but lose resistance-insulation. Unit for unit, it may be worth sacrificing heat through the enclosure to achieve a lower artificial lighting requirement, since fossil fuel for heat replacement can be over three times cheaper than electric power for lights. But, again, what will fuel-cost trends be in the future? Only recently, many energy strategists were calling for the minimum window area compatible with psychological needs – unsurprisingly, this was the view of the electricity authorities! But this extreme view is now modifying and window area, in moderation, is now being seen as a potential energy saver. Of course the argument is irrelevant to deep-plan factory buildings and so on, but it can be a telling one for the shallow plan. In fact there is now a detectable return towards the shallow plan for office buildings – artificial lighting in offices can account for 50 per cent of the energy use. A better information base for minimum standards framed in the building codes could help improve design guidelines, and again computer evaluation is surely a prospect for the longer term.

In artificial lighting – quantitatively – illumination levels are still needlessly high in many countries. This is a matter of fashion, which increased energy concern will do nothing to support. Qualitatively, it is curious

that, at the time of writing, the UK is the only country with a properly developed specification for glare-control. There is undoubted scope for improvement in this and other qualitative design. But the most significant improvements are coming in the lights themselves – virtually all the manufacturers seem to have been striving in this. Traditional tungsten lighting, producing around 12 lumens/watt, has long been supplanted, in commercial application, by fluorescent lighting producing perhaps 70 lumens/watt. Now, fluorescent types which are compact and have reasonably warm colour rendering are verging on the domestic market as well. Further, discharge lighting, such as high-pressure sodium producing 100 lumens/watt, is resolving both its colour-rendering problems and its mechanical problems in making the fittings strong enough to contain the high pressures required. Such increased efficiencies are doubly significant in producing more light per watt *and* less unwanted heat. There are imminent implications in the energy picture as a whole.

ACOUSTICS

Acoustics practice (lighting practice, too, in fact) is, at the moment, lying rather divided between US, UK and Continental standards, and acousticians are seeking some rationalisation here. More generally, in room acoustics, the criteria for accepted good listening conditions are starting to widen. Up to now, reverberation-time calculations have stood at the centre of auditorium design, with secondary thoughts for achieving diffuse conditions and for stressing beneficial reflections. But now there is increasing attention to other factors, such as the qualitative importance of the relationship between early and later reflected sound and – the subject of continuing research – the whole field of 'psychoacoustics', the subjective effect that sound quality has on the listener.

And there are the intriguing technical innovations. In the UK, research at Cambridge has investigated the potential for using scaled-down models to predict the acoustic response in buildings. And there is the computer potential in evaluating the various factors influencing a space's acoustical response. Electro-acoustical installations are, also, a current talking point, theoretically allowing reverberation time and other aspects of sound quality to be controlled and altered to suit different types of performance. They are to be welcomed, but cautiously. The idea of getting several auditoria acoustically for virtually the price of one is, of course, attractive. But many acousticians have misgivings that these 'stick-on' electrics may be wrongly seen as a short-cut substitute for, rather than a bonus to, good acoustical design. As with services, systems that look good on paper may not work so well in practice.

In noise control, required standards should continue to rise as part of our increased expectations of environmental quality – or indeed to protect it. Acoustic failure of party walls and floors in modern construction, multi-occupancy buildings has been a particular failure in recent years – and one not simple to solve after construction, remembering that mass and discontinuity are usually the main factors in partition transmission loss.

There are some novel possibilities in 'active' noise control. Electronically produced 'white noise' has had some success in commercial applications: it is a continuous low-volume sound, designed to soothe and to mask intermittent disturbance. Also, there have been experiments with firing out-of-phase sound waves back at a disturbing source, to 'cancel out' noise. This has possible applications in suppressing ventilation-system noise in ducts.

FIRE SAFETY

In fire safety, further refinement is inevitable in the ancillary technology, in extinguishing, detection, automatic door closure and so on. But the need with the greatest potential impact on design is for better basic escape provision and related constructional compartmentation in those places where provisions are so woefully inadequate – or absent. Electronic fire-security systems and sprinklers can help save lives but, on a 'cost per life saved' basis, it is the unenclosed stair, the unlockable final exit, the single-exit discothèque, that need urgent remedy. This is a matter for political and public awareness as much as for architecture.

But, looking at changes in the immediate future, it is actually possible that some other countries will come to be seen as having done too much. This is an argument that has to be very carefully and delicately couched. The amount society should spend on all forms of safety – and on welfare in the widest sense – is open to economic and philosophical discussion. But, whatever the amount, it is not without limit. Take the case of the UK. No country has done more towards fire safety. One informal estimate from the Fire Research Station has it that improved fire safety over the recent period may have cost more than £10 million per life saved. We need not elaborate here on how that figure was arrived at – the point is that even half that sum would buy an awful lot of road improvement or kidney dialysis machines. Fire officers say how hard it is to tell hospital boards to buy the full extent of the provisions required, against the fire which may never occur, when even a quarter of the amount spent on other medical facilities can be virtually guaranteed to save a predict-

able number of extra lives annually. It is not that there should be less focus on fire-safety principles and their enforcement, only that there should be an assessment on an actuarial basis of what the realistic, target, minimum provisions should be. The immediate future will, it is hoped, see increased provision in *most* places.

DATA AT THE DRAWING BOARD

The emerging point in all this technical review is that, in terms of the products and systems available to buildings, we now have a great deal of capability to be going on with. There are exceptions, but the capability is mostly here and the most important question is how we use it, how we make the best use of it in design. Product innovation will hardly cease, industry will go on innovating, both in response to demand-pull and because that is one of the ways to compete in the market place. But talking to architects, engineers and researchers, one finds few of them looking to radical new products – the hope is for a better ability to make the right choices within the multiplicity of choice, in materials, components, systems, and in strategies in the widest sense, as in energy.

We have spoken of architectural design being, at least, daunted by the fast-changing technology surrounding it. Added to this, there are the increasingly complex purposes we have for buildings and the increasing expectations in environmental quality and running economy. Whereas, once, the architect could hold all the information needed to design a building, large buildings now need a multi-disciplinary team and this poses all sorts of consequent challenges in team communication and systems interaction. This kind of thing has long been a recognised problem in other branches of technology – one gifted engineer could design a Second World War fighter, but the team required for something like a 747 has called for radically new design methodology. Architectural technology has come to similar problems. In closing, we might just touch on four areas commonly looked to for improving the technical data base in building design – namely, research, legislative standards, computers and education.

Relatively speaking, we do not spend so much on building research. It is curious to think that whereas a car costing £5000 could well have absorbed £500 million in its research, a building costing £500 000 would be lucky if its research absorbed £5000! We only live and work in buildings, that's all. True, the analogy is not entirely fair in that much building research is done generally by government and other agencies rather than on a per building basis, and also in that a building is a one-off product. But then its being one-off is part of the problem – architectural and engineering practices have little resource for research, considering the contingencies of getting a building up for the client in time and within budget. The manufacturing industries invest in research but this tends to be product rather than overall-strategy orientated: greater demand-pull on the strategies side could bring a change of emphasis here, producing products, products information and information-handling systems that acknowledged strategy in a more overall sense. But much of the needed research must continue to come from the public agencies. There may be difficulties, both in keeping this kind of effort married to the real practicalities of putting up buildings, and in getting the fruits of the research across to the practitioner and accepted by clients, developers and manufacturers – but they are difficulties to tackle.

Minimum legislative standards, informed by research, have considerable potential for improving the quality of technical information available to the designer. But what kind of standards are most effective? The standard makers complain that even the best standards meet unseemly reaction out in the field, while out in the field there is the complaint that standards are too much in the hands of academics and unrelated to real building problems. And there is the 'empirical' versus 'performance' standards argument. Empirical standards, spelling out in some detail the ways in which good performance can be achieved, are inclined to by many countries – the UK, for example. But others have it that they are a strait-jacket, inhibiting design, and that, in being necessarily comprehensive to cover *all* building cases, they tend to impose irrelevant restrictions on *particular* cases. The alternative, favoured in the USA for example, is to have performance standards establishing minimum performance criteria, but not stipulating how the criteria are to be met. This less specific approach, arguably, reduces the potential education value of standards and increases the difficulty for one professional to evaluate or interact with the work of another. And prescribed minimum performance is hard to measure – it is hard for the designer to specify the materials and construction that will give such and such a transmission loss or U value and, in every day building, away from the laboratory, it is obviously hard to measure whether the performance has been achieved. Also there is the question of the increased legal liabilities attracted by designers finding their own way towards acceptable performance. Architectural mistakes can cost dear and, indeed, this thought in itself can lead to over-specification and inhibition.

The best answer probably lies towards harmonised, empirical standards as *minimum* criteria – principally for their information value. This is provided they can be flexible in operation, responsive to change, and especially if they can be presented so that they demonstrate the rational principles from which they spring rather than just stating the quantities as symptoms of principles. But standard writing is difficult at the best of times.

Computers may not 'design' but they will have growing implications for design. The hardware is cheapening as the markets grow to a viable size. This also seems to be applying currently to the program writing, the software, but admittedly the effect may be temporary since intelligent programs are expensive to write and since rising performance demands are tending to increase program complexity. Certainly, given the growing demand for better and more simply available data – in architecture as everywhere else – the penetration into practice can only increase. Computers are already established tools in many of the architectural disciplines. They take the labour out of the iterative calculations in structural and environmental engineering. It is easy to see the quantity surveyor released from the repetitive chore of laying-off quantities and becoming a building economist of a more general sort. In large buildings, optimum planning in terms of circulation and space is susceptible to computer analysis. And, though the publicity here has perhaps been disproportionate, computers can now draft and throw up buildings and their internal systems in 3-D.

But what is so interesting now is the emerging capability of computers for evaluation and information-handling generally. There are already programs for obtaining optimum structural solutions and minimising servicing layouts and, at last, programs are now appearing for finding the most favourable energy strategies – we have touched on these things. One can speculate further. Given that designers can no longer rely solely on their experience and evolved traditions to guide them, the computer information bank could just help enlarge this 'experience' at the drawing board, it could help fill the knowledge gaps. It could improve the information flow on products, systems and minimum legislative standards. And, quite conceivably, computers could improve the information flow between the members of the design team itself. The idea of having a computer 'model' of the continually evolving design, with each discipline contributing and interacting, may be quite far off – but it is a highly seductive one. And how far might the computer come to *replace* the technical input of the professional in the design team? We will abide that question!

Finally, there is education. As the base from which so much springs, it makes a fitting point on which to close. Of course, a book like this must have its bias, and the introduction expressed a personal view. But certainly, there is now healthy debate on what ought to be the nature, and extent, of technical teaching in architectural schools. Different schools have different characters and this is valuable, some being more arts-orientated and some more technically-orientated. In general, the technical emphasis may have receded, or it may be just that the technical emphasis, in remaining about the same, has failed to respond to the growing complexity of buildings. Whichever it is, the misgivings expressed in parts of the architectural profession, and by the allied professions, are undeniable. Of course, schools as part of a sensitive and caring profession are becoming aware of this: there are clear signs of response, of an acceptance of the need for a firmer technical base interacting with design in the studio – though there is rather less consensus on how this should happen.

There has been some experiment with having a 'multi-disciplinary' approach in the early years, having architectural students attend engineering students lectures and *vice versa*. But this has some drawbacks. It can tend to separate the technology from the drawing board and this is extremely undesirable. Also, it is not wholly relevant for architectural students to know the skilled business of calculating multi-storey frames or service-plant capacity, nor for engineers to address all the contexts of building design. It depends on the particular course but, sometimes, each is learning in the right area but at the wrong depth.

There are questions as to how far technical courses ought to follow a standard syllabus. The argument against standardisation *per se*, is that, given the undoubted scope for different kinds of architect in practice, having different schools with different emphases is a good thing. The argument *for* standardisation is that it can help ensure that minimum standards are provided. Actually, the arguments are by no means irreconcilable. An attractive suggestion is to accept a degree of standardisation in a core technical course running parallel to the studio work over the early years – in many courses, over the first three years – and later allow greater specialisation in pure design, town planning, information technology or wherever particular students' inclinations and abilities may lie. The option of mid-course changes between colleges, inclining towards different specialisations, could be part of this. But, whatever the format, to see building technology as something to be underplayed, or as foreign to the real business of learning at the drawing board, is wrong.

BIBLIOGRAPHY

The following are recommended for further reading – many were valuable sources for this book. The sub-categorisations are only a guide and there are some overlaps within them. Short comments amplify the titles where necessary.

GENERAL REFERENCE

American Institute of Architects (1981) *Architectural Graphic Standards*, John Wiley, New York.
(Most essential single reference of US building industry. Comprehensive on methods and products. Expensive.)

Cowan, H. J. (1973) *Dictionary of Architectural Science*, Applied Science Publishers, London.
(Standard reference.)

Fleming, J., H. Honour, N. Pevsner (1972) *A Dictionary of Architecture*, Penguin Books, Harmondsworth, England.

Harrison, D. (editor) (1982) *Specification*, Vols 1 and 2, Architectural Press, London.
(Comprehensive, standard, annual reference on UK building methods and products. Expensive.)

Scott, J. S. (1974) *A Dictionary of Building*, Penguin Books, Harmondsworth, England.

Tutt, P., and **D. Adler** (1981) *New Metric Handbook*, Architectural Press, London.
(Multi-disciplinary; useful for theory as well as quantities.)

Uvarov, E. B., D. R. Chapman, and **A. Isaacs** (1979) *The Penguin Dictionary of Science*, Penguin Books, Harmondsworth, England.

STRUCTURE

Architects' Journal (1980) *Handbook of Building Structure*, Architectural Press, London.
(Panel of authors on theory, material and method – extensive reference.)

Cowan, H. J. (1976) *Architectural Structures*, Pitman, London and Marshfield, Mass.
(Theory and calculation.)

Gordon, J. E. (1976) *The New Science of Strong Materials*, Penguin Books, Harmondsworth, England.

Gordon, J. E. (1978) *Structures*, Penguin Books, Harmondsworth, England.

Logie, K. F. (1981) *Structures, basic theory with worked examples*, Nelson, London.

Salvadori, M. and **R. Heller** (1963) *Structure in Architecture*, Englewood Cliffs, New Jersey.
(Basic, illustrated concepts, so still relevant.)

Torroja, E. (1962) *Philosophy of Structures*, University of California Press.
(Comment as above)

STRUCTURE/CONSTRUCTION

Chudley, R. (1979) *Construction Technology Vols 1–4*, Longman, London.

Foster, J. S. and **R. Harington** (1975) *Structure and Fabric Pt. 2* (Mitchell's Building Construction), Batsford, London.

Greater London Council (1979) *Good Practice Details*, Architectural Press, London.

Handisyde, C. C. (1976) *Everyday Details*, Architectural Press, London.

McKay, W. B. (1970) *Building Construction Vols 1–4*, Longman, London.
(Becoming dated but still a standard reference.)

ENCLOSURE

Architects' Journal (1974) *Handbook of Building Enclosure*, Architectural Press, London.
(Valuable, comprehensive reference.)

Banham, R. (1969) *The Architecture of the Well-tempered Environment*, Architectural Press, London.
(Twentieth century historical account – reference to services also.)

Givoni, B. (1969) *Man, Climate and Architecture*,

Elsevier, New York.

Olgyay, V. and **A. Olgyay** (1963) *Design With Climate*, Princeton University Press, New Jersey.
(Dated but the accepted classic.)

Rostron, R. M. (1964) *Light Cladding of Buildings*, Architectural Press, London.
(Dated but still a useful reference.)

ENERGY-ORIENTATED

Brinkworth, B. J. (1974) *Solar Energy for Man*, Compton Press, Salisbury.

Burberry, P. (1979) *Building for Energy Conservation*, Architectural Press, London; John Wiley, New York.

Clarke, R. (1976) *Technological Self Sufficiency*, Faber & Faber, London.

Randell, J. E. (editor) (1978) *Ambient Energy and Building Design*, Construction Press, London, New York.
(16 authors on solar, wind, heat pumps etc.)

Szokolay, S. V. (1977) *Solar Energy and Building*, Architectural Press, London.

Vale, B. and **R. Vale** (1976) *The Autonomous House*, Thames and Hudson, London.

SERVICES

Burberry, P. (1979) *Environment and Services* (Mitchell's Building Construction) Batsford, London.

Burberry, P. and **A. Aldersey-Williams** (1978) *A Guide to Domestic Heating Installation and Controls*, Architectural Press, London.

Faber, O., J. R. Kell, and **P. L. Martin** (1979) *Heating and Air-Conditioning of Buildings*, Architectural Press, London.
(A standard text)

Hall, F. (1978) *Water Installation and Drainage Systems* (Construction Press) Longman, London.

Hall, F. (1981) *Building Services and Equipment, Vols 1–3*, Longman, London.

Lewis, M. L. (1979) *Electrical Installation Technology Vols 1 and 2*, Hutchinson, London.

McGuiness, W. J., B. Stein, and **J. S. Reynolds** (1980) *Mechanical and Electrical Equipment for Buildings*, John Wiley, New York.

McIntyre, D. A. (1980) *Indoor Climate*, Applied Science Publishers, Essex, England.
(Describes comfort conditions physiologically as well as how to achieve them.)

Say, M. G. (1976) *Electrical Engineer's Reference Book*, Butterworths, London

(Comprehensive for the practitioner.)

Wise, A. F. E. (1979) *Water, Sanitary and Waste Services for Buildings* (Mitchell's Building) Batsford, London.

LIGHTING

Durrent, D. W. (editor) (1980) *Interior Lighting Design*, Lighting Industry Federation, London.
Artificial lighting-quantity and quality

Gregory, R. L. (1977) *Eye and Brain,* Weidenfeld and Nicolson, London.
(Physiology and psychology of seeing).

Hopkinson, R. G. and **J. D. Kay** (1972) *The Lighting of Buildings*, Faber & Faber, London.
(Accepted, standard text.)

Hopkinson, R. G., P. Petherbridge, and **J. Longmore,** (1966) *Daylighting*, Heinemann, London.

The Illuminating Engineering Society, (1977) *The IES Code,* London.
(Artificial lighting – quantity and quality.)

Lynes, J. A. (1968) *Principles of Natural Lighting*, Applied Science Publishers, London.

ACOUSTICS

Knudsen, V. O. and **C. M. Harris** (1950) *Acoustical Designing in Architecture*, John Wiley, New York.
(Dated but an accepted classic.)

Kuttruff, H. (1979) *Room Acoustics*, Applied Science Publishers, Essex, England.

Moore, J. E. (1978) *Design for Good Acoustics and Noise Control*, Macmillan London, New York.

Parkin, P. H., H. R. Humphreys, and **J. R. Cowell,** (1979) *Acoustics, Noise and Buildings*, Faber & Faber, London, Boston.
(Accepted, standard text.)

FIRE SAFETY

Langdon-Thomas, G. J. (1972) *Fire Safety in Buildings – Principles and Practice*, A. & C. Black, London.

Lei, T. T. (1972) *Fire and Buildings*, Applied Science Publishers, Essex, England.

Malhotra, H. L. (1982) *Design of Fire Resisting Structures*, Blackie, Glasgow.

Taylor, J. and **G. Cooke** (1978) *The Fire Precautions Act in Practice*, Architectural Press, London.
(Orientated to UK regulations – but clear, illustrated theory.)

INDEX

Items not individually listed may be found under their generic heading, e.g. see: *dual-duct* (system) under *Air-conditioning*; *air-change criteria* under *Ventilation*. Where general subjects in the text (e.g. *drainage; fire-safety*) further apply to a specific type of building they are also indexed under that type of building (e.g. see under *Hotel* and *Hospital* and for both of the subjects exampled). Where an item is covered only, or mainly, in an illustration, rather than in the text, the illustration page number is given in *italic type*.

Absorbers, sound
  cavity, panel, 165
  porous, 164
Absorption, sound, 162–5
  coefficient of, 164
Acceleration, heating response, 83, 96
Acoustic ceiling, 165, 169
  in auditorium ventilation, 107
Acoustics, 160–77
  see also room acoustics, noise
    control
Adaptation
  comfort criteria, 79
  visual, 141, 147
    between adjacent areas, 156
Aerogenerators, 197
Aggregate, 12
Air-brick, 55
Air-conditioning
  controls, 108–9
  delivery and extract in space, 107–8
  distribution systems
    dual-duct, 105
    'packaged' and central plant,
      104–5
    variable volume, 105
    zoned, 105–6
  factors for choice, 102
  historical development, 102
  humidity control, *103*, 104
  induction units, 107

secondary (fan-coil) units, 106, 108
typical plant, 103–4
Alarm
  fire, 185
  security, 75
Alternator, 129
Amplification, sound, 169–70
Anaerobic bacteria, 126
Anodising, 60
Anti-gravity valve, 90
Anti-shear straps, 27
Anti-siphon trap, 119
Apartment block
  air-conditioning, 105
  cellular construction, 21
  drainage, 122
  drinking water supply, 116
  electrical supply, 134
  heating system, 94
  hot water distribution, 116
  refuse disposal, 127
  telecommunications, 136
  see also under multi-storey
Apostilb, 141
Apron wall *see* spandrel wall
Arch
  built-up, 7
  catenary, 19
  historical development, 6, 7
  in bending, 19–20
  masonry theory, 6–7
  parabolic, 20
Art gallery, lighting, 159
Arup, Ove & Ptnrs., 37
Aspect ratio, 102
Asphalt (roof), 71
Atriums, fire safety in, 187
Auditoria
  acoustical design, 162–70
  air-conditioning, 105, 107
  fire safety, 180, 184, 187
  reverberation criteria, 163
Auditory localisation, 162

Back boiler, 80
Back-filling, excavation, *13*, 73–4
Back-siphonage, 112

Balancing
  central heating circuit, 96
  mechanical ventilation, 100–1, 107
Balloon frame, timber, 14
Basement
  buoyancy raft, 25
  domestic, typical construction, 15
  larger buildings, 64–5, 73–4
Battens, roof, 53
Bead
  at plaster edge, *56*
  cladding, 71, 72
  glazing, 60, *61*
Beam
  beam types, 11–12
  frame types, 16–20
  in trabeated construction, 6
  theory
    bending moment, 9, 16–20
    bending stresses, 9–10
    internal moments, 11
    limiting deflection, 11
    shear forces, 11
Bending *see* beams
Beneficial reflections, sound, 168–9
Bituminous felt, 53, 54
Black body temperature, 153
Blind passage *see* dead-end corridor
Blinding, 13, *57*
Blinds *see* shading
Blockwork, 13–14
Boilers
  domestic, 83–4, 112–13
  larger buildings, 95–6, 117
Bonding, masonry, 4
Boundary-disturbance in shells, 36
Box beam, timber, 11
Box-deck bridges, 29
Break-cistern, tank
  fire extinguishing provision, 186
  water supply, 115, 116
Breather membrane, *55*, *57*, 70
Brick
  cavity construction, 13–14
  cladding, 63, *64*, 72
  permeability, 54–8
  resistance insulation, 47, 48
  see also masonry